Convex geometry is at once simple and amazingly rich. While the classical results go back many decades, during the last ten years the integral geometry of convex bodies has undergone a dramatic revitalization, brought about by the introduction of methods, results, and, most importantly, new viewpoints, from probability theory, harmonic analysis, and the geometry of finite-dimensional normed spaces.

This book is a collection of research and expository articles on convex geometry and probability, suitable for researchers and graduate students in several branches of mathematics coming under the broad heading of "Geometric Functional Analysis." It continues the Israel GAFA Seminar series, which is widely recognized as the most useful research source in the area.

The collection reflects the work done at the program in Convex Geometry and Geometric Analysis that took place at Mathematical Sciences Research Institute in 1996, emphasizing the links between the geometry of convex bodies, probability theory, harmonic analysis, and recent probabilistic methods in computation. It includes contributions from Christer Borell, Jean Bourgain, E. D. Gluskin, W. T. Gowers, Gil Kalai, Greg Kuperberg, Bernard Maurey, Vitali Milman, Alain Pajor, Gideon Schechtman, Michael Schmuckenschlager, Carsten Schütt, Gaoyong Zhang, and several of the most promising representatives of the new generation.

Mathematical Sciences Research Institute
Publications

34

Convex Geometric Analysis

Mathematical Sciences Research Institute
Publications

Volume 1	Freed and Uhlenbeck:	*Instantons and Four-Manifolds*, second edition
Volume 2	Chern (editor):	*Seminar on Nonlinear Partial Differential Equations*
Volume 3	Lepowsky, Mandelstam, and Singer (editors):	*Vertex Operators in Mathematics and Physics*
Volume 4	Kac (editor):	*Infinite Dimensional Groups with Applications*
Volume 5	Blackadar:	*K-Theory for Operator Algebras*
Volume 6	Moore (editor):	*Group Representations, Ergodic Theory, Operator Algebras, and Mathematical Physics*
Volume 7	Chorin and Majda (editors):	*Wave Motion: Theory, Modelling, and Computation*
Volume 8	Gersten (editor):	*Essays in Group Theory*
Volume 9	Moore and Schochet:	*Global Analysis on Foliated Spaces*
Volume 10	Drasin, Earle, Gehring, Kra, and Marden (editors):	*Holomorphic Functions and Moduli I*
Volume 11	Drasin, Earle, Gehring, Kra, and Marden (editors):	*Holomorphic Functions and Moduli II*
Volume 12	Ni, Peletier, and Serrin (editors):	*Nonlinear Diffusion Equations and Their Equilibrium States I*
Volume 13	Ni, Peletier, and Serrin (editors):	*Nonlinear Diffusion Equations and Their Equilibrium States II*
Volume 14	Goodman, de la Harpe, and Jones:	*Coxeter Graphs and Towers of Algerbras*
Volume 15	Hochster, Huneke, and Sally (editors):	*Commutative Algebra*
Volume 16	Ihara, Ribet, and Serre (editors):	*Galois Groups over Q*
Volume 17	Concus, Finn, and Hoffman (editors):	*Geometric Analysis and Computer Graphics*
Volume 18	Bryant, Chern, Gardner, Goldschmidt, and Griffiths:	*Exterior Differential Systems*
Volume 19	Alperin (editor):	*Arboreal Group Theory*
Volume 20	Dazord and Weinstein (editors):	*Symplectic Geometry, Groupoids, and Integrable Systems*
Volume 21	Moschovakis (editor):	*Logic from Computer Science*
Volume 22	Ratiu (editor):	*The Geometry of Hamiltonian Systems*
Volume 23	Baumslag and Miller (editors):	*Algorithms and Classification in Combinatorial Group Theory*
Volume 24	Montgomery and Small (editors):	*Noncommutative Rings*
Volume 25	Akbulut and King:	*Topology of Real Algebraic Sets*
Volume 26	Judah, Just, and Woodin (editors):	*Set Theory of the Continuum*
Volume 27	Carlsson, Cohen, Hsiang, and Jones (editors):	*Algebraic Topology and Its Applications*
Volume 28	Clemens and Kollar (editors):	*Current Topics in Complex Algebraic Geometry*
Volume 29	Nowakowski (editor):	*Games of No Chance*
Volume 30	Grove and Petersen (editors):	*Comparison Geometry*
Volume 31	Levy (editor):	*Flavors of Geometry*
Volume 32	Cecil and Chern (editors):	*Tight and Taut Submanifolds*
Volume 33	Axler, McCarthy, and Sarason (editors):	*Holomorphic Spaces*

Volumes 1–4 and 6–27 are available from Springer-Verlag

Convex Geometric Analysis

Edited by

Keith M. Ball
University College, London

Vitali Milman
Tel Aviv University

Keith M. Ball
Department of Mathematics
University College, London
London WC1E 6BT

Mathematical Sciences Research
 Institute
1000 Centennial Drive
Berkeley, CA 94720

Vitali Milman
Department of Mathematics
Tel Aviv University
Tel Aviv 69978
Israel

MSRI Editorial Committee
Hugo Rossi (chair)
Alexandre Chorin
Silvio Levy (series editor)
Jill Mesirov
Robert Osserman
Peter Sarnak

The Mathematical Sciences Research Institute wishes to acknowledge
support by the National Science Foundation.

Published by the Press Syndicate of the University of Cambridge
The Pitt Building, Trumpington Street, Cambridge CB2 1RP
40 West 20th Street, New York, NY 10011–4211, USA
10 Stamford Road, Oakleigh, Melbourne 3166, Australia

© Mathematical Sciences Research Institute 1999

Printed in the United States of America

Library of Congress cataloging-in-publication data is available.

A catalogue record for this book is available from the British Library.

ISBN 0-521-64259-0 hardback

QA
331.7
.C674
1999

Contents

Introduction: The Convex Geometry and Geometric Analysis Program, MSRI, Spring 1996	ix
Program Seminars	xi
GAFA Seminars 1994–1996	xix
Integrals of Smooth and Analytic Functions over Minkowski's Sums of Convex Sets SEMYON ALESKER	1
Localization Technique on the Sphere and the Gromov–Milman Theorem on the Concentration Phenomenon on Uniformly Convex Sphere SEMYON ALESKER	17
Geometric Inequalities in Option Pricing CHRISTER BORELL	29
Random Points in Isotropic Convex Sets JEAN BOURGAIN	53
Threshold Intervals under Group Symmetries JEAN BOURGAIN AND GIL KALAI	59
On a Generalization of the Busemann–Petty Problem JEAN BOURGAIN AND GAOYONG ZHANG	65
Isotropic Constants of Schatten Class Spaces SEAN DAR	77
On the Stability of the Volume Radius EFIM D. GLUSKIN	81
Polytope Approximations of the Unit Ball of ℓ_p^n W. TIMOTHY GOWERS	89
A Remark about the Scalar-Plus-Compact Problem W. TIMOTHY GOWERS	111

Another Low-Technology Estimate in Convex Geometry   117
    GREG KUPERBERG

On the Equivalence Between Geometric and Arithmetic Means for
Log-Concave Measures   123
    RAFAŁ LATAŁA

On the Constant in the Reverse Brunn–Minkowski Inequality for p-Convex
Balls   129
    ALEXANDER E. LITVAK

The Extension of the Finite-Dimensional Version of Krivine's Theorem to
Quasi-Normed Spaces   139
    ALEXANDER E. LITVAK

A Note on Gowers' Dichotomy Theorem   149
    BERNARD MAUREY

An "Isomorphic" Version of Dvoretzky's Theorem, II   159
    VITALI MILMAN AND GIDEON SCHECHTMAN

Asymptotic Versions of Operators and Operator Ideals   165
    VITALI MILMAN AND ROY WAGNER

Metric Entropy of the Grassmann Manifold   181
    ALAIN PAJOR

Curvature of Nonlocal Markov Generators   189
    MICHAEL SCHMUCKENSCHLÄGER

An Extremal Property of the Regular Simplex   199
    MICHAEL SCHMUCKENSCHLÄGER

Floating Body, Illumination Body, and Polytopal Approximation   203
    CARSTEN SCHÜTT

A Note on the M^*-Limiting Convolution Body   231
    ANTONIS TSOLOMITIS

The Convex Geometry and Geometric Analysis Program MSRI, Spring 1996

During the last ten years the integral geometry of convex bodies has undergone a dramatic revitalisation, brought about by the introduction of methods, results and, most importantly, new viewpoints, from probability theory, harmonic analysis and the geometry of finite-dimensional normed spaces. The principal goal of this program was to bring together researchers from several different fields, Classical Convex Geometry, Geometric Functional Analysis, Computational Geometry and related areas of Harmonic Analysis. The main reason for doing so was that research in these areas has found considerable overlap in recent years. Several problems and classes of problems have been come upon independently from different directions, and techniques from some areas have been found important in others. This goal was achieved beyond even our most optimistic expectations.

As well as an introductory workshop, consisting of four lecture series with an educational format, the program included one full-scale research workshop and two concentrations of visitors, in addition to the regular activity during the principal five months. About 190 mathematicians attended the program in some capacity or other and there were over 150 lectures and seminars during the period. These were of several types. There was a regular educational seminar, two or three times a week which enabled participants to become acquainted with material from other fields. Three or four lectures a week dealt with recent research by members, and there was a "young research seminar" (roughly once a week) which gave postdocs and students a chance to describe their work in an informal atmosphere.

MSRI provided an unique environment in which to bring together such a group of people. The spectacular scenery served as a fitting backdrop to the immensely invigorating mathematical atmosphere.

The articles in this collection present recent research in the areas covered by the program. All this research was either completed at MSRI or presented in lectures during the program. The full list of lectures is included in the next few pages, in order to give some indication of the scope of the program. Since this volume takes the place of the regular GAFA seminar series for the year 1995–96, we have also included a list of lectures given in that seminar.

We would like to express our sincere thanks to Silvio Levy for his careful preparation of the manuscript of these proceedings.

Keith Ball
Vitali Milman

MSRI PROGRAM SEMINARS

Introductory Workshop on Convex Geometry and Geometric Functional Analysis

The contents of this introductory workshop were published in the form of two sections, by Ball and Bollobás respectively, in the book *Flavors of Geometry*, Silvio Levy (editor), MSRI Publications **31**, Cambridge University Press, 1997.

January 29, 1996
K. Ball, Basic notions in convex geometry

January 30, 1996
K. Ball, Fritz John's theorem
G. Schechtman, The central limit theorem and large deviation inequalities

January 31, 1996
J. Lindenstrauss, Volume ratios and their uses

February 1, 1996
K. Ball, The Brunn–Minkowski theorem
G. Schechtman, Concentration of measure in geometry

February 2, 1996
J. Lindenstrauss, Embedding Euclidean spaces into l_1

February 5, 1996
B. Bollobás, Rapid mixing and volume estimation, part I
K. Ball, Convolutions and volume ratios

February 6, 1996
G. Schechtman, Dvoretzky's theorem

February 7, 1996
B. Bollobás, Rapid mixing and volume estimation, part II
K. Ball, The slicing problem (Informal talk)

February 8, 1996
J. Lindenstrauss, Distributing points uniformly on spheres

February 9, 1996
B. Bollobás, Rapid mixing and volume estimation, part III

Regular Seminars

January 17, 1996
K. Ball, A new lower bound for lattice packing density in high dimensions

February 2, 1996
S. Kwapien, An inductive approach to moment estimates on product spaces

February 5, 1996

K. Ball, Volume in $\mathbb{R}^n$ and its relationship to linear structure (MSRI–UC Berkeley Lecture)

February 6, 1996

A. Dembo, Information inequalities and concentration of measure

February 7, 1996

S. Kwapien, An inductive approach to moment inequalities, part II

February 8, 1996

S. Dar, The isotropic constant of non-symmetric convex bodies

February 12, 1996

D. Klain, Valuations and Hadwiger characterization theorem

G. Schechtman, Alexandrov–Fenchel inequalities, part I

February 13, 1996

S. Szarek, A 'restricted' Brunn–Minkowski inequality

A. Giannopoulos, On some vector balancing problems

February 14, 1996

G. Schechtman, Alexandrov–Fenchel inequalities, part II

February 15, 1996

J. Mount, Sampling contingency tables

N Tomczak–Jaegermann, Complexity and higher order Schreier families

February 16, 1996

S. Alesker, Hilbert polynomial and number of points in the sum of finite sets (after Khovanskii)

Concentration in Infinite Dimensional Convex Geometry

February 20, 1996

N. Kalton, Complements of Sidon sets

E. Odell, Proximity to l_1 and distortion in asymptotic l_1 spaces

J. Lindenstrauss, The uniform classification of Banach spaces

B. Randriantoanina, 1-complemented subspaces in complex sequence spaces

N. Randriantoanina, Absolutely summing operators on non-commutative C^*-algebras

M. Girardi, Completely continuous operators

February 21, 1996

P. Wojtaszczyk, Wavelets for Banach spacers

T. Gamelin, Hankel operators on bounded analytic functions

H. Rosenthal, Invariants for differences of bounded semi-continuous functions, with applications to Banach space theory

S. Dilworth, Differentiability of the Pettis integral and weak and scalar convergence almost everywhere

A. Koldobsky, The Levy representation of norms and inequalities for Gaussian expectations

D. Speegle, A construction of indicator function wavelets

P. Habala, A Banach space whose subspaces do not have the GL-property

February 22, 1996

M. Ostrovskii, Classes of Banach spaces stable with respect to the opening

V. Fonf, Countable proximal sets in infinite dimensional Banach spaces

S. Argyros, Convex unconditionality and summability of weakly null sequences

T. Schlumprecht, A Banach space with hereditarily huge asymptotic structure

R. Wagner, Gowers dichotomy for asymptotic structure

G. Androulakis, Distorting mixed Tsirelson spaces

R. Judd, Calculating the l_1 index of certain Banach spaces

February 23, 1996

B. Maurey, Banach spaces with small spaces of operators

G. Godefroy, Progress on the approximation properties

P. Casazza, Complemented unconditional basic sequences in Banach lattices

I. Deliyanni, Examples of asymptotic l_1 Banach spaces

D. Kutzarova, On some asymptotic l_1 spaces

M. Robdera, On the analytic complete continuity property

P. Saab, On convolution operators associated with vector measures

S. Saccone, Tight uniform algebras

Regular Seminars

February 26, 1996

J. Lindenstrauss, Zonoids whose polars are zonoids

February 27, 1996

M. Wodzicki, A fresh look at Banach spaces

M. Girardi, Strongly measurable Banach-space valued functions: examples and results

February 28, 1996

M. Rudelson, Direct construction of majorizing measures

February 29, 1996

J. M. Rojas, Affine space and toric varieties, part II

March 4, 1996

Y. Gordon, Volume computation for quotients of L_p spaces and applications

March 5, 1996

R. Schneider, From areas in Minkowski spaces to zonoids

K. Swanepoel, Collapsing conditions for sets of vectors in Mikowski spaces

March 6, 1996

A. Arias, Pisier's example of a polynomially bounded operator which is not similar to a contraction (after G. Pisier)

A. Petrunin, Introduction to Alexandrov spaces

March 8, 1996

V. Milman, Some problems in local theory

E. Grinberg, Microlocal analysis of convex surfaces and Funk's characterization of the sphere

Workshop in Random Methods in Convex Geometry

March 11, 1996

V. Milman, Global versus local views in high dimensional convexity

M. Talagrand, A functional point of view for concentration of measure

A. Zee, Universal correlation and other results in random matrix theory

B. Bollobás, Random partial orders

S. Szarek, Komlos conjecture, Sidak's inequality and local Lovász lemma

S. Alesker, Integrals of analytic and smooth functions over Minkowski sum of convex bodies

R. Latała, Estimates of moments of sums of independent real random variables

March 12, 1996

R. Kannan, Logarithmic Sobolev inequalities and geometric random walks

M. Simonovits, Randomized volume algorithms

H. König, Isometric embeddings of Euclidean spaces into l_p^N spaces and cubature formulas on spheres

M. Junge, Mixed volumes for l_p sums of convex bodies

A. Tsolomitis, Limiting convolution bodies

M. Schmückenschlager, Hu's inequality

J. Wenzel, Sequences of ideal norms and the UMD-property

March 13, 1996

R. Schneider, Determination of convex bodies from projection functions

J. Pach, On uniformly distributed distances — a geometric application of Janson's inequality

M. Rudelson, Contact points and applications

B. Bollobás, Volume estimates and rapid mixing (MSRI–UC Berkeley Lecture)

March 14, 1996

I. Barany, Affine perimeter and limit shape

W. Banaszczyk, The width of lattice-point-free convex bodies

P. Mankiewicz, Groups of operators acting on random quotients of l_1^m

A. Giannopoulos, Low M^* estimates for coordinate subspaces

T. Schlumprecht, The Gaussian correlation problem for ellipsoids

S. Dar, Slicing problem for trace classes

March 15, 1996

D. Welsh, Randomized approximation of geometrical Tutte invariants

S. Kwapien, Differential inequalities and comparison of moments

F. Chung, Logarithmic Harnack inequalities

G. Zhang, Ellipsoidal decompositions of centered bodies

W. Weil, Section and projection means of convex bodies

D. Klain, A continuous analogue of Sperner's theorem

R. Vitale, The Wills functional and Gaussian processes

Regular Seminars

March 18, 1996

R. Vitale, The Wills functional and Gaussian processes II

March 19, 1996

A. Tsolomitis, Convolution bodies

F. Barthe, Extremal sections of the unit ball of l_p^n

March 20, 1996

V. Klee, Some unsolved problems in convex geometry: seven favorites

W. Banaszczyk, Some inequalities for polar reciprocal n-dimensional lattices and convex bodies (transference theorems in the geometry of numbers)

March 22, 1996

W. Weil, Local formulae in integral geometry

March 25, 1996

D. Klain, Blaschke sums and mixed bodies,

March 26, 1996

N. Tomczak–Jaegermann, Spaces of type p containing arbitrarily distortable subspaces,

March 27, 1996

E. Lutwak, The Brunn–Minkowski–Firey theory,

March 28, 1996

G. Zhang, Volume inequalities for sections of convex bodies,

March 29, 1996

I. Bárány, On the number of convex lattice polytopes,

April 1, 1996

M. Simonovits, Localization lemmas and isoperimetric inequalities, part I

April 2, 1996

K. M. Ball, Bang's Lemma and the De Leeuw–Kahane–Katznelson Theorem (d'après Nazarov)

V. Ferenczi, Several properties of hereditarily indecomposable Banach spaces

April 3, 1996

A. Pajor, Kolmogorov's entropy in convex geometry

April 4, 1996

W. B. Johnson, Extensions of c_o: an addendum to a paper by Kalton and Pelczynski

April 5, 1996

M. Simonovits, Localization lemmas and isoperimetric inequalities, part II

A. Petrunin, Alexandrov spaces, part II

April 9, 1996

A. Petrunin, Alexandrov spaces, part III

P. Habala, A Banach space whose subspaces fail the Gordon–Lewis property

April 1, 1996

B. Maurey, Factorization of linear operators

April 12, 1996

W. T. Gowers, A lower bound of tower type for Szemerédi's uniformity lemma

A. Khovanskii, Connections between algebraic and convex geometry

April 15, 1996

A. Khovanskii, Connections between algebraic and convex geometry, part II

April 16, 1996

N. Ghoussoub, Improved Moser–Aubin–Onofri's inequalities on S^2

R. Latała, On the equivalence between arithmetic and geometric mean for logarithmically concave measures

April 17, 1996

J. M. Rojas, Mixed subdivisions and some practical results on mixed volume computation

S. J. Szarek, Free probability, part I

April 19, 1996

S. J. Szarek, Free probability, part II

April 22, 1996

P. M Gruber, Some aspects of approximation of convex bodies by polytopes

April 24, 1996

G. Pisier, Quadratic forms in unitary operators

April 26, 1996

M. Loss, The Best Constant in the Hardy–Littlewood–Sobolev Inequality

April 29, 1996

S. Szarek, Free probability and random matrices

April 30, 1996

K. Oleszkiewicz, On the discrete version of the antipodal theorem

S. Dar, A Brunn–Minkowski type inequality

May 1, 1996

J. Bourgain, Influence of variables and threshold intervals

May 2, 1996

E. Carlen, Logarithmic Sobolev inequalities and sharp estimates for smoothing and decay for the 2-d Navier–Stokes equation

May 3, 1996

J. Bourgain, Random points in isotropic convex sets

T. Oikhberg, Exact operator spaces with exact dual

G. Pisier, Similarity problems and completely bounded maps

May 6, 1996

S. Bates, Nonlinear surjections of Banach spaces

May 7, 1996

A. Pelczynski, The Dunford–Pettis property of L_1 on cosidon sets

Sharp Inequalities in Harmonic Analysis and Convex Geometry

May 8, 1996

K. Ball, Convolution inequalities and convex geometry

K. Oleszkiewicz, Best constant in the Khinchine–Kahane inequality

B. Ruskai, Contraction of relative entropy in information theory and quantum theory

May 9, 1996

G. Pisier, Various inequalities for vector valued C_p spaces

R. Ambartzumian, Hilbert's fourth problem: parametric versions

C. Morpurgo, Zeta functions and fractional integral inequalities

E. Grinberg, The cosine transform of higher order

May 10, 1996

D. Burkholder, Some sharp inequalities in stochastic analysis

A. Barvinok, Computing mixed discriminant, mixed volumes, and permanents

D. Jerison, Variational problems for the capacity and for the first eigenvalue of convex bodies

May 13, 1996

A. Burchard, The Riesz rearrangement inequality

C. Borell, On Brunn–Minkowski inequalities in option theory

T. Bisztriczky, A proof of Hadwiger's conjecture for dual cyclic polytopes

S. Alesker, Polynomial rotation invariant valuations on convex sets

May 14, 1996

E. Lieb, Inequalities related to stability of matter: overview

R. Laugesen, Extremals for zeta functions (and more) on Laplacians under conformal mapping

May 15, 1996

M. Loss, Inequalities related to stability of matter: matter and fields

Regular Seminars

May 14, 1996

A. Litvak, An extension of reverse Brunn–Minkowski inequality to non-convex case

May 16, 1996

R. Wagner, Asymptotic versions of operators and operator ideals

May 17, 1996

E. Lieb, Gaussian kernels and some of their applications

May 20, 1996

G. Schechtman, Three problems in geometric functional analysis

May 22, 1996

L. Lovasz, Stopping rules for random walks in convex bodies

May 23, 1996

L. Tzafrizi, Legendre and Jacobi polynomials in Banach Space theory

May 24, 1996

A. Giannopoulos, On the diameter of proportional sections of a symmetric convex body

May 28, 1996

M. Rudelson, Bringing a body into an isotropic position

May 29, 1996

E. Gluskin, On Kashin approach to correction theorems

M. Loss, On the paper of Elliott Lieb — Gaussian kernels have Gaussian optimizers

May 30, 1996

H. Groemer, Half-sections and half-girths of convex bodies

GAFA SEMINARS 1994–1996

November 25, 1994

Yaki Sternfeld (Haifa), Recent problems and methods in dimensional theory

M. Dubiner (Tel Aviv), The equivalence of two polynomially defined metrics on a general convex subset of $\mathbb{R}^n$

December 9, 1994

M. Krivelevich (Tel Aviv), Probabilistic techniques in Ramsey theory

S. Reisner (Haifa), Constructing a polytope to approximate a convex body (joint work with Y. Gordon (Haifa) and Y. Meyer (France))

M. Rudelson (Jerusalem), Approximate John's decomposition

December 23, 1994

S. Dar (Tel Aviv), Isotropic capacity of convex non-symmetric bodies

Joel Zinn (Texas), Hypercontractivity and a Gaussian correlation inequality

A. Leiderman (Beer Sheva), The Kolmogorov superposition theorem and the free locally convex space on the unit interval

April 7, 1995

G. Schechtman (Rehovot), Isomorphic version of Dvoretzky's theorem (joint work with V. Milman (Tel Aviv))

G. Schechtman (Rehovot), A peculiar rearrangement of the Haar system (joint work with P. Muller)

R. Wagner (Tel Aviv), Asymptotic constants of Tzirelson type spaces (joint work with E. Odell and N. Tomczak)

November 24, 1995

V. Milman (Tel Aviv), Global vs. local results in asymptotic theory of normed spaces (joint work with G. Schechtman (Rehovot))

H. Hofer (Zurich), From periodic orbits to symplectic homology

A. Shnirelman (Tel Aviv), Geometry and dynamics on the group of area-preserving diffeomorphisms

December 8, 1995

S. Dar (Tel Aviv), Gradient map and Legendre transform (after M. Gromov)

S. Alesker (Tel Aviv), Minkowski type theorems for smooth and analytical weights

J. Lindenstrauss (Jerusalem), Differentiability of Lipschitz maps between Banach spaces

December 29, 1995

I. Benjamini (Rehovot), On the support of harmonic measures for random walks

Y. Gordon (Haifa), Variants of the Cauchy–Binet formula, with applications to volume estimates (joint work with M. Junge)

M. Braverman (Beer Sheva), Rosenthal's inequality and characterization of L_p-spaces

January 12, 1996

Y. Benjamini (Haifa), Linear approximation of quasi-isometries on $\mathbb{R}^n$ (work of Fritz John)

A. Dembo (Haifa), Information inequalities and concentration of measure

Integrals of Smooth and Analytic Functions over Minkowski's Sums of Convex Sets

SEMYON ALESKER

1. Introduction and Statement of Main Results

Let $\bar{K} = (K_1, K_2, \ldots, K_s)$ be an s-tuple of compact convex subsets of $\mathbb{R}^n$. For any continuous function $F : \mathbb{R}^n \longrightarrow \mathbb{C}$, consider the function

$$M_{\bar{K}} F : \mathbb{R}^s_+ \longrightarrow \mathbb{C}, \text{ where } \mathbb{R}^s_+ = \{(\lambda_1, \ldots, \lambda_s) \mid \lambda_j \geq 0\},$$

defined by

$$(M_{\bar{K}} F)(\lambda_1, \ldots, \lambda_s) = \int_{\sum_{i=1}^s \lambda_i K_i} F(x)\, dx. \qquad (*)$$

This defines an operator $M_{\bar{K}}$, which we will call a Minkowski operator. Denote by $\mathcal{A}(\mathbb{C}^n)$ the Frechet space of entire functions in n variables with the usual topology of the uniform convergence on compact sets in $\mathbb{C}^n$, and $C^r(\mathbb{R}^n)$ the Frechet space of r times differentiable functions on $\mathbb{R}^n$ with the topology of the uniform convergence on compact sets in $\mathbb{R}^n$ of all partial derivatives up to the order r ($1 \leq r \leq \infty$).

The main results of this work are Theorems 1 and 3 below.

THEOREM 1.
($\imath$) *If* $F \in \mathcal{A}(\mathbb{C}^n)$, *then* $M_{\bar{K}} F$ *has a (unique) extension to an entire function on* $\mathbb{C}^s$ *and defines a continuous operator from* $\mathcal{A}(\mathbb{C}^n)$ *to* $\mathcal{A}(\mathbb{C}^s)$ *(see Theorem 3 below).*
($\imath\imath$) *If* $F \in C^r(\mathbb{R}^n)$, *then* $M_{\bar{K}} F \in C^r(\mathbb{R}^s_+)$ *(it is smooth up to the boundary) and again* $M_{\bar{K}}$ *defines a continuous operator from* $C^r(\mathbb{R}^n)$ *to* $C^r(\mathbb{R}^s_+)$.

COROLLARY 2. *If* F *is a polynomial of degree* d, *then* $M_{\bar{K}} F$ *is a polynomial of degree at most* $d + n$.

Indeed, we can assume F to be homogeneous of degree d. Then $M_{\bar{K}}$ is an entire function, which is homogeneous of degree $d + n$, hence it is a polynomial.

In fact, this corollary is well known and it is a particular case of the Pukhlikov–Khovanskii Theorem ([P-Kh]; see another proof below).

THEOREM 3. *Assume that a sequence $F^{(m)} \in \mathcal{A}(\mathbb{C}^n)$ (or $C^r(\mathbb{R}^n)$, respectively), $m \in \mathbb{N}$ is such that*

$$F^{(m)} \longrightarrow F \text{ in } \mathcal{A}(\mathbb{C}^n) \text{ (or } C^r(\mathbb{R}^n)).$$

Let $K_i^{(m)}$, K_i, $i = 1, 2, \ldots, s$, $m \in \mathbb{N}$ be convex compact sets in $\mathbb{R}^n$, and suppose $K_i^{(m)} \longrightarrow K_i$ in the Hausdorff metric for every $i = 1, \ldots, s$. Then

$$M_{\bar{K}^{(m)}} F^{(m)} \longrightarrow M_{\bar{K}} F$$

in $\mathcal{A}(\mathbb{C}^s)$ (or $C^r(\mathbb{R}^s_+)$).

REMARKS. 1. It follows from Theorem 1 that, if K is a compact convex set, D is the standard Euclidean ball and γ_n is the standard Gaussian measure in $\mathbb{R}^n$, then $\gamma_n(K + \varepsilon \cdot D)$ is an entire function of ε and the coefficients of the corresponding power expansion are rotation invariant continuous valuations on the family of compact convex sets (see the related definitions in Section 4).

2. There is a different simpler proof of Theorem 1 in the case when all the K_i are convex polytopes. However, the standard approximation argument cannot be applied automatically, since Theorem 3 on the continuity does not follow from that simpler construction even for polytopes.

In Section 4 we present another proof of the Pukhlikov–Khovanskii Theorem.

2. Preliminaries

Before proving these theorems, let us recall some facts, which are probably quite classical, but we will follow Gromov's work [G] (see also [R]).

A function $f : \mathbb{R}^n \longrightarrow \mathbb{R}$ is called convex if for all $x, y \in \mathbb{R}^n$ and $\mu \in [0, 1]$,

$$f(\mu x + (1-\mu)y) \leq \mu f(x) + (1-\mu)f(y);$$

f is called strictly convex if

$$f(\mu x + (1-\mu)y) < \mu f(x) + (1-\mu)f(y)$$

whenever $x \neq y$ and $\mu \in (0, 1)$. Define a Legendre transform of the convex function f (which is also called a conjugate function of f)

$$Lf(y) := \sup_{x \in \mathbb{R}^n} ((y, x) - f(x)).$$

Then Lf is a convex function and $-\infty < Lf \leq +\infty$. A set $K_f := \{y \in \mathbb{R}^n \mid Lf(y) < +\infty\}$ is called the effective domain of Lf. Obviously, K_f is a convex set. For any convex set K, we will denote the relative interior of K by $\operatorname{Int} K$.

LEMMA 4.

(ı) Let $f : \mathbb{R}^n \longrightarrow \mathbb{R}$ be a strictly convex C^2-function. Then K_f is a convex set and the gradient map $\nabla f : \mathbb{R}^n \longrightarrow \mathbb{R}^n$ is a one-to-one map of $\mathbb{R}^n$ onto $\operatorname{Int} K_f$.
(ıı) If f_1, f_2 are as in (ı), then for all λ_1, $\lambda_2 > 0$,

$$\operatorname{Im}(\nabla(\lambda_1 f_1 + \lambda_2 f_2)) = \lambda_1 \operatorname{Im}(\nabla f_1) + \lambda_2 \operatorname{Im}(\nabla f_2).$$

PROOF. (ı) The injectivity of ∇f immediately follows from the strict convexity of f.

For any $x_0, x \in \mathbb{R}^n$,

$$f(x) \geq f(x_0) + (\nabla f(x_0), x - x_0).$$

Hence $Lf(\nabla f(x_0)) = (\nabla f(x_0), x_0) - f(x_0) < \infty$ and $\operatorname{Im}(\nabla f) \subset K_f$. In order to check that $\operatorname{Im}(\nabla f) \subset \operatorname{Int} K_f$, let us choose any $a \in \partial K_f$ and assume that there exists $b \in \mathbb{R}^n$ such that $\nabla f(b) = a$. Without loss of generality, one may assume that $a = 0 = b$ and $f(0) = 0$. Then $f(x) \geq 0$ for all $x \in \mathbb{R}^n$.

Since K_f is convex and $0 \in \partial K_f$, one can find a unit vector $u \in \mathbb{R}^n$ such that $\lambda u \notin K_f$ for all $\lambda > 0$. Consider a new convex function on $\mathbb{R}^1$

$$\phi(t) := \inf \{f(y + tu) \mid y \perp u\}.$$

Clearly, $\phi(t) \geq 0$ everywhere and $\phi(0) = 0$.

Case 1. Assume that there exists $t_0 > 0$ such that $\phi(t_0) > 0$. Then, by the convexity of ϕ, $\phi(t) \geq \frac{\phi(t_0)}{t_0} t$ for $t \geq t_0$ and for $t \leq 0$. Hence

$$L\phi\left(\frac{\phi(t_0)}{t_0}\right) \leq \sup \left\{\frac{\phi(t_0)}{t_0} t - \phi(t) \;\Big|\; t \in [0, t_0]\right\} < \infty.$$

But for the Legendre transform of f, one has

$$Lf\left(\frac{\phi(t_0)}{t_0} u\right) = \sup_{x \in \mathbb{R}^n} \left(\left(\frac{\phi(t_0)}{t_0} u, x\right) - f(x)\right)$$

$$= \sup_{s \in \mathbb{R},\, y \perp u} \left(\left(\frac{\phi(t_0)}{t_0} u, su + y\right) - f(su + y)\right)$$

$$= \sup_{s \in \mathbb{R}} \left(\frac{\phi(t_0)}{t_0} s - \phi(s)\right) = L\phi\left(\frac{\phi(t_0)}{t_0}\right) < \infty.$$

Thus $\frac{\phi(t_0)}{t_0} u \in K_f$, and this contradicts the choice of u.

Case 2. Assume that $\phi(t) = 0$ for all $t \geq 0$. Let us show that this case is impossible (this will finish the proof of part (ı) of Lemma 4). It would follow from the fact that $f(x) \longrightarrow \infty$ as $x \longrightarrow \infty$.

If the last statement is false, then there exists a sequence of vectors $x_k \longrightarrow \infty$ such that $|f(x_k)| \leq C$ (where C is some constant). Passing to a subsequence, we may assume that

$$\frac{x_k}{|x_k|} \longrightarrow u \in \mathbb{R}^n,$$

where $|\cdot|$ denotes the Euclidean norm in $\mathbb{R}^n$. Since f is strictly convex, $f(0) = 0$ and $\nabla f(0) = 0$ by assumption, then $f(u) > 0$. Also, for all $x \in \mathbb{R}^n$,

$$f(x) \geq f(u) + (\nabla f(u), x - u).$$

Substituting $x = 0$ or $x = x_k$, we obtain

$$(\nabla f(u), u) \geq f(u) > 0,$$

$$f(x_k) \geq f(u) + (\nabla f(u), x_k - u).$$

The last inequality can be rewritten

$$f(x_k) \geq f(u) - (\nabla f(u), u) + |x_k| \left(\nabla f(u), \frac{x_k}{|x_k|} \right).$$

But $(\nabla f(u), \frac{x_k}{|x_k|}) \longrightarrow (\nabla f(u), u) > 0$, hence $f(x_k) \longrightarrow \infty$, which contradicts our assumptions.

(ii) Under conditions of the lemma $\lambda_1 f_1 + \lambda_2 f_2$ is also a strictly convex function. By the part (i),

$$\mathrm{Im}(\nabla f_i) = \mathrm{Int}\, K_{f_i} \text{ for } i = 1, 2.$$

Then easily

$$\mathrm{Im}(\lambda_1 \nabla f_1 + \lambda_2 \nabla f_2) \subset \lambda_1 \, \mathrm{Int}\, K_{f_1} + \lambda_2 \, \mathrm{Int}\, K_{f_2} \subset$$

$$\mathrm{Int}(\lambda_1 K_{f_1} + \lambda_2 K_{f_2}) \subset \mathrm{Int}(K_{\lambda_1 f_1 + \lambda_2 f_2}) = \mathrm{Im}(\nabla(\lambda_1 f_1 + \lambda_2 f_2)).$$

Hence all the sets in the above sequence of inclusions coincide. $\square$

LEMMA 5. [G] *Let $K \subset \mathbb{R}^n$ be an open bounded convex set, let μ be the Lebesgue measure in $\mathbb{R}^n$. Define*

$$f(x) := \log \int_K \exp(x, y) \, d\mu(y). \tag{1}$$

Then f is a strictly convex C^∞-function and $\mathrm{Im}(\nabla f) = K$.

Now let K_i, $1 \leq i \leq s$ be compact convex subsets of $\mathbb{R}^n$. For every i, fix a point $a_i \in K_i$. Let μ_i denote ($\dim K_i$)-dimensional Lebesgue measure supported on $\mathrm{span}(K_i - a_i)$. Define

$$f_i(x) := (x, a_i) + \int_{K_i - a_i} \exp(x, y) \, d\mu_i(y).$$

For every i, $f_i(x)$ depends only on the orthogonal projection of x on $\mathrm{span}(K_i - a_i)$. Moreover, f_i is a convex function on $\mathbb{R}^n$ and strictly convex on $\mathrm{span}(K_i - a_i)$. Then it is easy to see that $K_{f_i} \subset a_i + \mathrm{span}(K_i - a_i)$. Thus, by Lemmas 5 and 4,

$$\mathrm{Im}\, \nabla f_i = \mathrm{Int}\, K_i = \mathrm{Int}\, K_{f_i}.$$

COROLLARY 6. *Let K_i, f_i, $1 \leq i \leq s$ be as above, $\lambda_i > 0$. Then*

$$\operatorname{Im}\left(\nabla\left(\sum_{i=1}^{s} \lambda_i f_i\right)\right) = \sum_{i=1}^{s} \lambda_i \operatorname{Int} K_i.$$

PROOF. It is sufficient to consider all $\lambda_i = 1$. Set $L := \operatorname{span}\left(\sum_i (K_i - a_i)\right)$. Without loss of generality, we may assume that $L = \mathbb{R}^n$. Then obviously the function $f := \sum f_i$ is strictly convex on $\mathbb{R}^n$, and by Lemma 4, $\operatorname{Im} \nabla f = \operatorname{Int} K_f$ is an open and convex set. Clearly,

$$\operatorname{Im} \nabla f \subset \sum \operatorname{Im} \nabla f_i = \sum \operatorname{Int} K_i = \sum \operatorname{Int} K_{f_i} = \operatorname{Int}\left(\sum K_{f_i}\right)$$

(the last equality holds for general convex bounded subsets of $\mathbb{R}^n$). One can easily see that $\sum K_{f_i} \subset K_f$, hence

$$\operatorname{Im} \nabla f \subset \sum \operatorname{Int} K_i \subset \operatorname{Int} K_f = \operatorname{Im} \nabla f. \qquad \square$$

3. Proofs of Theorems 1 and 3

PROOF OF THEOREM 1. For every K_i, choose $f_i : \mathbb{R}^n \longrightarrow \mathbb{R}$ as above. Then $\nabla f_i = (\frac{\partial f_i}{\partial y_1}, \ldots, \frac{\partial f_i}{\partial y_n})$, and the Jacobian of the gradient map equals the Hessian of f_i,

$$H(f_i) = \left(\frac{\partial^2 f_i}{\partial y_p \partial y_q}\right)_{p,q=1}^n,$$

which is a non-negative definite matrix, since f_i is convex.

We have for $\lambda_i > 0$,

$$\int_{\sum \lambda_i K_i} F(x)\, dx = \int_{\mathbb{R}^n} F\left(\sum \lambda_i \nabla f_i(y)\right) \det\left(H\left(\sum \lambda_i f_i(y)\right)\right) dy. \qquad (2)$$

Write for simplicity $H_i(y) = H(f_i(y))$, so that the last expression is

$$\int_{\mathbb{R}^n} F\left(\sum \lambda_i \nabla f_i(y)\right) \det\left(\sum \lambda_i H_i(y)\right) dy$$

$$= \sum_{j_1, \ldots, j_n} \lambda_{j_1} \ldots \lambda_{j_n} \int_{\mathbb{R}^n} F\left(\sum \lambda_i \nabla f_i(y)\right) D(H_{j_1}(y), \ldots, H_{j_n}(y))\, dy, \qquad (3)$$

where $D(H_{j_1}(y), \ldots, H_{j_n}(y))$ denotes the mixed discriminant of non-negative definite symmetric matrices $H_{j_1}(y), \ldots, H_{j_n}(y)$. But it is well known that the mixed discriminant of such matrices is nonnegative (see, e.g., [Al]).

Let us substitute $F \equiv 1$ into (2). We obtain

$$\operatorname{vol}\left(\sum \lambda_i K_i\right) = \sum_{j_1, \ldots, j_n} \lambda_{j_1} \ldots \lambda_{j_n} \int_{\mathbb{R}^n} D(H_{j_1}(y), \ldots, H_{j_n}(y))\, dy.$$

Hence $\int_{\mathbb{R}^n} D(H_{j_1}(y), \ldots, H_{j_n}(y))\, dy = V(K_{j_1}, \ldots, K_{j_n})$ (the right hand side denotes the mixed volume of $K_{j_1}, \ldots, K_{j_n}$; see, e.g., [B-Z], [Sch]).

Observe that the integrand in (3) makes sense also for $\lambda_i < 0$, if $F \in C^r(\mathbb{R}^n)$ and for all complex λ_i, if $F \in \mathcal{A}(\mathbb{C}^n)$. We only have to check the convergence of the integral for such λ_i and its convergence after taking partial derivatives with respect to λ_i. Then Theorem 1 (i) and(ii) will be proved.

Let us show that for the integral in (3), and the same proof works for the partial derivatives with respect to the λ_i.

Since $\text{Im}(\nabla f_i) \subset K_i$, there exists a constant C, such that $\|\sum \lambda_i \nabla f_i(y)\| \leq C \cdot \sum |\lambda_i|$ for all $y \in \mathbb{R}^n$, where $\|\cdot\|$ is some norm in $\mathbb{C}^n$ (or in $\mathbb{R}^n$). By the continuity of F, $F(\sum \lambda_i \nabla f_i)$ is bounded by some constant $K(R)$ if $\sum |\lambda_i| \leq R$ and $y \in \mathbb{R}^n$. Hence

$$\int |F(\sum \lambda_i \nabla f_i(y)) D(H_{j_1}(y), \ldots, H_{j_n}(y))| \, dy$$

$$\leq K(R) \int_{\mathbb{R}^n} D(H_{j_1}(y), \ldots, H_{j_n}(y)) \, dy$$

$$= K(R) V(K_{j_1}, \ldots, K_{j_n}) < \infty. \qquad \square$$

REMARK. We have actually shown that, if $F \in C^r(\mathbb{R}^n)$, then the equality (2) gives us a smooth extension of $M_{\bar{K}} F(\lambda_1 \ldots, \lambda_n)$ from $\mathbb{R}_+^s$ to $\mathbb{R}^s$. It turns out that this extension is natural in some sense, i.e. it does not depend on the choice of the functions f_i. Indeed, assume that we have two such extensions $M_{\bar{K}} F$ and $M'_{\bar{K}} F$ corresponding to f_i and f'_i. Choose a sequence of polynomials $\{P_m\}$ approximating F uniformly on compact sets in $\mathbb{R}^n$. Then for corresponding extensions, we have $M_{\bar{K}} P_m \longrightarrow M_{\bar{K}} F$ and $M'_{\bar{K}} P_m \longrightarrow M'_{\bar{K}} F$ uniformly on compact sets in $\mathbb{R}^s$. By Corollary 2, $M_{\bar{K}} P_m$ and $M'_{\bar{K}} P_m$ are polynomials and since they coincide on $\mathbb{R}_+^s$, they coincide everywhere on $\mathbb{R}^s$. Hence $M_{\bar{K}} P_m \equiv M'_{\bar{K}} P_m$ on $\mathbb{R}^s$ and $M_{\bar{K}} F \equiv M'_{\bar{K}} F$.

PROOF OF THEOREM 3. *Step 1.* It is sufficient to prove the continuity of $M_{\bar{K}} F$ separately with respect to F and $\bar{K} = (K_1, \ldots, K_s)$, because $M_{\bar{K}} F = M(F; K_1, \ldots, K_s)$ can be considered as a map $M : L_1 \times \mathcal{K}^s \longrightarrow L_2$, where L_1 and L_2 are Frechet spaces, $L_1 = \mathcal{A}(\mathbb{C}^n)$ or $C^r(\mathbb{R}^n)$, $L_2 = \mathcal{A}(\mathbb{C}^s)$ or $C^r(\mathbb{R}^s)$, and $\mathcal{K}$ is the space of compact convex subsets of $\mathbb{R}^n$ with the Hausdorff metric. Since M is linear with respect to the first argument $F \in L_1$, and $\mathcal{K}^s$ is locally compact (by the Blaschke's selection theorem), then M is continuous as a function of two arguments (it is an easy and well known consequence of the Banach–Steinhaus theorem, which says that if L_1, L_2 are Frechet spaces, T is a locally compact topological space and $M : L_1 \times T \longrightarrow L_2$ is linear with respect to the first argument and continuous with respect to each argument separately, then M is continuous as a function of two variables).

Step 2. Let $K_1, \ldots, K_s$ be fixed, $F^{(m)} \longrightarrow F$. Using formula (2) and simple estimates as in the proof of Theorem 1, one can easily see that $M_{\bar{K}} F^{(m)} \longrightarrow M_{\bar{K}} F$.

Step 3. Now suppose $F \in \mathcal{A}(\mathbb{C}^n)$ (respectively, $C^r(\mathbb{R}^n)$) is fixed, $K_i^{(m)} \longrightarrow K_i$ as $m \longrightarrow \infty$ for all $i = 1, \ldots, s$. Let us choose $a_i^{(m)} \in K_i^{(m)}$, $a_i \in K_i$. Define

$$f_i(y) = (a_i, x) + \log \int_{K_i - a_i} \exp(x, y) \, d\mu_i(x),$$

$$f_i^{(m)}(y) = (a_i^{(m)}, x) + \log \int_{K_i^{(m)} - a_i^{(m)}} \exp(x, y) d\mu_i^{(m)}(x),$$

where μ_i, $\mu_i^{(m)}$ are measures as in the discussion after Lemma 5 with K_i, $K_i^{(m)}$ instead of K. By (3), $M_{\bar{K}^{(m)}} F(\lambda_1, \ldots, \lambda_s) =$

$$\sum_{j_1, \ldots, j_n} \lambda_{j_1} \ldots \lambda_{j_n} \int_{\mathbb{R}^n} F\left(\sum \lambda_i \nabla f_i^{(m)}(y)\right) D(H_{j_1}^{(m)}(y), \ldots, H_{j_n}^{(m)}(y)) \, dy$$

and $M_{\bar{K}} F(\lambda_1, \ldots, \lambda_s) =$

$$\sum_{j_1, \ldots, j_n} \lambda_{j_1} \ldots \lambda_{j_n} \int_{\mathbb{R}^n} F\left(\sum \lambda_i \nabla f_i(y)\right) D(H_{j_1}(y), \ldots, H_{j_n}(y)) \, dy.$$

Since all K_i, $K_i^{(m)}$ are uniformly bounded, there exists a large Euclidean ball U containing all these sets. As in the proof of Theorem 1, if $\sum |\lambda_i| \le R$, then

$$|M_{\bar{K}^{(m)}} F(\lambda_1, \ldots, \lambda_s)| \le \sum_{j_1, \ldots, j_n} R^n \max_{x \in R \cdot U} |F(x)| \cdot V(K_{j_1}^{(m)}, \ldots, K_{j_n}^{(m)})$$

$$\le \left(\max_{x \in R \cdot U} |F(x)|\right) \cdot \left(\sum_{j_1, \ldots, j_n} R^n \operatorname{vol}(U)\right)$$

$$\le K(R) \cdot \max_{x \in R \cdot U} |F(x)|,$$

where $K(R)$ is some constant depending on R.

The same estimate holds for $M_{\bar{K}} F$. Hence, for every $\varepsilon > 0$, one can choose a polynomial P_ε approximating F on the set $R \cdot U$, such that for all i, m, λ_i with $\sum |\lambda_i| \le R$ we have

$$|M_{\bar{K}^{(m)}}(F - P_\varepsilon)(\lambda_1, \ldots, \lambda_s)| < \varepsilon, \tag{5}$$

$$|M_{\bar{K}}(F - P_\varepsilon)(\lambda_1, \ldots, \lambda_s)| < \varepsilon. \tag{6}$$

But by Corollary 2, the degrees of $M_{\bar{K}^{(m)}} P_\varepsilon$ and $M_{\bar{K}} P_\varepsilon$ are independent of m. Obviously, by the definition (*) in the Introduction, $M_{\bar{K}^{(m)}} P_\varepsilon$ converges to $M_{\bar{K}} P_\varepsilon$ uniformly on compact sets in the non-negative orthant $\mathbb{R}_+^s$. Hence because of the boundedness of their degrees, $M_{\bar{K}^{(m)}} P_\varepsilon \longrightarrow M_{\bar{K}} P_\varepsilon$ in $\mathbb{R}^s$ (respectively, $\mathbb{C}^s$). This and (5) and (6) imply that, for large m,

$$|M_{\bar{K}^{(m)}} F(\lambda_1, \ldots, \lambda_s) - M_{\bar{K}} F(\lambda_1, \ldots, \lambda_s)| < 3\varepsilon$$

whenever $\sum |\lambda_i| \le R$.

A similar argument can be applied to prove uniform convergence of partial derivatives of $M_{\bar{K}^{(m)}} F$ on compact sets. $\square$

4. Polynomial Valuations

We are now going to present another proof of the Pukhlikov–Khovanskii Theorem. They introduced in [P-Kh] the notion of the polynomial valuation, generalizing the classical translation invariant and translation covariant valuations.

Let Λ be an additive subgroup of $\mathbb{R}^n$. Denote by $\mathcal{P}(\Lambda)$ the set of all polytopes with vertices in Λ. We will assume that span $\Lambda = \mathbb{R}^n$.

DEFINITION. (a) A function $\phi : \mathcal{P}(\Lambda) \longrightarrow \mathbb{R}$ is called a valuation, if for all $P_1, P_2 \in \mathcal{P}(\Lambda)$, such that $P_1 \cup P_2$ and $P_1 \cap P_2$ belong to $\mathcal{P}(\Lambda)$ we have

$$\phi(P_1 \cup P_2) + \phi(P_1 \cap P_2) = \phi(P_1) + \phi(P_2). \tag{7}$$

(b) The valuation ϕ is called fully additive if, for every finite family of polytopes $P_1, \ldots, P_k$ in $\mathcal{P}(\Lambda)$ such that the intersection $\bigcap_{i \in \sigma} P_i$ over every nonempty subset $\sigma \subset \{1 \ldots k\}$ and their union $\bigcup_{i=1}^k P_i$ lie in $\mathcal{P}(\Lambda)$, the following equation holds:

$$\phi\left(\bigcup_{i=1}^k P_i\right) = \sum_{\sigma \subset \{1,\ldots,k\}, \sigma \neq \varnothing} (-1)^{|\sigma|+1} \phi\left(\bigcap_{i \in \sigma} P_i\right), \tag{8}$$

where $|\sigma|$ is the cardinality of σ.

Obviously, for $k = 2$, (8) is equivalent to (7). We will consider only fully additive valuations; however it is true that, if $\Lambda = \mathbb{R}^n$, then every valuation on $\mathcal{P}(\Lambda)$ is fully additive (see [V], [P-S]). But it is not known to the author whether this holds in the general case. In the definitions (a) and (b) one can replace $\mathcal{P}(\Lambda)$ by the set of all convex compact sets $\mathcal{K}$. If ϕ is continuous with respect to the Hausdorff metric on $\mathcal{K}$, then (a) implies (b) [Gr].

(c) The valuation $\phi : \mathcal{P}(\Lambda) \longrightarrow \mathbb{R}$ is called polynomial of degree at most d, if for every fixed $K \in \mathcal{P}(\Lambda)$, $\phi(K + x)$ is a polynomial of degree at most d with respect to $x \in \Lambda$.

EXAMPLES. 1. Let μ be any signed locally finite measure on $\mathbb{R}^n$. Then $\phi(K) := \mu(K)$ is a fully additive valuation.

2. The mixed volume

$$\phi(K) = V(K[j], A_1, \ldots, A_{n-j}),$$

where $K[j]$ means that K occurs j times, and A_l are fixed convex compact sets, is known to be a fully additive translation invariant continuous valuation.

3. Let $\Lambda = \mathbb{Z}^n \subset \mathbb{R}^n$ be an integer lattice, and let f be a polynomial of degree d. Then for $K \in \mathcal{P}(\Lambda)$,

$$\phi(K) := \sum_{x \in K \cap \mathbb{Z}^n} f(x)$$

is a fully additive polynomial valuation of degree d.

4. Let $\Lambda = \mathbb{Z}^n$, let Ω be a subset of $\mathbb{R}^n$, which is invariant with respect to translations to vectors in $\mathbb{Z}^n$, and let K and f be as in example 3. Then $\phi(K) := \int_{K \cap \Omega} f(x)\, dx$ is also a fully additive polynomial valuation of degree d.

For more information about valuations, especially those which are translation invariant and translation covariant, see the surveys [Mc-Sch] and [Mc2].

THEOREM 6. [P-Kh] *Let $\phi : \mathcal{P}(\Lambda) \longrightarrow \mathbb{R}$ be a fully additive polynomial valuation of degree d. Fix $K_1, \ldots, K_s \in \mathcal{P}(\Lambda)$. Then $\phi(\sum_i \lambda_i K_i)$ is a polynomial in $\lambda_i \in \mathbb{Z}_+$ of degree at most $d+n$. Moreover, if $\mathbb{Q} \cdot \mathcal{P}(\Lambda) = \mathcal{P}(\Lambda)$, then it is a polynomial in $\lambda_i \in \mathbb{Q}_+$.*

REMARK. For translation invariant valuations this theorem was proved in [Mc1], and our proof uses some constructions of that work.

LEMMA 7. *(Well known; see, e.g., [GKZ, p. 215].) Let $P \subset \mathbb{R}^n$ be a polytope. Then there exists a family of k-simplices $\{S_\alpha\}_{\alpha \in I}$, $0 \leq k \leq n$, such that*
(i) $P = \bigcup_{\alpha \in I} S_\alpha$;
(ii) each vertex of each S_α is a vertex of P;
(iii) every two S_β and S_γ intersect in a common face;
(iv) for all β and γ, $S_\beta \cap S_\gamma \in \{S_\alpha\}_{\alpha \in I}$.

LEMMA 8. *Let $K_1, \ldots, K_s$ be polytopes in $\mathbb{R}^n$. Then for all $\lambda_i \geq 0$, $1 \leq i \leq s$, the set $K(\bar{\lambda}) := \sum_i \lambda_i K_i$ has a decomposition*
$$K(\bar{\lambda}) = \bigcup_{\alpha \in I} S_\alpha(\bar{\lambda}),$$
where $S_\alpha(\bar{\lambda})$ are polytopes (not necessarily simplices) such that
(i) they satisfy (i) – (iv) in Lemma 7;
(ii) if for some $\bar{\lambda}^0$, $\lambda_i^0 > 0$, and $\alpha, \beta, \gamma \in I$, $S_\alpha(\bar{\lambda}^0) \cap S_\beta(\bar{\lambda}^0) = S_\gamma(\bar{\lambda}^0)$, then for all $\bar{\lambda} = (\lambda_i)$, $\lambda_i \geq 0$ we have $S_\alpha(\bar{\lambda}) \cap S_\beta(\bar{\lambda}) = S_\gamma(\bar{\lambda})$;
(iii) each $S_\alpha(\bar{\lambda})$ has the form
$$S_\alpha(\bar{\lambda}) = \sum_i \lambda_i S_{i,\alpha}$$
where $S_{i,\alpha}$ are simplices with vertices in K_i, independent of $\bar{\lambda}$ and $\dim S_\alpha(\bar{\lambda}) = \sum_i \dim(\lambda_i S_{i,\alpha})$ (note that $\dim(\lambda_i S_{i,\alpha}) = \dim S_{i,\alpha}$ for $\lambda_i > 0$).

PROOF. Because of the homogeneity it is sufficient to prove the lemma only for $\lambda_i \geq 0$, $\sum \lambda_i = 1$. Consider in $\mathbb{R}^s \oplus \mathbb{R}^n$ a convex polytope
$$P := \left\{ (\mu_1, \ldots, \mu_s; x) \mid \mu_i \geq 0,\ \sum_{i=1}^s \mu_i = 1,\ x \in \sum_i \mu_i K_i \right\}$$
Now apply Lemma 7 to $P = \bigcup_\alpha S_\alpha$. Set
$$S_\alpha(\bar{\lambda}) := S_\alpha \cap \{(\mu_1, \ldots, \mu_s; x) \mid \mu_i = \lambda_i \text{ for all } i\}.$$
One can easily check that $S_\alpha(\bar{\lambda})$ satisfy all the properties in Lemma 8. □

LEMMA 9. *Let $P \in \mathcal{P}(\Lambda)$. Then $\phi(N \cdot P)$ is a polynomial in $N \in \mathbb{Z}_+$ of degree at most $d + n$.*

PROOF. By Lemma 7, $P = \bigcup_{\alpha \in I} S_\alpha$, where the S_α are simplices. Hence

$$\phi(N \cdot P) = \sum_{\sigma \subset I, \sigma \neq \emptyset} (-1)^{|\sigma|-1} \phi \left(N \cdot \left(\bigcap_{\alpha \in \sigma} S_\alpha \right) \right)$$

But for fixed σ, there exists $\gamma \in I$ such that $\bigcap_{\alpha \in \sigma} S_\alpha = S_\gamma$. So we have to show that for every simplex $\Delta \in \mathcal{P}(\Lambda)$, $\phi(N \cdot \Delta)$ is a polynomial of degree at most $d + n$.

Fix Δ and write $k = \dim \Delta$. The proof will be by induction in k. If $k = 0$, then $\Delta = \{v\}$ is a point and $\phi(N \cdot \{v\}) = \phi(\{0\} + Nv)$ is a polynomial of degree at most d by the definition of the polynomial valuation.

Let $k > 0$. For simplicity of notation we will assume that $k = n$. In an appropriate coordinate system Δ has the form $\Delta = a + \tilde{\Delta}$, where $a \in \mathbb{R}^n$, $\tilde{\Delta} = \{(x_1, \ldots, x_n) \mid 0 \leq x_1 \leq \ldots \leq x_n \leq 1\}$. Thus

$$N \cdot \tilde{\Delta} = \{(x_1, \ldots, x_n) \mid 0 \leq x_1 \leq \ldots \leq x_n \leq N\}.$$

$N \cdot \tilde{\Delta}$ can be represented as a disjoint union

$$N \cdot \tilde{\Delta} = \bigcup_{z \in \mathbb{Z}^n \cap ((N-1) \cdot \tilde{\Delta})} \left((z + \tilde{Q}) \cap (N \cdot \tilde{\Delta}) \right) \bigcup (N \cdot \Delta'), \quad (9)$$

where $\tilde{Q} := \{(x_1, \ldots, x_n) \mid 0 \leq x_i < 1 \text{ for all } i\}$ and

$$\Delta' = \{(x_1, \ldots, x_n) \mid 0 \leq x_1 \leq \ldots \leq x_{n-1} \leq x_n = 1\}.$$

Of course, $(z + \tilde{Q}) \cap (N \cdot \tilde{\Delta})$ is not a compact polytope, so ϕ is not defined on it. But we can define ϕ on this set in the following natural way. First, for $\tau \subset \{1, \ldots, n\}$, denote $F_\tau :=$

$$\{(x_1, \ldots, x_n) \mid 0 \leq x_i \leq 1 \text{ for all } i \in \{1, \ldots, n\}, \text{ and } x_j = 1 \text{ for all } j \in \tau\}$$

Clearly, F_τ is an $(n - |\tau|)$-dimensional face of the unit cube $[0, 1]^n$. Now define

$$\phi \left((z + \tilde{Q}) \cap (N \cdot \tilde{\Delta}) \right) := \sum_{\tau \subset \{1, \ldots, n\}} (-1)^{|\tau|} \phi \left((z + F_\tau) \cap (N \cdot \tilde{\Delta}) \right).$$

Since in (9) we have a disjoint union,

$$\phi(N \cdot \Delta) = \phi(N \cdot a + N \cdot \Delta') + \sum_{z \in \mathbb{Z}^n \cap ((N-1) \cdot \tilde{\Delta})} \phi \left(N \cdot a + (z + \tilde{Q}) \cap N \cdot \tilde{\Delta} \right). \quad (10)$$

Every $z \in \mathbb{Z}^n \cap ((N - 1) \cdot \tilde{\Delta})$ has the form $z = (z_i)_{i=1}^n$, where

$$z_1 = \cdots = z_{j_1} < z_{j_1+1} = \cdots = z_{j_2} < \cdots < z_{j_{l-1}+1} = \cdots = z_{j_l} \leq N - 1, \quad (11)$$

and $j_l = n$.

Set for $1 \leq i \leq j \leq n$, $T_{i,j} :=$

$\{(x_1, \ldots, x_n) \mid 0 \leq x_i \leq x_{i+1} \leq \ldots \leq x_j \leq 1 \text{ and } x_l = 0 \text{ for } l < i \text{ or } l > j\}$.

For a sequence $0 < j_1 < \cdots < j_{l-1} < n$, denote (as in [Mc1]) $T_{j_1\ldots j_{l-1}} := T_{0j_1} + \cdots + T_{l-1,n}$. Now let $\tilde{T}_{j_1\ldots j_{l-1}} := T_{j_1\ldots j_{l-1}} \cap \tilde{Q}$. So if z belongs to $\mathbb{Z}^n \cap (N-1)\cdot\tilde{\Delta}$ and satisfies (11), then obviously $(z + \tilde{Q}) \cap N\cdot\tilde{\Delta} = z + \tilde{T}_{j_1\ldots j_{l-1}}$. Define $S_{j_1\ldots j_{l-1}}(N) = \{z \in \mathbb{Z}^n \mid z \text{ satisfies (11)}\}$. Then (10) can be rewritten:

$$\phi(N\cdot\Delta) = \phi(N\cdot a + N\cdot\Delta')$$
$$+ \sum_{0<j_1<\cdots<j_{j-1}<n}\left(\sum_{z \in S_{j_1\ldots j_{l-1}}(N)} \phi(N\cdot a + z + \tilde{T}_{j_1\ldots j_{l-1}})\right). \quad (12)$$

By the inductive hypothesis, $\phi(N\cdot a + N\cdot\Delta')$ is a polynomial in $N \in \mathbb{Z}_+$ of degree at most $d+n$. Now fix $0 < j_1 < \cdots < j_{l-1} < n$. Then $\phi(x + \tilde{T}_{j_1\ldots j_{l-1}})$ is a polynomial in x of degree at most d, let us denote it $q(x)$. It is sufficient to show that $\sum_{z \in S_{j_1\ldots j_{l-1}}(N)} q(N\cdot a + z)$ is a polynomial in $N \in \mathbb{Z}_+$ of degree at most $d+n$.

We can write $q(N\cdot a + z) = \sum_{t=0}^{d} N^t q_t(z)$, where $q_t(z)$ is a polynomial of degree at most $d - t$. Recall that for any $z \in S_{j_1\ldots j_{l-1}}(N)$ and $m = 1, \ldots l-1$, $z_{j_{m-1}+1} = \ldots = z_{j_m}$. So set $w_m := z_{j_m}$. We have $0 \leq w_1 < w_2 < \ldots < w_l \leq N-1$. Actually, $q_t(z)$ is a polynomial in the vector $w = (w_1, \ldots, w_l) \in \mathbb{R}^l$. We will show that

$$f(N) := \sum_{0 \leq w_1 < w_2 < \cdots < w_l \leq N-1} q_t(w)$$

is a polynomial in $N \in \mathbb{Z}_+$ of degree at most $\deg q_t + l$ (note that, if $N \leq l-1$ the sum is extended over an empty set and for such an N, we just define $f(N) := 0$). This and (12) will imply that $\phi(N\cdot\Delta)$ is a polynomial of degree at most $d+n$.

In order to prove that $f(N)$ is a polynomial of degree g, it is sufficient to show that $f(N+1) - f(N)$ is a polynomial of degree $g - 1$.

Let us apply induction in l. If $l = 1$,

$$f(N+1) - f(N) = q_t(N) \text{ for } N \geq 0, \quad (13)$$

and the lemma follows.

Assume that $l > 0$. We have

$$f(N+1) - f(N) = \sum_{0 \leq w_1 < \cdots < w_{l-1} < w_l = N} q_t(w).$$

We may assume q_t to be a monomial $q_t(w) = w_1^{\alpha_1} \ldots w_l^{\alpha_l}$, $\alpha_j \geq 0$. Hence

$$f(N+1) - f(N) = N^{\alpha_l} \cdot \sum_{0 \leq w_1 < \cdots < w_{l-1} \leq N-1} w_1^{\alpha_1} \ldots w_{l-1}^{\alpha_{l-1}}.$$

By the inductive hypothesis, the last sum is a polynomial of degree at most $l - 1 + \sum_1^{n-1} \alpha_j$. Hence $f(N)$ is a polynomial of degree at most $l + \sum_1^n \alpha_j$. □

PROOF OF THEOREM 6. Using the same notation as previously, we have to show that $\phi(K(\bar{\lambda}))$ is a polynomial in $\lambda_i \in \mathbb{Z}_+$ of degree at most $d+n$. By Lemma 8 and the full additivity of ϕ,

$$\phi(K(\bar{\lambda})) = \phi\left(\bigcup_{\alpha \in I} S_\alpha(\bar{\lambda})\right) = \sum_{\sigma \subset I, \sigma \neq \emptyset} (-1)^{|\sigma|+1} \phi\left(\bigcap_{\beta \in \sigma} S_\beta(\bar{\lambda})\right).$$

Fix some $\sigma \subset I, \sigma \neq \emptyset$. By Lemma 8 (ii) there exists $\gamma \in I$, such that $\bigcap_{\beta \in \sigma} S_\beta(\bar{\lambda}) = S_\gamma(\bar{\lambda})$ for every vector $\bar{\lambda}$ with nonnegative coordinates.

So it is sufficient to show that for any γ, $\phi(S_\gamma(\bar{\lambda}))$ is a polynomial. But $S_\gamma(\bar{\lambda}) = \sum_{i=1}^s \lambda_i \cdot S_{\gamma,i}$ as in Lemma 8 (iii).

Suppose that for $1 \leq i \leq p$, $\dim S_{\gamma,i} > 0$ and for $i > p$, $\dim S_{\gamma,i} = 0$, i.e. $S_{\gamma,i} = \{v_i\}$ is a point for $i > p$.

Define $\Delta_i := S_{\gamma,i} - v_{\gamma,i}$, where $v_{\gamma,i}$ is some vertex of $S_{\gamma,i}$. So $S_\gamma(\bar{\lambda}) = \sum_{i=1}^p \lambda_i \Delta_i + \sum_{i=1}^p \lambda_i v_{\gamma,i} + \sum_{i>p} \lambda_i u_i$. By Lemma 8 (iii),

$$\dim S_\gamma(\bar{\lambda}) = \sum_i \dim(\lambda_i \Delta_i).$$

This implies that $\sum \lambda_i \Delta_i$ is, in fact, a direct sum of the $\lambda_i \Delta_i$. So we have to check that

$$\phi\left(\bigoplus_{i=1}^s (\lambda_i \Delta_i) + \sum_{j=1}^l \mu_j u_j\right)$$

is a polynomial in $\lambda_i, \mu_j \in \mathbb{Z}_+$ of degree at most $d+n$, where the u_j are fixed integer vectors.

Let $L_1 = \bigoplus_{j=1}^{s-1} \operatorname{span} \Delta_j$, $L_2 = \operatorname{span} \Delta_s$. For any polytopes $K_1, K_2, K_i \subset L_i, i = 1, 2$, consider the polynomial $\phi(K_1 \oplus K_2 + x)$, which we will denote by $W_{K_1 \oplus K_2}(x)$. Obviously, all the previous definitions of the valuation and the polynomial valuation can be formulated not only for the real valued functions on $\mathcal{P}(\Lambda)$, but also for the vector valued functions with values in a linear space (and even in an abelian semigroup). The proofs of all the previous lemmas of Section 4 will work without any change.

Then obviously $W_{K_1 \oplus K_2}(x)$ is a fully additive polynomial valuation with respect to each argument K_1 and K_2, with values in the linear space of polynomials in x (here we use the fact that the sum of K_1 and K_2 is direct). Hence by Lemma 9 (applied in the vector valued case), $W_{K_1 \oplus N \cdot K_2}(x)$ is a polynomial in N (at the moment we are not interested in its degree). In particular, this implies that $\phi(K_1 \oplus N \cdot K_2 + x)$ is a polynomial in N and x, where $N \in \mathbb{Z}_+, x \in \Lambda$. Then obviously if we decompose $W_{K_1 \oplus N \cdot K_2}$ with respect to the powers in N, then its coefficients will be polynomial valued fully additive polynomial valuations with respect to K_1 (now K_2 is fixed). Applying an inductive argument in s, we see that

$$\phi\left(\bigoplus_{i=1}^s (\lambda_i \Delta_i) + \sum_{j=1}^l \mu_j u_j\right) \tag{14}$$

is a polynomial in $\lambda_i, \mu_j \in \mathbb{Z}_+$.

Let us estimate its degree. If λ_i and μ_j are fixed,

$$\phi\left(\bigoplus_{i=1}^{s}(t\cdot\lambda_i\Delta_i) + \sum_{j=1}^{l} t\cdot\mu_j u_j\right)$$

is a polynomial in $t \in \mathbb{Z}_+$ of degree at most $d+n$ by Lemma 9. Hence the degree of (14) cannot be bigger than $d+n$.

Now consider the case $\mathbb{Q}\cdot\mathcal{P}(\Lambda) = \mathcal{P}(\Lambda)$. Let $K_1,\ldots,K_s \in \mathcal{P}(\Lambda)$. For any natural number m, $\phi(\sum_{i=1}^{s}\lambda_i(\frac{1}{m}K_i))$ is a polynomial in $\lambda_i \in \mathbb{Z}_+$, hence $\phi(\sum_{i=1}^{s}\lambda_i K_i)$ is a polynomial in $\lambda_i \in \frac{1}{m}\cdot\mathbb{Z}_+$ for any $m \in \mathbb{N}$. Consequently, it is a polynomial in $\lambda_i \in \mathbb{Q}_+$. □

REMARKS. 1. The valuation ϕ can be defined not only on polytopes, but on the family of all convex compact sets. If ϕ is continuous with respect to the Hausdorff metric, then it is called a continuous valuation (this implies its full additivity, see [Gr]). If a continuous valuation is polynomial of degree at most d, then for all convex compact sets $K_1, \ldots, K_s$, the function $\phi(\sum_i \lambda_i K_i)$ is a polynomial in $\lambda_i \in \mathbb{R}_+$ of degree at most $d+n$. This can be deduced immediately from Theorem 6 using approximation by polytopes.

2. We would like to recall here some results in the same spirit due to Khovanskii [Kh1, Kh2].

Let A and B be finite subsets of an abelian semigroup G. Denote by $N * A$ the sum of N copies of the set A. Let $\chi : G \longrightarrow \mathbb{C}$ be a multiplicative character, i.e. $\chi(x+y) = \chi(x)\cdot\chi(y)$. Let $f(N)$ denote the sum of values of the character χ over all elements of the set $B + N * A$.

THEOREM 10. [Kh2] *For sufficiently large N, the function $f(N)$ is a quasi-polynomial in N, i.e. for large N, the function $f(N) = \sum q_i^N P_i(N)$, where q_i are values of the character χ on the set A, and P_i are polynomials of degree strictly less than the number of points in A, in which the value of χ is equal to q_i.*

Now let A and B be finite subsets of an abelian group G. Denote by $G(A)$ the subgroup of the group G consisting of the elements of the form $\sum n_i a_i$, where $a_i \in G, n_i \in \mathbb{Z}$ and $\sum n_i = 0$. Now take $\chi \equiv 1$, then $f(N)$ is equal to the cardinality of the set $B + N * A$.

THEOREM 11. [Kh1] *Let G be the lattice $\mathbb{Z}^n \subset \mathbb{R}^n$ and assume that $G(A) = \mathbb{Z}^n$. Then for large N, the function $f(N)$ is a polynomial of degree at most n and the coefficient of N^n is equal to the volume of the convex hull of A.*

The methods of [Kh1] and [Kh2] in fact imply the following more general versions of these theorems:

THEOREM 10'. *Let G and χ be as in Theorem 10, and let $B, A_1, \ldots, A_s$ be finite subsets of G. Let $f(N_1, \ldots, N_s)$ be the sum of values of the character χ over all the elements of the set $B + N_1 * A_1 + \cdots + N_s * A_s$. Then if all the N_i, $1 \leq i \leq s$, are sufficiently large, $f(N_1, \ldots, N_s)$ is a quasi-polynomial.*

THEOREM 11'. *Let B, A_i, $1 \leq i \leq s$ be finite subsets of the lattice $\mathbb{Z}^n \subset \mathbb{R}^n$ and $G(\bigcup A_i) = \mathbb{Z}^n$. Then, if all the N_i, $1 \leq i \leq s$ are sufficiently large, the cardinality of $\sum N_i * A_i$ is a polynomial of degree at most n, whose homogeneous component of degree n is equal to the polynomial $\text{vol}(N_1 \cdot \text{conv} A_1 + \cdots + N_s \cdot \text{conv} A_s)$.*

As we were informed by Prof. Khovanskii, these facts were known to him (unpublished).

Acknowledgements

I am grateful to Prof. V. D. Milman for his guidance in this work and to the referee for very important remarks. I would also like to thank S. Dar, through whose talk on GAFA Seminar I became familiar with Gromov's work used in the paper.

References

[Al] Alexandrov, A. D.; Die gemischte Diskriminanten und die gemischte Volumina, Math. Sbornik **3** (1938), 227–251.

[B-Z] Burago, Yu. D.; Zalgaller, V. A.; Geometric inequalities. Translated from the Russian by A. B. Sosinskiĭ. Grundlehren der Mathematischen Wissenschaften, 285. Springer Series in Soviet Mathematics. Springer, Berlin and New York, 1988.

[GKZ] Gelfand, I. M.; Kapranov, M. M.; Zelevinsky, A. V.; Discriminants, resultants and multidimensional determinants. Birkhäuser, Boston, 1994.

[Gr] Groemer, H. On the extension of additive functionals on classes of convex sets. Pacific J. Math. **75**:2 (1978), 397–410.

[G] Gromov, M; Convex sets and Kähler manifolds. Advances in differential geometry and topology, 1–38, World Sci. Publishing, Teaneck, NJ, 1990.

[Kh1] Khovanskii, A. G.; The Newton polytope, the Hilbert polynomial and sums of finite sets. Funct. Anal. Appl., **26**:4 (1992), 276–281.

[Kh2] Khovanskii, A. G.; Sums of finite sets, orbits of commutative semigroups and Hilbert functions. Funct. Anal. Appl., **29**:2 (1995), 102–112.

[Mc1] McMullen, P.; Valuations and Euler-type relations on certain classes of convex polytopes. Proc. London Math. Soc. (3) **35**:1 (1977), 113–135.

[Mc2] McMullen, P.; Valuations and dissections; Handbook of convex geometry, edited by P. M. Gruber and J. M. Wills, 1993.

[Mc-Sch] McMullen, P.; Schneider, R.; Valuations on convex bodies. In: Convexity and its Applications, 170–247, Birkhäuser, Basel and Boston, 1983.

[P-S] Perles, M. A.; Sallee, G. T.; Cell complexes, valuations, and the Euler relation. Canad. J. Math. **22** (1970), 235–241.

[P-Kh] Pukhlikov, A. V.; Khovanskii, A. G.; Finitely additive measures of virtual polyhedra (in Russian). Algebra i Analiz 4:2 (1992), 161–185; translation in St. Petersburg Math. J. 4:2 (1993), 337–356.

[R] Rockafellar R. T.; Convex Analysis. Princeton University Press, 1970.

[Sch] Schneider, R.; Convex Bodies: the Brunn–Minkowski Theory. Encyclopedia of Mathematics and its Applications **44**. Cambridge University Press, Cambridge, 1993.

[V] Volland, W.; Ein Fortsetzungssatz für additive Eipolyederfunktionale im euklidischen Raum. (German) Arch. Math. **8** (1957), 144–149.

SEMYON ALESKER
DEPARTMENT OF MATHEMATICS
TEL-AVIV UNIVERSITY
RAMAT-AVIV
ISRAEL
semyon@math.tau.ac.il

Localization Technique on the Sphere and the Gromov–Milman Theorem on the Concentration Phenomenon on Uniformly Convex Sphere

SEMYON ALESKER

ABSTRACT. We give a simpler proof of the Gromov–Milman theorem on concentration phenomenon on uniformly convex sphere. We also outline Rohlin's theory of measurable partitions used in the proof.

The purpose of this note is to present a localization technique for the sphere S^n on an example of the Gromov–Milman theorem [Gr-M] about the concentration phenomenon on uniformly convex spheres. This result was obtained in [Gr-M] in a some more general setting. Our approach follows the same general reasoning, but is simpler and more direct than the original approach. We also outline Rohlin's theory of measurable partitions, which is used in the proof. Note that the terminology of "localization" was introduced for $\mathbb{R}^n$ by L. Lovász and M. Simonovits [L-S1, L-S2]. [Gr-M] did not use such terminology and also did not put the scheme of localization explicitly.

NOTE. K. Ball has informed us recently that he, jointly with R. Villa, found an extremely short proof of the Gromov–Milman theorem for uniformly convex sphere as an application of the Prekopa–Leindler inequality (see, e.g., [P]).

1. Related Definitions and Formulation of the Gromov–Milman Theorem

DEFINITION 1.1. Let us say that a finite dimensional normed space $X = (\mathbb{R}^{n+1}, \|\cdot\|)$ has modulus of convexity at least $\delta(\varepsilon) > 0$ for $\varepsilon > 0$, if for all vectors $x, y \in X$ such that $\|x\| = \|y\| = 1$ and $\|x - y\| \geq \varepsilon$ we have $\|\frac{x+y}{2}\| \leq 1 - \delta(\varepsilon)$.

We may assume $\delta(\varepsilon)$ to be a monotone increasing function of positive ε. Denote by $K(X) := \{x \in X : \|x\| \leq 1\}$ the unit ball of X and by $S(X) := \{x \in X : \|x\| = 1\}$ the unit sphere of X.

Work partially supported by a BSF Grant.

For any subset $A \subset S(X)$, let us denote $\hat{A} := \bigcup_{0 \leq t \leq 1} t \cdot A$. Now define a probability measure $\hat{\mu}$ on $S(X)$ induced by the standard Lebesgue measure vol_{n+1} on $\mathbb{R}^{n+1}$: for any Borel subset $A \subset S(X)$, let

$$\hat{\mu}(A) := \mathrm{vol}_{n+1}(\hat{A})/\mathrm{vol}_{n+1} K(X).$$

We will prove the following theorem, due to Gromov and Milman [Gr-M].

THEOREM 1.1. *Let $\delta(\varepsilon)$ be the modulus of convexity of the normed $(n+1)$-dimensional space X and let $\hat{\mu}$ be the probability measure on $S(X)$ as above. Then, for every Borel set $A \subset S(X)$ such that $\hat{\mu}(A) \geq \frac{1}{2}$, and every $\varepsilon > 0$,*

$$\hat{\mu}(A_\varepsilon) \geq 1 - \exp(-a(\varepsilon)n),$$

where $A_\varepsilon := \{x \in S(X) : \mathrm{dist}(x, A) \leq \varepsilon\}$, $\mathrm{dist}(x, A) := \inf_{y \in A} \|x - y\|$, $a(\varepsilon) := \delta((\varepsilon/8) - \theta_n)$, where θ_n is such that $\delta(\theta_n) = 1 - (1/2)^{1/(n-1)} \approx \frac{\log 2}{n-1}$.

2. Rohlin's Theory

Following [Gr-M], we will use some results of Rohlin's theory [R]. Let (M, Ω_ν, ν) be a complete measure space, i.e. M is a set, Ω_ν is a σ-algebra of subsets of M, and ν is a complete probability measure on Ω_ν.

Let ζ be some partition of M into pairwise disjoint subsets, whose union is equal to M.

DEFINITION 2.1. *A partition ζ of M is called measurable, if there exists a countable family $\Sigma = \{S_\alpha\}_{\alpha=1}^\infty$ of measurable subsets of M such that each element $C \in \zeta$ has the form $C = \bigcap_{\alpha=1}^\infty R_\alpha$, where for all α either $R_\alpha = S_\alpha$ or $R_\alpha = \bar{S}_\alpha$, where $\bar{S}_\alpha$ denotes the complement of S_α.*

Obviously, each element of a measurable partition is measurable.

Denote by H_ζ the canonical homomorphism from M onto the factor set M/ζ. Then M/ζ turns out to be a complete measure space, if we introduce a measure ν_ζ by setting a subset $X \subset M/\zeta$ to be measurable in M/ζ iff $H_\zeta^{-1}(X)$ is measurable in M and $\nu_\zeta(X) := \nu(H_\zeta^{-1}(X))$.

We will need the following theorem due to Rohlin:

THEOREM 2.2 [R] *Let M be a metric separable complete space, ν be a complete Borel probability measure on M and ζ be a measurable partition of M generated by a countable family $\Sigma = \{S_\alpha\}_{\alpha=1}^\infty$ (in the sense of Definition 2.1). Then there exists a canonical family of complete Borel probability measures $\{\nu_C\}_{C \in M/\zeta}$ on M satisfying these conditions:*

(1) *For ν_ζ-a.e. element $C \in M/\zeta$, ν_C is concentrated on $C \subset M$, i.e. $\nu_C(C) = 1$ (here we denote both the element C of M/ζ and its preimage $H_\zeta^{-1}(C)$ in M by the same letter C).*
(2) *For every ν-measurable subset $A \subset M$, $\nu_C(A)$ is a ν_ζ-measurable function of $C \in M/\zeta$ and*

(3) $\nu(A) = \int_{M/\zeta} \nu_C(A \cap C) \, d\nu_\zeta(C)$.

(4) *The canonical family* $\{\nu_C\}$ *is unique, i.e. if* $\{\nu'_C\}$ *satisfies* (1)–(3), *then* $\nu_C = \nu'_C$ *for* ν_ζ-*a.e.* C.

(5) *Furthermore, the family* Σ', *which is an image of* Σ *under* H_ζ, *generates the* σ-*algebra of* ν_ζ-*measurable subsets of* M/ζ.

COROLLARY 2.3. *Let* M, ν, ζ *be as in Theorem* 2.2. *Let* $f \in L_1(M, \nu)$ *be an integrable function.*

Then the integral $\int_M f \, d\nu_C = \int_C f \, d\nu_C$ *is a* ν_ζ- *integrable function of* $C \in M/\zeta$ *and*

$$\int_M f \, d\nu = \int_{M/\zeta} \left(\int_C f \, d\nu_C \right) d\nu_\zeta(C).$$

PROOF (standard). This corollary is obvious for the step functions. In general, we may assume $f \geq 0$.

For $k, j \in \mathbb{N} \cup \{0\}$, define

$$A_{kj} := \left\{ x \in M : \frac{j}{2^k} \leq f(x) < \min\left(\frac{j+1}{2^k}, k \right) \right\}$$

(obviously, $A_{kj} = \varnothing$ for $j \geq k2^k$) and

$$f_k := \sum_{j=0}^{\infty} \frac{j}{2^k} \chi_{A_{kj}},$$

where $\chi_{A_{kj}}$ are characteristic functions of A_{kj}. Clearly, f_k are step functions, $0 \leq f_k(x) \leq f(x)$ for every $x \in M$, the sequence $\{f_k(x)\}_{k \in \mathbb{N}}$ is nondecreasing, and $f_k \longrightarrow f$ everywhere on M and in $L_1(M, \nu)$. For f_k we have:

$$\int_{M/\zeta} \left(\int_C f_k \, d\nu_C \right) d\nu_\zeta(C) = \int_M f_k \, d\nu \leq \int_M f \, d\nu.$$

Set $\phi_k(C) = \int_C f_k \, d\nu_C$. It is well defined for ν_ζ-a.e. $C \in M/\zeta$. Clearly, $\{\phi_k(C)\}$ is nondecreasing and $\sup_k \int_{M/\zeta} \phi_k(C) \, d\nu_\zeta(C) \leq const < \infty$.

Hence by B. Levy's theorem $\{\phi_k\}$ converges ν_ζ-a.e. and in $L_1(M/\zeta, \nu_\zeta)$ to some function $\phi(C) \in L_1(M/\zeta, \nu_\zeta)$, and $\phi_k(C) \leq \phi(C)$. Then for ν_ζ-a.e. C,

$$\int_C f \, d\nu_C = \lim_{k \to \infty} \int_C f_k \, d\nu_C = \phi(C)$$

again, by B. Levy's theorem applied to the measure ν_C.

Thus we obtain

$$\int_{M/\zeta} \left(\int_C f \, d\nu_C \right) d\nu_\zeta = \int_{M/\zeta} \phi(C) \, d\nu_\zeta = \lim_{k \to \infty} \int_{M/\zeta} \left(\int_C f_k \, d\nu_C \right) d\nu_\zeta$$

$$= \lim_{k \to \infty} \int_M f_k \, d\nu = \int_M f \, d\nu. \quad \square$$

If the partition ζ is generated by the family $\Sigma = \{S_\alpha\}_{\alpha=1}^\infty$, denote by $\mathcal{F}_N$ a finite (σ-) algebra of sets generated by $\{S_\alpha\}_{\alpha=1}^N$, and let $\bar{\mathcal{F}}_N$ be its image in M/ζ under H_ζ. So $\bar{\mathcal{F}}_1 \subset \bar{\mathcal{F}}_2 \subset \cdots \subset \bar{\mathcal{F}}_N \subset \cdots$. Let $\bar{\mathcal{F}}_\infty$ be the minimal complete σ-algebra containing $\bigcup_{n=1}^\infty \bar{\mathcal{F}}_N$. By Theorem 2.2 (5), $\bar{\mathcal{F}}_\infty$ coincides with the σ-algebra of ν_ζ-measurable subsets of M/ζ.

For every element $C \in M/\zeta$ and every $N \in \mathbb{N}$, denote by $\bar{\Phi}_N(C)$ the unique minimal element of $\bar{\mathcal{F}}_N$, which contains C (clearly, $\bar{\Phi}_N(C) = H_\zeta(\bigcap_{\alpha=1}^N R_\alpha)$, where $R_\alpha = S_\alpha$ or $\bar{S}_\alpha$). Denote its preimage in M by $\Phi_N(C)$.

COROLLARY 2.4. *Let M, ν, ζ, f be as in Corollary 2.3. Then, for ν_ζ-a.e. $C \in M/\zeta$,*

$$\int_C f\, d\nu_C = \lim_{N \to \infty} \frac{1}{\nu(\Phi_N(C))} \int_{\Phi_N(C)} f\, d\nu.$$

PROOF. The function $\phi(C) = \int_C f\, d\nu_C$ is $\bar{\mathcal{F}}_\infty$-measurable by Corollary 2.3. Then, by the classical P. Levy martingale convergence theorem (see, e.g., [L-Sh]),

$$\phi \stackrel{\nu_\zeta-a.e.}{=} \lim_{N \to \infty} \mathbb{E}\left(\phi\,|\,\bar{\mathcal{F}}_N\right).$$

But

$$\mathbb{E}\left(\phi\,|\,\bar{\mathcal{F}}_N\right)(C) = \frac{1}{\nu_\zeta(\bar{\Phi}_N(C))} \int_{\bar{\Phi}_N(C)} \phi(C_1)\, d\nu_\zeta(C_1).$$

By the definition of ν_ζ, $\nu_\zeta(\bar{\Phi}_N(C)) = \nu(\Phi_N(C))$. Using Corollary 2.3, we easily check that

$$\int_{\bar{\Phi}_N(C)} \phi(C_1)\, d\nu_\zeta(C_1) = \int_{\Phi_N(C)} f\, d\nu.$$

So $\mathbb{E}\left(\phi\,|\,\bar{\mathcal{F}}_N\right)(C) = \frac{1}{\nu(\Phi_N(C))} \int_{\Phi_N(C)} f\, d\nu$ and the corollary is proved. □

3. Convex Restrictions of Measures

Let K be a convex bounded (not necessarily compact) subset of $\mathbb{R}^N$.

DEFINITION 3.1. A function $\gamma : K \longrightarrow \mathbb{R}_+$ is called α-concave ($\alpha > 0$), if $\gamma^{1/\alpha}$ is concave.

Assume that $K \subset \mathbb{R}^k \subset \mathbb{R}^N$ and $\dim K = k$. Let μ be a nonnegative Borel measure on $\mathbb{R}^N$, which is absolutely continuous with respect to the standard Lebesgue measure m_N, and let $g := \frac{d\mu}{dm_N}$.

DEFINITION 3.2. A measure ν on K is called a convex restriction of the measure μ, if there exists an $(n-k)$-concave function γ on K such that $d\nu = g \cdot \gamma \cdot dm_k$, where m_k is the Lebesgue measure on $\mathbb{R}^k$.

REMARK. Our definition of the convex restriction of measures is different from that given in [Gr-M], but both definitions are equivalent.

LEMMA 3.3. *Assume that $K_2 \subset K_1 \subset \mathbb{R}^N$ and $\dim K_i = k_i$. Let a measure ν_1 on K_1 be a convex restriction of a measure μ. Let a measure ν_2 on K_2 be a convex restriction of ν_1.*

Then ν_2 is a convex restriction of μ.

PROOF. If $d\mu = g\, dm_N$, then $d\nu_1 = g\, \gamma_1\, dm_{k_1}$ and $d\nu_2 = g\, \gamma_1\, \gamma_2\, dm_{k_2}$, where γ_1 is an $(N - k_1)$-concave function on K_1, γ_2 is a $(k_1 - k_2)$-concave on K_2. Set $\alpha = N - k_1$ and $\beta = k_1 - k_2$.

It is sufficient to show that $\gamma_1 \cdot \gamma_2$ is an $(\alpha + \beta)$-concave on K_2 [Gr-M, Appendix, Lemma 1]. Indeed, using the Hölder inequality with $p = (\alpha + \beta)/\alpha$ and $q = (\alpha + \beta)/\beta$, we obtain for every $x, y \in K_2$ and every $0 < \theta < 1$,

$$\theta \cdot [\gamma_1(x)\, \gamma_2(x)]^{1/(\alpha+\beta)} + (1-\theta) \cdot [\gamma_1(y)\, \gamma_2(y)]^{1/(\alpha+\beta)}$$
$$\leq [\theta \cdot \gamma_1(x)^{1/\alpha} + (1-\theta) \cdot \gamma_1(y)^{1/\alpha}]^{\alpha/(\alpha+\beta)} \cdot [\theta \cdot \gamma_2(x)^{1/\beta} + (1-\theta) \cdot \gamma_2(y)^{1/\beta}]^{\beta/(\alpha+\beta)}$$
$$\leq \gamma_1(\theta x + (1-\theta) y)^{1/(\alpha+\beta)} \cdot \gamma_2(\theta x + (1-\theta) y)^{1/(\alpha+\beta)}. \quad \square$$

Later we will need the following result:

LEMMA 3.4. *Let a measure μ on $\mathbb{R}^N$ is such that $d\mu = f \cdot dm_N$, where f is continuous, $f > 0$ m_N-a.e., and suppose we are given a decreasing sequence of convex compact sets $K_1 \supset K_2 \supset \cdots \supset K_n \supset \cdots$ of full dimension N. Let $K := \bigcap_{n=1}^{\infty} K_n$, $k := \dim K$. Define a sequence of probability measures $\{\lambda_n\}$ such that for any Borel subset $A \subset \mathbb{R}^N$ $\lambda_n(A) := \frac{\mu(A \cap K_n)}{\mu(K_n)}$ (note that our assumptions imply that $\mu(K_n) \neq 0$).*

Then one can choose a subsequence $\{n_l\}$ such that $\{\lambda_{n_l}\}$ converges weakly to a measure concentrated on K, which is a convex restriction of μ.

PROOF. Let E be the affine hull of K, and put $k = \dim E$.

Consider new convex sets

$$\tilde{K}_n := \{(x, y) \in E \oplus E^\perp : (x, \operatorname{vol}(K_n)^{1/(N-k)} \cdot y) \in K_n\}.$$

By the Cavalieri principle, $\operatorname{vol}(\tilde{K}_n) = 1$. Replace $\tilde{K}_n$ by its $(N-k)$-dimensional Schwarz symmetrization K'_n with respect to E. Then K'_n are also convex compact bodies, $\operatorname{vol}(K'_n) = 1$ and $K'_n \supset K$. This and their rotation invariance imply easily that K'_n are uniformly bounded. Hence by the Blaschke selection theorem one can choose a subsequence $\{n_l\}$ such that K'_{n_l} converges to some convex compact set M with respect to the Hausdorff metric. Obviously, M is also invariant with respect to rotations around E, $\operatorname{vol}(M) = 1$, $M \supset K$, and $M \cap E = K$, because $\bigcap_l K_{n_l} = K$.

Consider a function γ on K:

$$\gamma(x) = \operatorname{vol}_{N-k}\left(M \cap (x + E^\perp)\right).$$

Then γ is $(N-k)$-concave by Brunn's theorem. We will show that for every continuous function u on $\mathbb{R}^N$

$$\frac{1}{\mu(K_{n_l})} \int_{K_{n_l}} u(x) f(x) dm_N(x) \longrightarrow \int_K u(x) f(x) \gamma(x) dm_k(x).$$

This will prove the lemma.

Denote $v := u \cdot f$ and consider a function $v'(x) := v(Pr_E \, x)$, where Pr_E is the orthogonal projection onto E. Since $v \equiv v'$ on K, for any $\varepsilon > 0$ there exists an open neighborhood U of K such that $|v - v'| < \varepsilon$ on U. But $K_n \subset U$ for large n, hence

$$\frac{1}{\mu(K_n)} \int_{K_n} (v - v') \, dm_N \longrightarrow 0, \, n \longrightarrow \infty.$$

By the Fubini theorem,

$$\frac{1}{\mu(K_{n_l})} \int_{K_{n_l}} v'(x) \, dm_N(x) = \int_{K'_{n_l}} v'(x) \, dm_N(x) \longrightarrow$$

$$\int_M v'(x) dm_N(x) = \int_K v'(x) \gamma(x) dm_k(x) = \int_K u(x) f(x) \gamma(x) \, dm_k(x). \quad \square$$

4. Convex Partitions

Assume that $M \subset \mathbb{R}^N$ is a convex compact body, $\dim M = N \geq 3$, $M \ni 0$. Let μ be a probability measure on M, which is absolutely continuous with respect to the Lebesgue measure m_N, $d\mu = f dm_N$, where f is continuous and $f > 0$ m_N-a.e. on M.

Fix A_1, A_2 disjoint closed subsets of ∂M such that $\hat{A}_i := \bigcup_{0 \leq t \leq 1} t \cdot A_i$, $i = 1, 2$ have nonzero measure μ. Set $\lambda := \frac{\mu(\hat{A}_1)}{\mu(\hat{A}_2)}$.

Using the idea of [Gr-M], we will construct a measurable (cf. Definition 2.1) partition ζ of the convex set M satisfying the following properties (in the notation of Section 2):

(4.1) Every element $C \in \zeta$ of this partition is a convex subset of M and has the form $C = \bigcup_{0 \leq t \leq 1} t \cdot (C \cap \partial M)$.
(4.2) ν_C is the convex restriction of μ to C for ν_ζ-a.e. $C \in \zeta$.
(4.3) $\nu_C(\hat{A}_1) = \lambda \nu_C(\hat{A}_2)$ for ν_ζ-a.e. $C \in \zeta$.
(4.4) Moreover, if the measure μ is homogeneous of degree $\alpha > 0$, i.e. for every Borel subset $T \subset M$ and every $t \in [0,1]$ $\mu(t \cdot T) = t^\alpha \cdot \mu(T)$, then ν_C is also homogeneous of degree α for ν_ζ-a.e. $C \in \zeta$.

The construction of such partition uses the Borsuk–Ulam theorem.

Let S^{N-1} be the Euclidean sphere in $\mathbb{R}^N$. For $x \in S^{N-1}$, denote $H_x^+ := \{y \in \mathbb{R}^N : (y,x) \geq 0\}$ the closed half-space. So $H_x^- := \mathbb{R}^N - H_x^+$ is an open half-space. Then $M^+ := M \cap H_x^+$ and $M^- := M \cap H_x^-$ are convex sets.

LOCALIZATION AND THE GROMOV–MILMAN THEOREM   23

It will be more convenient to consider M^- as a compact set. Namely, replace M by a new set, where the hyperplane $H_x = \{y \mid \langle y, x \rangle = 0\}$ is considered as a "double" set, that is, one copy of it belongs to M^+ and another to M^- (this is similar to the situation where, if we consider the dyadic points of the unit interval as "double" points, we obtain the Cantor set). In the steps that follow, each hyperplane we construct will be considered as "double". This will not change M and its factor set by the partition constructed below, since these spaces are Lebesgue spaces in the sense of [R].

Consider a map $\phi : S^{N-1} \longrightarrow \mathbb{R}^2$ such that

$$\phi(x) = \left(\mu(\hat{A}_1 \cap H_x^+), \mu(\hat{A}_2 \cap H_x^+) \right).$$

Since ϕ is continuous and $N \geq 3$, we can apply the Borsuk–Ulam theorem and find $x \in S^{N-1}$ such that $\mu(\hat{A}_i \cap H_x^\pm) = \frac{1}{2}\mu(\hat{A}_i)$, for $i = 1, 2$. Now apply the same argument to M^+ and M^- separately, replacing A_i by $A_i^+ := A_i \cap H_x^+ \subset M^+$ and setting $A_i^- := A_i \cap H_x^- \subset M^-$ correspondingly. So after the second use of the Borsuk–Ulam theorem we obtain a partition of M into four disjoint convex subsets $M^{++}, M^{+-}, M^{-+}, M^{--}$. By construction $\mu(\hat{A}_1 \cap M^{++}) = \lambda \mu(\hat{A}_2 \cap M^{++})$, and this holds for all the other elements of the partition.

Repeating this procedure infinitely, we obtain a partition ζ of M, which is obviously measurable and satisfies (4.1) by construction. The property (4.2) follows immediately from Corollary 2.4 and Lemma 3.4. Corollary 2.4 implies also (4.3).

In order to prove (4.4), recall that the Borel σ-algebra of subsets of $\mathbb{R}^N$ is generated by a countable number of sets $\{T_j\}_{j=1}^\infty$. Since for ν_ζ-a.e. C ν_C is the convex restriction of μ, ν_C is absolutely continuous with respect to the Lebesgue measure on C; hence it is sufficient to check (4.4) only for $t \in \mathbb{Q}$. So we have to prove (4.4) for fixed T and t. And this again follows from Corollary 2.4.

By Theorem 2.1, $\mu(\hat{A}_1) = \int_{M/\zeta} \nu_C(\hat{A}_1) \, d\nu_\zeta(C)$. Hence we can choose C such that $\nu_C(\hat{A}_1 \cap C) = \nu_C(\hat{A}_1) \geq \mu(\hat{A}_1)$, and C satisfies (4.1)-(4.4). Let us show that $\dim C < N$. Indeed, $C = \bigcap_{k=1}^\infty V_k$, where V_k denotes the unique element of the partition of M constructed on the k-th step as above, which contains C. All V_k are convex, hence if $\dim C = N$, then $\dim V_k = N$. By Corollary 2.4 and the construction,

$$\nu_C(\hat{A}_1) = \lim_{k \to \infty} \frac{1}{\mu(V_k)} \mu(\hat{A}_1 \cap V_k) = \lim_{k \to \infty} \frac{1}{\mu(V_k)} \cdot \frac{1}{2^k} \mu(\hat{A}_1).$$

Since we have assumed that $\frac{d\mu}{dm_N} > 0$ m_N-a.e., $\mu(V_k) \geq \mu(C) > 0$. So the right hand limit is equal to 0, contradicting the choice of C.

Let us fix such a C and denote it by M_1. Denote also ν_C by μ_1. Now we come back to the situation where $M = K(X)$ is the unit ball of $X = (\mathbb{R}^{n+1}, \|\cdot\|)$, μ is the normalized Lebesgue measure on M, and $A_1, A_2 \subset \partial M = S(X)$ are compact and disjoint. Thus μ_1 is a convex restriction of the Lebesgue measure, and it satisfies (4.4) with $\alpha = n + 1$.

Since $\mu_1(\hat{A}_1 \cap M_1) = \lambda \mu_1(\hat{A}_2 \cap M_1) > 0$ (recall that $\lambda = \frac{\mu(\hat{A}_1)}{\mu(\hat{A}_2)} = \frac{m_{n+1}(\hat{A}_1)}{m_{n+1}(\hat{A}_2)}$), we have $\dim M_1 \geq 2$. Obviously, M_1 is convex and compact.

If $\dim M_1 \geq 3$, the use of the Borsuk–Ulam theorem is possible and by the same procedure we construct a convex compact subset $M_2 \subset M_1$ and a convex restriction μ_2 of the measure μ_1 satisfying (4.1)-(4.4) with $\alpha = n+1$ and $\lambda \mu_2(\hat{A}_2 \cap M_2) = \mu_2(\hat{A}_1 \cap M_2) \geq \mu_1(\hat{A}_1 \cap M_1) \geq \mu(\hat{A}_1) = m_{n+1}(\hat{A}_1)$.

By Lemma 3.3, μ_2 is a convex restriction of m_{n+1}. Repeating this argument, after at most $n-1$ steps we obtain a 2-dimensional convex compact set $N \subset M$ and a measure ν on N such that:

(4.5) $N = \bigcup_{0 \leq t \leq 1} t \cdot (N \cap S(X))$ and N is contained in some half-plane (by construction).

(4.6) There exists an $(n-1)$-concave function γ on N such that $d\nu = \gamma \, dm_2$ (where m_2 is the Lebesgue measure on $\mathbb{R}^2$).

(4.7) $\lambda \nu(\hat{A}_2 \cap N) = \nu(\hat{A}_1 \cap N) \geq m_{n+1}(\hat{A}_1) \, (= \hat{\mu}(A_1))$, where $\lambda = \frac{m_{n+1}(\hat{A}_1)}{m_{n+1}(\hat{A}_2)}$ as above.

(4.8) ν is homogeneous of degree $n+1$, i.e. for every Borel subset $T \subset \mathbb{R}^2$ and every $t \in [0, 1]$,
$$\nu(t \cdot T) = t^{n+1} \cdot \nu(T).$$

Note that (4.6) and (4.8) immediately imply

(4.9) γ is homogeneous of degree $n-1$, i.e. $\gamma(t \cdot x) = t^{n-1} \gamma(x)$ for every $x \in N$, $t \in [0, 1]$.

Clearly, by (4.5) $N \cap S(X)$ is a spherical segment. Denote it by $I = [a, b]$. Since the Banach–Mazur distance between any 2-dimensional normed space and the Euclidean ball is at most $\sqrt{2}$, we can find a Euclidean norm $|\cdot|$ on $\mathrm{span}\, N$ such that

(4.10) $$\frac{1}{\sqrt{2}} |x| \leq \|x\| \leq |x|, \ \forall x \in \mathrm{span}\, N.$$

For every two points $x, y \in I$, denote by $\rho(x, y)$ the length of the segment $[x, y] \subset I$ with respect to $\|\cdot\|$, i.e. if $[x, y]$ is parameterized by some parameter $\tau \in [0, 1]$, then
$$\rho(x, y) := \sup_{0 \leq \tau_1 < \cdots < \tau_k \leq 1} \sum_{j=1}^{k-1} \|\tau_{j+1} - \tau_j\|.$$

Similarly, denote by $d(x, y)$ the length of $[x, y]$ with respect to $|\cdot|$.

By a result of [S],
$$\|x - y\| \leq \rho(x, y) \leq 2 \|x - y\|.$$

Thus we obtain

(4.11) $\qquad |x - y| \leq d(x, y) \leq \sqrt{2} \rho(x, y) \leq 2\sqrt{2} \|x - y\| \leq 2\sqrt{2} |x - y|.$

On I we have a measure $\hat{\nu}$ such that $\hat{\nu}(A) := \nu(\hat{A})$, where $A \subset I$ is any Borel subset and $\hat{A} = \bigcup_{0 \leq t \leq 1} t \cdot A$. Then $d\hat{\nu} = f_\nu \, dt$, where dt is an element of the Euclidean length and f_ν is a continuous function. By (4.6) and (4.9),

$$\nu(\hat{A}) = \int_{\hat{A}} \gamma \, dm_2 = \frac{1}{n+1} \int_A \gamma(t) \, dt.$$

So $f_\nu = \frac{1}{n+1} \gamma$. The rest of the paper closely follows [Gr-M].

For $x, y \in I$, (4.6) implies

(4.12) $$\gamma^{1/(n-1)}\left(\frac{x+y}{2}\right) \geq \frac{\gamma^{1/(n-1)}(x) + \gamma^{1/(n-1)}(y)}{2}.$$

Set $z = \frac{x+y}{2} / \|\frac{x+y}{2}\| \in [x, y]$. By (4.9) and the inequality

$$\|\frac{x+y}{2}\| \leq 1 - \delta(\|x - y\|),$$

we have

(4.13) $$\gamma\left(\frac{x+y}{2}\right) \leq (1 - \delta(\|x - y\|))^{n-1} \gamma(z).$$

It easily follows from the inequality (4.11) that for some absolute constant $\alpha \in \left(0, \frac{1}{2}\right)$,

(4.14) $$d(z, x) \geq \alpha \, d(x, y), \text{ and } d(z, y) \geq \alpha \, d(x, y).$$

Let us parameterize the segment $I = [a, b]$ by the Euclidean length of the segment $[a, x]$, namely if x corresponds to t_1, it means $d(x, a) = t_1$. Let y corresponds to $t_2 > t_1$, then $d(x, y) = t_2 - t_1$. Clearly, (4.12)–(4.14) imply

(4.15) $$\frac{f_\nu^{1/(n-1)}(t_1) + f_\nu^{1/(n-1)}(t_2)}{2}$$
$$\leq (1 - \delta(\|x - y\|)) \cdot \max_{z \in [t_1 + \alpha(t_2 - t_1), \, t_2 - \alpha(t_2 - t_1)]} f_\nu(z)^{1/(n-1)}.$$

Then easily f_ν has no local minima and at most one local maximum inside I (this local maximum must be global). Denote the global maximum of f_ν by $t_0 \in [0, l]$ (where $l = d(a, b)$). Then obviously f_ν increases on $[0, t_0]$ and decreases on $[t_0, l]$.

For any $t \in [0, t_0]$ and any θ such that $0 \leq t - \theta < t \leq t_0$, (4.15) and the monotonicity of f_ν on $[0, t_0]$ imply

$$f_\nu(t - \theta) \leq (1 - \delta(\|x - y\|))^{n-1} f_\nu(t),$$

where x corresponds to $t - \theta$, and y corresponds to t. But by (4.11) $\|x - y\| \geq \frac{1}{2\sqrt{2}} d(x, y) = \frac{\theta}{2\sqrt{2}}$, and we obtain

(4.16) $$f_\nu(t - \theta) \leq \left(1 - \delta\left(\frac{\theta}{2\sqrt{2}}\right)\right)^{n-1} f_\nu(t).$$

Similarly, if $t_0 \leq t < t + \theta \leq l$, then

$$(4.17) \qquad f_\nu(t+\theta) \leq \left(1 - \delta\left(\frac{\theta}{\sqrt{2}}\right)\right)^{n-1} f_\nu(t).$$

Hence, for $t_0 - 2\theta \geq 0$,

$$\hat{\nu}([0, t_0 - 2\theta]) = \int_0^{t_0 - 2\theta} f_\nu(t)\, dt \leq \left(1 - \delta\left(\frac{\theta}{2\sqrt{2}}\right)\right)^{n-1} \int_\theta^{t_0 - \theta} f_\nu(t)\, dt$$

$$\leq \left(1 - \delta\left(\frac{\theta}{2\sqrt{2}}\right)\right)^{n-1} \left(\hat{\nu}([0, t_0 - 2\theta]) + \hat{\nu}([t_0 - 2\theta, t_0 - \theta])\right).$$

Thus

$$(4.18) \qquad \hat{\nu}([0, t_0 - 2\theta]) \leq \frac{\left(1 - \delta\left(\frac{\theta}{2\sqrt{2}}\right)\right)^{n-1}}{1 - \left(1 - \delta\left(\frac{\theta}{2\sqrt{2}}\right)\right)^{n-1}} \hat{\nu}([t_0 - 2\theta, t_0]).$$

In the same way, for $t_0 + 2\theta \leq l$ we have

$$(4.19) \qquad \hat{\nu}([t_0 + 2\theta, l]) \leq \frac{\left(1 - \delta\left(\frac{\theta}{2\sqrt{2}}\right)\right)^{n-1}}{1 - \left(1 - \delta\left(\frac{\theta}{2\sqrt{2}}\right)\right)^{n-1}} \hat{\nu}([t_0, t_0 + 2\theta]).$$

Adding (4.18) and (4.19) and using $\hat{\nu}([0, l]) = 1$, we obtain:

LEMMA 4.20. $\hat{\nu}(I - [t_0 - 2\theta, t_0 + 2\theta]) \leq \left(1 - \delta\left(\frac{\theta}{2\sqrt{2}}\right)\right)^{n-1} \approx e^{-\delta\left(\frac{\theta}{2\sqrt{2}}(n-1)\right)}.$

5. Proof of Theorem 1.1

(We repeat the argument of [Gr-M].)

Let $A \subset S(X)$, $\hat{\mu}(A) \geq \frac{1}{2}$ (the measure $\hat{\mu}$ was defined in Section 1). Fix $\varepsilon \in (0, 1)$. Set $A_1 := A$, $A_2 := S(X) - A_\varepsilon$. Hence we can find a compact convex 2-dimensional set N with a probability measure ν satisfying (4.5)–(4.9). Let c be the point on I with the maximal density of $\hat{\nu}$.

If θ_n is such that $\delta(\theta_n) = 1 - (\frac{1}{2})^{1/(n-1)} \approx \frac{\log 2}{n-1}$, then $\hat{\nu}\{x \in I : d(x, c) \leq 4\sqrt{2}\,\theta_n\} \geq \frac{1}{2}$. By (4.7), $\hat{\nu}(A_1 \cap I) \geq \frac{1}{2}$; hence there exists $x' \in A_1 \cap I$ such that $\|x' - c\| \leq d(x', c) \leq 4\sqrt{2}\,\theta_n$. Now let us take θ such that $\varepsilon = 4\sqrt{2}\,(\theta + \theta_n)$. For an ε-neighborhood of $\{x'\}$ (with respect to the original norm $\|\cdot\|$), we have $\{x'\}_\varepsilon \supset \{c\}_{4\sqrt{2}\,\theta}$ and $\{x'\}_\varepsilon \cap A_2 = \emptyset$. Therefore, again by Lemma 4.20 and (4.11)

$$\hat{\nu}(A_2 \cap I) \leq \hat{\nu}(I - \{x : d(x, c) \leq 4\sqrt{2}\,\theta\}) \leq (1 - \delta(\theta))^{n-1}$$

$$\approx \exp(-\delta(\theta)(n - 1)) = \exp\left(-\delta\left(\frac{\varepsilon}{4\sqrt{2}} - \theta_n\right)(n - 1)\right).$$

By (4.7),

$$\mu(A_2) = \frac{\hat{\nu}(A_2 \cap I)}{\hat{\nu}(A_1 \cap I)} \mu(A_1) \leq \hat{\nu}(A_2 \cap I) \leq \exp\left(-\delta\left(\frac{\varepsilon}{4\sqrt{2}} - \theta_n\right)(n-1)\right). \quad \square$$

Acknowledgements

I would like to thank Prof. V. D. Milman for his guidance in this work.

References

[Gr-M] Gromov, M.; Milman, V. D.; Generalization of the spherical isoperimetric inequality to uniformly convex Banach spaces. Compositio Mathematica **62** (1987), 263–282.

[L-S1] Lovász, L.; Simonovits M.; Mixing rate of Markov chaines, an isoperimetric inequality, and computing the volume. Proc. 31st Ann. Symp. on Found. of Comput. Sci., IEEE Computer Soc., 1990, 346–355.

[L-S2] Lovász, L.; Simonovits M.; Random walks in a convex body and an improved volume algorithm. Random Structures and Algorithms 4 (1993), 359–412.

[L-Sh] Liptser, R. S.; Shiryayev, A. N.; Statistics of Random Processes I, General Theory; Springer, New York, Heidelberg, Berlin, 1977.

[P] Pisier, G.; The Volume of Convex Bodies and Banach Space Geometry. Cambridge University Press, Cambridge 1989.

[R] Rohlin, V. A.; On the fundamental ideas of measure theory; Mat. Sbornik, N.S. 25 (67), (1949), 107–150 (Russian). Translation in Amer. Math. Soc. Translations **71**, 55 pp. (1952).

[S] Schäffer, J. J.; Inner diameter, perimeter, and girth of spheres. Math. Ann. **173** (1967), 59–82.

SEMYON ALESKER
DEPARTMENT OF MATHEMATICS
TEL-AVIV UNIVERSITY
RAMAT-AVIV
ISRAEL
semyon@math.tau.ac.il

Convex Geometric Analysis
MSRI Publications
Volume **34**, 1998

Geometric Inequalities in Option Pricing

CHRISTER BORELL

ABSTRACT. This paper discusses various geometric inequalities in option pricing assuming that the underlying stock prices are governed by a joint geometric Brownian motion. In particular, inequalities of isoperimetric type are proved for different classes of derivative securities. Moreover, the paper discusses the option on the minimum of several assets and, among other things, proves a log-concavity property of its price.

1. Introduction

The purpose of this paper is to prove various geometric inequalities in option pricing using familiar inequalities of the Brunn-Minkowski type in Gauss space.

To begin with, recall that a European (American) call [put] option is defined as the right to buy [sell] one share of stock at a specified price on (or before) a specified date. The specified price is referred to as the exercise price and the terminal date of the contract is called the expiration date or maturity date. In fact, already the early paper [20] by Merton treats a variety of convexity properties of puts and calls, sometimes without any distributional assumptions on the underlying stock prices. Here, however, it will always be assumed that the price process $X(t) = (X_1(t), \ldots, X_m(t))$, $t \geq 0$, of the underlying risky assets $\mathcal{X}_1, \ldots, \mathcal{X}_m$ is governed by a so called joint geometric Brownian motion. Furthermore, all options will be of European type and so, from now on, option will always mean option of European type.

Now suppose $f : \mathbb{R}_+^m \to [0, +\infty[$ is a continuous function such that

$$f(x) \leq A\left(1 + \sum_{i=1}^m |x_i|\right)^a \quad \text{for } x = (x_1, ..., x_m) \in \mathbb{R}_+^m,$$

1991 *Mathematics Subject Classification.* primary 52A40, 90A09; secondary 60G15, 60H05.

Key words and phrases. put, call, derivative security, isoperimetric inequality, log-concave, Gaussian random variable.

Research supported by the Swedish Natural Science Research Council.

for appropriate constants $a, A \geq 0$, and suppose a certain derivative security $\mathcal{U}_f^T$ pays $f(X(T))$ at the maturity time T. Here f is termed a payoff function. If t is a time point prior to T, set $\tau = T - t$, and denote by $u_f(\tau, X(t))$ the (theoretic) price of $\mathcal{U}_f^T$ at time t. If $u_f(\tau, x) = u_f(\tau, x_1, \ldots, x_m)$ is positive and $i \in \{1, \ldots, m\}$ is fixed, the quantity

$$\psi_f^i(\tau, x) = \frac{x_i}{u_f(\tau, x)} \frac{\partial u_f(\tau, x)}{\partial x_i}$$

is called the elasticity of the price $u_f(\tau, x)$ relative to the price x_i. The quantities $\psi_f^1(\tau, x), \ldots, \psi_f^m(\tau, x)$ enter quite naturally in option pricing in connection with so called hedging against the contingent claim $\mathcal{U}_f^T$. Actually, we will below occasionally consider a slightly larger class of payoff functions than stated here.

Now let the function $f(x)$, $x \in \mathbb{R}_+^m$, be a log-concave function of the log-price vector $\ln x = (\ln x_1, \ldots, \ln x_m)$. In Section 3, we prove, among other things, that the function $\tau^{m/2} u_f(\tau, x)$ is a log-concave function of $(\tau, \ln x)$. In particular, if f is not identically equal to zero, then for any fixed $i \in \{1, \ldots, m\}$ and $\tau > 0$ the elasticity function $\psi_f^i(\tau, x)$ is a non-increasing function of x_i when the other prices $x_1, \ldots, x_{i-1}, x_{i+1}, \ldots, x_m$ are held fixed. Note that these results apply to the payoff function

$$f(x) = \min_{i=1,\ldots,m} x_i \qquad (1)$$

which is of interest in connection with the cheapest to deliver option. The derivative security corresponding to the payoff function in (1) is sometimes referred to as the quality option (see e.g. Boyle [11]).

The main concern in this paper is to prove certain inequalities of isoperimetric type. More explicitly, consider the same risky assets as above and suppose $a > 0$ is given. We shall write $f \in \mathcal{C}_a$ if $f : \mathbb{R}_+^m \to [0, +\infty[$ is a locally Lipschitz continuous function such that

$$\sum_{i=1}^m x_i \left| \frac{\partial f}{\partial x_i} \right| \leq a + f(x) \text{ a.e.}$$

with respect to Lebesgue measure in $\mathbb{R}^m$. The class $\mathcal{C}_a$ is convex and contains the zero payoff function. Moreover, the class $\mathcal{C}_a$ contains the payoff functions of all puts and calls on the $\mathcal{X}_i, i = 1, \ldots, m$, with exercise prices less than or equal to a. In addition, if $f, g \in \mathcal{C}_a$, then $\max(f, g) \in \mathcal{C}_a$ and $\min(f, g) \in \mathcal{C}_a$. In particular, the function in (1) as well as the function

$$f(x) = \max_{i=1,\ldots,m} x_i \qquad (2)$$

belong to the class $\mathcal{C}_a$ for all $a > 0$.

In Section 4 we discuss, among other things, the Monte Carlo method for computing the option price $u_f(\tau, x)$ when f is as in (2). Let $\mathcal{X}_m$ be the most volatile asset of the risky assets $\mathcal{X}_1, \ldots, \mathcal{X}_m$ and let σ_m be the volatility of $\mathcal{X}_m$.

The (crude) Monte Carlo method then gives us a certain unbiased estimator $\overline{Z}_N$ of the option price $u_f(\tau, x)$ and we prove that

$$\mathbb{P}\left[\left|\frac{\overline{Z}_N - u_f(\tau, x)}{u_f(\tau, x)}\right| \geq \varepsilon\right] \leq \frac{e^{\sigma_m^2 \tau} - 1}{\varepsilon^2 N}, \quad \varepsilon > 0. \tag{3}$$

Note that the right-hand side of (3) is independent of the option price $u_f(\tau, x)$.

In Section 4 we also prove the following property of the class $\mathcal{C}_a$ for fixed $a > 0$. Suppose v is the expected exercise value of a call on $\mathcal{X}_m$ with the maturity date T and exercise price a. Then, amongst all derivative securities $\mathcal{U}_f^T$ with $f \in \mathcal{C}_a$ and with the expected payoff v at time T, the payoff at time T has maximal variance for the call on $\mathcal{X}_m$ with the exercise price a.

Finally, in Section 5 we discuss inequalities of isoperimetric type for other classes of payoff functions than those considered above.

2. Notation and Basic Results

Throughout this paper $\mathcal{X}_i, i = 1, \ldots, m$, stand for m risky assets with a joint price process $X(t) = (X_1(t), \ldots, X_m(t))$, $t \geq 0$, governed by an m-dimensional geometric Brownian motion. Stated more explicitly, there are linearly independent unit vectors c_i, $i = 1, \ldots, m$, in $\mathbb{R}^n$ and a normalized Brownian motion $(W(t))$ in $\mathbb{R}^n$ such that

$$\frac{dX_i(t)}{X_i(t)} = (\mu_i + \sigma_i^2/2)dt + \sigma_i dW_i(t), \quad i = 1, \ldots, m$$

for suitable $\mu_1, \ldots, \mu_m \in \mathbb{R}$ and $\sigma_1 > 0, \ldots, \sigma_m > 0$, where

$$W_i(t) = \langle c_i, W(t) \rangle, \quad i = 1, \ldots, m.$$

Here, $\langle \cdot, \cdot \rangle = \langle \cdot, \cdot \rangle_{\mathbb{R}^n}$ denotes the standard scalar product in $\mathbb{R}^n$.

In what follows, $t < T$ and we set

$$M_{\sigma_i}^{W_i}(\tau) = e^{-\frac{\sigma_i^2}{2}\tau + \sigma_i W_i(\tau)} \quad \text{for } i = 1, \ldots, m$$

and

$$M_\sigma^W(\tau) = (M_{\sigma_1}^{W_1}(\tau), \ldots, M_{\sigma_m}^{W_m}(\tau)),$$

where $\tau = T - t$. Moreover, if $\xi = (\xi_1, \ldots, \xi_m)$, $\eta = (\eta_1, \ldots, \eta_m) \in \mathbb{R}^m$, we will make frequent use of the following notation:

$$|\xi| = (|\xi_1|, \ldots, |\xi_m|)$$

$$\|\xi\|_1 = \sum_1^m |\xi_i|$$

$$\|\xi\|_2 = \sqrt{\langle \xi, \xi \rangle_{\mathbb{R}^m}}$$

$$\|\xi\|_\infty = \max_{i=1,\ldots,m} |\xi_i|$$

$$e^\xi = (e^{\xi_1}, \ldots, e^{\xi_m})$$

$$\ln e^\xi = \xi$$

and

$$\xi\eta = (\xi_1\eta_1, \ldots, \xi_m\eta_m).$$

Now consider a derivative security $\mathcal{U}_f^T$ with the payoff $f(X(T))$ at time T. Below, for technical reasons, it will be assumed that $f : \mathbb{R}_+^m \to \mathbb{R}$ is a continuous function such that

$$\mathbb{E}[|f(xM_\sigma^W(\tau))|^p] < +\infty$$

for all $x \in \mathbb{R}_+^m$, $\tau > 0$, and $p > 0$ and a function f satisfying these assumptions will be called a payoff function. If r denotes the risk-free interest rate and if $u_f(\tau, X(t))$ denotes the value of the derivative security $\mathcal{U}_f^T$ at time $t \in [0, T[$, we have

$$u_f(\tau, x) = \mathbb{E}[e^{-r\tau} f(xe^{r\tau} M_\sigma^W(\tau))]. \tag{4}$$

A proof this equation is given e.g. in Duffie's book [14] or in the basic paper [16] by Harrison and Pliska. If f is a payoff function it is readily seen that $u_f(\tau, x)$ is a payoff function as a function of x for fixed τ. We now define $u_f(\tau, x)$ for all $\tau > 0$ by the equation (4) and set $(S_\tau f)(x) = u_f(\tau, x)$ if $\tau > 0$. Then the family $(S_\tau)_{\tau>0}$ becomes a semi-group, the so called option semi-group of the underlying risky assets $\mathcal{X}_1, \ldots, \mathcal{X}_m$.

Throughout this paper, if $\xi \in \mathbb{R}$, we let $\xi^+ = \max(0, \xi)$ and $\xi^- = (-\xi)^+$. Moreover, given $a > 0$ and $i \in \{1, \ldots, m\}$, let

$$c_{a,i}(x) = (x_i - a)^+$$

and

$$p_{a,i}(x) = (x_i - a)^- = (a - x_i)^+.$$

Here the derivative security $\mathcal{U}_{c_{a,i}}^T$ is called a call on $\mathcal{X}_i$ with exercise price a and maturity time T and the derivative security $\mathcal{U}_{p_{a,i}}^T$ is called a put on $\mathcal{X}_i$ with exercise price a and maturity time T. From (4) we have the following famous formula by Black and Scholes, viz.

$$u_{c_{a,i}}(x, \tau) = x\Phi\left(\frac{\ln\frac{x_i}{a} + (r + \frac{\sigma_i^2}{2})\tau}{\sigma_i\sqrt{\tau}}\right) - ae^{-r\tau}\Phi\left(\frac{\ln\frac{x_i}{a} + (r - \frac{\sigma_i^2}{2})\tau}{\sigma_i\sqrt{\tau}}\right)$$

where

$$\Phi(\xi) = \int_{-\infty}^\xi e^{-\frac{\eta^2}{2}} \frac{d\eta}{\sqrt{2\pi}}$$

is the distribution function of a real-valued Gaussian random variable with unit variance and expectation zero. Moreover, by the put-call parity relation we have

$$u_{p_{a,i}}(\tau, x) = ae^{-r\tau} + u_{c_{a,i}}(\tau, x) - x_i.$$

In the following $\mathcal{X}_0$ denotes a bond with the value $X_0(t) = e^{-r\tau}$ at time t. Furthermore, we set

$$\phi_f^i(\tau, x) = \frac{\partial u_f}{\partial x_i}(\tau, x), \quad i = 1, \ldots, m \tag{5}$$

and

$$\phi_f^0(\tau, x) = e^{r\tau}(u_f(\tau, x) - \sum_1^m x_i \phi_f^i(\tau, x)).$$

A portfolio consisting of $\phi_f^j(T - s, X(s))$ units of $\mathcal{X}_j$ for all $j = 0, \ldots, m$ at any time $s \in [t, T[$ has the value $u_f(\tau, X(t))$ at time t and

$$f(X(T)) = u_f(\tau, X(t)) + \sum_{j=0}^m \int_t^T \phi_f^j(T - s, X_j(s)) \, dX_j(s).$$

This so called self-financing trading strategy in the $\mathcal{X}_j$, $j = 0, \ldots, m$, is basic to the theory of option pricing and much more details may be found in [14] and [16]. The portfolio $\phi_f = (\phi_f^0, \phi_f^1, \ldots, \phi_f^m)$ is often called a hedge against the contingent claim $\mathcal{U}_f^T$. If $u_f(\tau, x)$ is positive for all x, the corresponding relative portfolio $\psi_f = (\psi_f^0, \psi_f^1, \ldots, \psi_f^m)$ is defined by

$$\psi_f^i(\tau, x) = x_i \phi_f^i(\tau, x) / u_f(\tau, x), \quad i = 1, \ldots, m$$

and

$$\psi_f^0 = 1 - \sum_{i=1}^m \psi_f^i.$$

Given $i \in \{1, \ldots, m\}$ the quantity $\psi_f^i(\tau, x)$ is called the elasticity of the price $u_f(\tau, x)$ relative to the price x_i.

A payoff function f is said to be homogeneous if

$$f(\alpha x) = \alpha f(x), \quad \alpha > 0, \ x \in \mathbb{R}_+^m$$

and for such functions, $\phi_f^0 = 0$ and u_f is independent of r. Typical examples of homogeneous payoff functions are

$$f_{\min}(x) = \min_{i=1,\ldots,m} x_i$$

and

$$f_{\max}(x) = \max_{i=1,\ldots,m} x_i.$$

Finally, for future reference recall that a real-valued random variable is said to have a $N(0; 1)$-distribution if its distribution function equals Φ.

3. Derivative Securities with Log-Concave Payoff Functions

Recall that a non-negative function h defined on a convex subset D of a vector space is log-concave if

$$h(\theta\xi + (1-\theta)\eta) \geq h(\xi)^\theta h(\eta)^{1-\theta} \qquad (6)$$

for all $\xi, \eta \in D$ and all $\theta \in {]}0,1{[}$. If the inequality in (6) is reversed, then h is said to be log-convex. It is well known and simple to prove that the class of all log-convex functions on a convex set is closed under addition. However, the class of all log-concave functions defined on a convex set containing more than one point is not closed under addition.

The options on the minimum and maximum of several assets have been treated by Stulz [22], Johnson [18], Boyle and Tse [12] and others. The important cheapest to deliver option involves the consideration of options on the minimum of several assets i.e., the so called quality option (for more details see e.g. Boyle [11]). In fact, options on the minimum and maximum of two assets already appear implicitly in the Margrabe paper [19], which considers the option to exchange one asset for another. We will comment more on the Margrabe option below. Note that the payoff function $f_{\min}(x)$ is a log-concave function of the asset price vector $x = (x_1, \ldots, x_m)$ as well as of the asset log-price vector $\ln x = (\ln x_1, \ldots, \ln x_m)$. Moreover, the payoff function $f_{\max}(x)$ is a log-convex function of the asset log-price vector $\ln x$. If a payoff function f is concave (convex), then the security price $u_f(\tau, x)$ is a concave (convex) function of x for fixed τ as is readily seen from equation (4) (cf. [20]). If the payoff function $f(x)$ is a log-convex function of the log-price vector $\ln x$ and f is not identically equal to zero, then it follows from the equation (4) that the option price $u_f(\tau, x)$ is a log-convex and positive function of $\ln x$ for fixed τ. In particular, for any fixed $i = 1, \ldots, m$, the function

$$x_i \to \psi_f^i(\tau, x_1, \ldots, x_{i-1}, x_i, x_{i+1}, \ldots, x_m) \qquad (7)$$

is non-decreasing since

$$\psi_f^i(\tau, x) = \frac{\partial \ln u_f(\tau, e^\xi)}{\partial \xi_i}, \qquad x = e^\xi. \qquad (8)$$

The main purpose of this section is to prove that the function in (7) is non-increasing if the payoff function $f(x)$ is a log-concave function of the log-price vector $\ln x$ and f is not identically equal to zero. To this end we will make use of a very nice property of log-concave functions, first proved in a general setting by Prékopa [21] and which reads as follows:

If the function $f(\xi, \eta_1, \ldots, \eta_n)$ is a log-concave function of $(\xi, \eta_1, \ldots, \eta_n) \in D \times \mathbb{R}^n$, where D is convex, then the integral

$$\int_{\mathbb{R}^n} f(\xi, \eta_1, \ldots, \eta_n) \, d\eta_1 \ldots d\eta_n$$

is a log-concave function of $\xi \in D$.

Below, this result will be referred to as Prékopa's theorem. Note that the Davidovič, Korenbljum, and Hacet early paper [13] treats an important special case of Prékopa's theorem.

THEOREM 3.1. (a) *If the payoff function $f(x)$ is a log-concave function of the log-price vector $\ln x$, then the function $\tau^{m/2} u_f(\tau, x)$ is a log-concave function of $(\tau, \ln x)$. In particular, if f is not identically equal to zero, then for any $i = 1, \ldots, m$ and $\tau > 0$, the function in (7) is non-increasing.*
(b) *If the payoff function $f(x)$ is homogeneous and a log-concave function of the log-price vector $\ln x$, then the function $\tau^{(m-1)/2} u_f(\tau, x)$ is a log-concave function of $(\tau, \ln x)$.*

To prove Theorem 3.1, we need the following result:

LEMMA 3.1. *If $g : \mathbb{R}_+^m \to \mathbb{R}$ is a homogeneous payoff function and $m \geq 2$, then*

$$\mathbb{E}[g(M_\sigma^W(\tau))] = \mathbb{E}[g(M_{\sigma_1^*}^{W_1^*}(\tau), \ldots, M_{\sigma_{m-1}^*}^{W_{m-1}^*}(\tau), 1)]$$

where

$$\sigma_i^* = \sqrt{\sigma_i^2 - 2\langle c_i, c_j \rangle \sigma_i \sigma_m + \sigma_m^2}$$

and

$$W_i^* = (\sigma_i W_i - \sigma_m W_m)/\sigma_i^*$$

for $i = 1, \ldots, m-1$.

In the special case $m = 2$, Lemma 3.1 is implicit in [20] (with a proof different from the one below).

PROOF. We have that

$$\mathbb{E}[g(M_\sigma^W(\tau))] = \mathbb{E}[g(e^{a_1 + \sigma_1^* W_1^*(\tau)}, \ldots, e^{a_{m-1} + \sigma_{m-1}^* W_{m-1}^*(\tau)}, 1) e^{a_m + \sigma_m W_m(\tau)}]$$

for appropriate constants $a_1, \ldots, a_m$ independent of g. By conditioning on $W^*(\tau) = (W_1^*(\tau), \ldots, W_{m-1}^*(\tau))$ the right-hand side equals

$$\mathbb{E}[g(e^{a_1 + \sigma_1^* W_1^*(\tau)}, \ldots, e^{a_{m-1} + \sigma_{m-1}^* W_{m-1}^*(\tau)}, 1) e^{a_m' + \langle b', W^*(\tau) \rangle}]$$

for appropriate $a_m' \in \mathbb{R}$ and $b' \in \mathbb{R}^{m-1}$. Therefore, by translating the probability law of $W^*(\tau)$, we get

$$\mathbb{E}[g(M_\sigma^W(\tau))] = \mathbb{E}[g(M_{\sigma_1^*}^{W_1^*}(\tau), \ldots, M_{\sigma_{m-1}^*}^{W_{m-1}^*}(\tau), 1) e^{a + \langle b, W^*(\tau) \rangle}] \quad (9)$$

for suitable $a \in \mathbb{R}$ and $b \in \mathbb{R}^{m-1}$. Now let C denote the covariance matrix of $W^*(1)$ and let $e_1, \ldots, e_{m-1}$ be the standard basis in $\mathbb{R}^{m-1}$. Then by choosing $g(x) = x_i$ for $i = 1, \ldots, m$, we have

$$\begin{cases} \langle b + \sigma_i^* e_i, C(b + \sigma_i^* e_i) \rangle + 2a - \sigma_i^{*2} = 0 & \text{for } i = 1, \ldots, m-1, \\ \langle b, Cb \rangle + 2a = 0. \end{cases}$$

From this we get $\langle e_i, Cb \rangle = 0$ for $i = 1, \ldots, m-1$, and it follows that $b = 0$ since C is invertible. Hence also $a = 0$. In view of (9), Lemma 3.1 is thereby completely proved. □

PROOF OF THEOREM 3.1. We first prove Part (b). However, for the sake of simplicity we restrict ourselves to the special case $m = 2$; the general case is proved in a similar way. To prove Part (b) with $m = 2$ note that Lemma 3.1 yields

$$u_f(\tau, x) = \int_{\mathbb{R}} f(x_1 e^{-\frac{\sigma_1^{*2}}{2}\tau + \sigma_1^* \sqrt{\tau}\zeta}, x_2) e^{-\frac{\zeta^2}{2}} \frac{d\zeta}{\sqrt{2\pi}}$$

and hence

$$\sqrt{\tau} u_f(\tau, x) = \int_{\mathbb{R}} f(x_1 e^{-\frac{\sigma_1^{*2}}{2}\tau + \sigma_1^* \zeta}, x_2) e^{-\frac{\zeta^2}{2\tau}} \frac{d\zeta}{\sqrt{2\pi}}.$$

We next introduce the new vector variable $\xi = \ln x$ and have

$$\sqrt{\tau} u_f(\tau, x) = \int_{\mathbb{R}} g(\tau, \xi, \zeta) \, d\zeta$$

where

$$g(\tau, \xi, \zeta) = f(e^{\xi_1 - \frac{\sigma_1^{*2}}{2}\tau + \sigma_1^* \zeta}, e^{\xi_2}) \frac{e^{-\frac{\zeta^2}{2\tau}}}{\sqrt{2\pi}}.$$

Since the function

$$\frac{\zeta^2}{\tau}, \quad \zeta \in \mathbb{R}, \quad \tau > 0$$

is convex we conclude that the function $g(\tau, \xi, \zeta)$ is a log-concave function of (τ, ξ, ζ). The Prékopa theorem now implies that the integral

$$\int_{\mathbb{R}} g(\tau, \xi, \zeta) \, d\zeta$$

is a log-concave function of (τ, ξ). This proves Part (b) of Theorem 3.1. The first statement in Part (a) of Theorem 3.1 is proved in a similar way as Part (b) of Theorem 3.1. Moreover, the last statement in Part (a) of Theorem 3.1 now follows from (8). This concludes our proof of Theorem 3.1. □

EXAMPLE 3.1. Set $f_0(x) = \min(x_1, x_2)$ and suppose $\alpha \in]0, +\infty[$. Then, in view of Theorem 3.1, the function $\tau^\alpha u_{f_0}(\tau, x)$ is a log-concave function of $(\tau, \ln x)$ if $\alpha \geq \frac{1}{2}$. We now claim that the condition $\alpha \geq \frac{1}{2}$ is necessary for this conclusion. To see this first note the equation

$$u_{f_0}(\tau, x) = x_1 \Phi\left(\frac{\ln \frac{x_2}{x_1} - \frac{\sigma_1^{*2}\tau}{2}}{\sigma_1^* \sqrt{\tau}}\right) + x_2 \Phi\left(\frac{\ln \frac{x_1}{x_2} - \frac{\sigma_1^{*2}\tau}{2}}{\sigma_1^* \sqrt{\tau}}\right) \qquad (10)$$

which is implicit in the Margrabe paper [19] (here σ_1^* is as in Lemma 3.1 with $m = 2$). In fact, Margrabe determines u_{f_1} when $f_1(x) = \max(0, x_2 - x_1)$ and, since $u_{f_0}(\tau, x) = x_2 - u_{f_1}(\tau, x)$, equation (10) is an immediate consequence of his paper. A direct derivation of (10) is also simple using Lemma 3.1. To see this, set $a = \sigma_1^*\sqrt{\tau}$ and let G be a real-valued centered Gaussian random variable with unit variance. Now using Lemma 3.1 it follows that

$$u_{f_0}(\tau, x) = \mathbb{E}[\min(x_1 e^{-\frac{a^2}{2}+aG}, x_2)]$$

so that

$$u_{f_0}(\tau,x) = x_1 \mathbb{E}\left[e^{-\frac{a^2}{2}+aG}; G \leq \frac{1}{a}\left(\ln\frac{x_2}{x_1}+\frac{a^2}{2}\right)\right] + x_2 \mathbb{P}\left[G > \frac{1}{a}\left(\ln\frac{x_2}{x_1}+\frac{a^2}{2}\right)\right].$$

By applying the translation formula of Gaussian measures, we get

$$u_{f_0}(\tau,x) = x_1 \mathbb{P}\left[G \leq \frac{1}{a}\left(\ln\frac{x_2}{x_1}-\frac{a^2}{2}\right)\right] + x_2 \mathbb{P}\left[-G < \frac{1}{a}\left(\ln\frac{x_1}{x_2}-\frac{a^2}{2}\right)\right]$$

and (10) follows. In particular, we have

$$u_{f_0}(\tau, (x_1, x_1)) = 2x_1 \Phi(-\frac{\sigma_1^*}{2}\sqrt{\tau}).$$

Now set

$$g(\tau) = \alpha \ln \tau + \ln(\Phi(-\sqrt{\tau})), \quad \tau > 0.$$

The claim above follows if we prove that g is not concave for any $\alpha \in \,]0, \frac{1}{2}[$. To this end, set $\varphi = \Phi'$ so that

$$g'(\tau) = \frac{\alpha}{\tau} - \frac{\varphi(-\sqrt{\tau})}{2\sqrt{\tau}\Phi(-\sqrt{\tau})}.$$

The function g' is non-increasing if and only if the function

$$h(s) = \frac{\alpha}{s^2} - \frac{\varphi(s)}{2s\Phi(-s)}, \quad s > 0$$

is non-increasing. Now

$$h'(s) = -\frac{2\alpha}{s^3} + \frac{\varphi(s)}{2\Phi(-s)}\left(1 + \frac{1}{s^2} - \frac{\varphi(s)}{s\Phi(-s)}\right)$$

and, by using the Laplace-Feller inequality (see e.g. Tong [24]),

$$\Phi(-s) = \varphi(s)\left(\frac{1}{s} - \frac{1}{s^3} + \frac{3}{s^5} + O\left(\frac{1}{s^7}\right)\right), \quad \text{as } s \to +\infty.$$

From this

$$\frac{\varphi(s)}{\Phi(-s)} = \frac{s}{1 - \left(\frac{1}{s^2} - \frac{3}{s^4} + O\left(\frac{1}{s^6}\right)\right)}, \quad \text{as } s \to +\infty,$$

and we get

$$h'(s) = -\frac{2\alpha}{s^3} + \frac{\varphi(s)}{2\Phi(-s)}\left[1 + \frac{1}{s^2} - \left(1 + \frac{1}{s^2} - \frac{3}{s^4} + \left(\frac{1}{s^2} - \frac{3}{s^4}\right)^2 + O\left(\frac{1}{s^6}\right)\right)\right],$$

as $s \to +\infty$. Thus

$$h'(s) = -\frac{2\alpha}{s^3} + \frac{\varphi(s)}{2\Phi(-s)}\frac{2}{s^4}\left(1 + O\left(\frac{1}{s^2}\right)\right), \quad \text{as } s \to +\infty,$$

and, finally,

$$h'(s) = -\frac{2\alpha}{s^3} + \frac{1}{1+O(\frac{1}{s^2})}\frac{1}{s^3}\left(1+O\left(\frac{1}{s^2}\right)\right), \quad \text{as } s \to +\infty.$$

Therefore, if $h'(s) \leq 0$ for all $s > 0$, then necessarily $\alpha \geq \frac{1}{2}$. This proves that the function g is not concave for any $\alpha \in \,]0, \frac{1}{2}[$ and, hence, that the function $\tau^\alpha u_{f_0}(\tau, x)$ is not a log-concave function of $(\tau, \ln x)$ for any $\alpha \in \,]0, \frac{1}{2}[$. □

THEOREM 3.2. *If $f_0 = f_{\min}$, then the functions $\tau^{(m-1)/2}\phi^i_{f_0}(\tau,x)$, $i = 1,\ldots,m$, are log-concave functions of $(\tau, \ln x)$.*

PROOF. Using (5) with $f = f_0$, we have from Lemma 3.1 that

$$\phi^m_{f_0}(\tau, x) = \mathbb{E}[h(x_1 M^{W_1^*}_{\sigma_1^*}(\tau), \ldots, x_{m-1} M^{W^*_{m-1}}_{\sigma^*_{m-1}}(\tau), x_m)]$$

where

$$h(x) = \begin{cases} 1 & \text{if } x_m < f_0(x_1,\ldots,x_{m-1},1), \\ 0 & \text{if } x_m \geq f_0(x_1,\ldots,x_{m-1},1). \end{cases}$$

Since h is a log-concave function of $\ln x$, as in the proof of Theorem 3.1, the Prékopa theorem implies that the function $\tau^{(m-1)/2}\phi^m_{f_0}(\tau, x)$ is a log-concave function of $(\tau, \ln x)$. In a similar way we conclude that the functions

$$\tau^{(m-1)/2}\phi^i_{f_0}(\tau, x), \quad i = 1, \ldots, m-1,$$

are log-concave functions of $(\tau, \ln x)$. This completes the proof of Theorem 3.2. □

4. Extremal Properties of Calls

In this section we are going to prove an inequality of the so called Berwald's type (cf. [3]) for a certain class of option prices. To begin with we therefore review the Berwald inequality as well as some other closely related results due to the author [6], [8].

A real-valued function ψ is said to be convex with respect to another real-valued function φ if there exists a convex continuous function κ such that $\psi = \kappa \circ \varphi$. We shall write $\psi \in \mathcal{V}_0(\varphi)$ if the function $\psi : [0, +\infty[\to \mathbb{R}$ is convex with respect to the non-decreasing continuous function $\varphi : [0, +\infty[\to \mathbb{R}$.

Now let K be a convex body in $\mathbb{R}^n$ with volume $|K|$ and suppose $f : K \to \,]0, +\infty[$ is a given concave function. Moreover, suppose $\psi \in \mathcal{V}_0(\varphi)$ and

$$\frac{1}{|K|}\int_K \varphi(f(x))\,dx = n\int_0^1 \varphi(\xi t)(1-t)^{n-1}\,dt$$

where ξ is a suitable positive number. Under these premises Berwald [3] proves that

$$\frac{1}{|K|}\int_K \psi(f(x))\,dx \leq n\int_0^1 \psi(\xi t)(1-t)^{n-1}\,dt.$$

In [6] the same inequality is established for so called dome functions on K (i.e. functions on K which are possible to represent as the supremum of a suitable family of uniformly bounded and positive concave functions on K). Clearly, the Berwald inequality then also remains true for all functions on K which are equimeasurable with appropriate dome functions on K, a class of functions, which is optimal in connection with the Berwald inequality [6]. All these results depend on the standard Brunn-Minkowski inequality for volume measure in $\mathbb{R}^n$. In our paper [8] we proved an inequality of the Berwald type for certain sublinear functions using the so called isoperimetric inequality in Gauss space. Here, again, we will apply the isoperimetric inequality in Gauss space but this time to a class of functions different from the one in [8].

Throughout the remaining part of this paper we assume that $G = (G_1, \ldots, G_n)$ is the standard Gaussian random vector in $\mathbb{R}^n$ with stochastically independent and $N(0; 1)$-distributed components. The isoperimetric inequality for the random vector $G = (G_1, \ldots, G_n)$, independently discovered by Sudakov and Tsyrelson [23] and the author [7], reads as follows:

If $A \subseteq \mathbb{R}^n$ is a Borel set and $\mathbb{P}[G \in A] = \mathbb{P}[G_n \leq \alpha]$ for an appropriate $\alpha \in [-\infty, +\infty]$, then $\mathbb{P}[G \in A + \bar{B}(0;\varepsilon)] \geq \mathbb{P}[G_n \leq \alpha + \varepsilon]$ for $\varepsilon > 0$, where $\bar{B}(0;\varepsilon) = \{\xi \in \mathbb{R}^n; \|\xi\|_2 \leq \varepsilon\}$.

For new proofs of the isoperimetric inequality in Gauss space, see Bakry and Ledoux [1] and Bobkov [5]. Before we apply isoperimetry in Gauss space to option pricing we have to discuss some properties of so called Lipschitz functions.

A real-valued function g defined on an open subset V of $\mathbb{R}^m$ belongs to the Lipschitz class $\mathrm{Lip}_\infty(V; C)$, if $C > 0$ and

$$|g(\xi) - g(\eta)| \leq C\|\xi - \eta\|_\infty, \quad \xi, \eta \in V.$$

By a theorem of Rademacher (see e.g. Federer [15]), any function g of Lipschitz class $\mathrm{Lip}_\infty(V; C)$ is differentiable a.e. with respect to Lebesgue measure and

$$\|\nabla g(\xi)\|_1 \leq C \quad \text{a.e.}$$

Furthermore, if $0 < C_0 \leq C$ and

$$\|\nabla g(\xi)\|_1 \leq C_0 \quad \text{a.e.}$$

then $g \in \mathrm{Lip}_\infty(V; C_0)$. Given an open set $U \subseteq \mathbb{R}^m$, we will write $g \in \mathrm{Lip}_{\mathrm{loc}}(U)$, if to any relatively compact open subset V of U, the restriction of g to V belongs to the class $\mathrm{Lip}_\infty(V; C)$ for an appropriate $C > 0$.

A function $f \in \mathrm{Lip}_{\mathrm{loc}}(\mathbb{R}^m_+)$ is said to belong to the class $\mathcal{C}$ if $f > 0$, that is, $f(x) > 0, x \in \mathbb{R}^m_+$, and

$$\langle x, |\nabla f(x)| \rangle \leq f(x) \quad \text{a.e.} \tag{11}$$

Given $a > 0$, we define

$$\mathcal{C}_a = (\mathcal{C} - a)^+.$$

Stated more explicitly, a function f belongs to the class $\mathcal{C}_a$ if and only if $f \in \text{Lip}_{\text{loc}}(\mathbb{R}^m_+)$, f is non-negative, and

$$\langle x, |\nabla f(x)| \rangle \leq a + f(x) \quad \text{a.e.}$$

THEOREM 4.1. *Suppose $f : \mathbb{R}^m_+ \to [0, +\infty[$ and let $a > 0$. Then $f \in \mathcal{C}$ if and only if $f > 0$ and*

$$f(xe^\xi) \leq f(x) e^{\|\xi\|_\infty}, \quad x \in \mathbb{R}^m_+, \ \xi \in \mathbb{R}^m. \tag{12}$$

Moreover, $f \in \mathcal{C}_a$ if and only if

$$a + f(xe^\xi) \leq (a + f(x)) e^{\|\xi\|_\infty}, \quad x \in \mathbb{R}^m_+, \ \xi \in \mathbb{R}^m.$$

In particular, any $f \in \mathcal{C}_a$ is a payoff function.

PROOF. Suppose first that $f > 0$ and set

$$g(\xi) = \ln f(e^\xi), \quad \xi \in \mathbb{R}^m.$$

Clearly, the inequality (12) just means that $g \in \text{Lip}_\infty(\mathbb{R}^m; 1)$.

Now let $f \in \mathcal{C}$. Then $g \in \text{Lip}_{\text{loc}}(\mathbb{R}^m)$ and

$$\nabla g(\xi) = \frac{e^\xi \nabla f(e^\xi)}{f(e^\xi)} \quad \text{a.e.} \tag{13}$$

Moreover, $\|\nabla g(\xi)\|_1 \leq 1$, a.e. and, hence, $g \in \text{Lip}_\infty(\mathbb{R}^m; 1)$.

Conversely, suppose $g \in \text{Lip}_\infty(\mathbb{R}^m; 1)$. Then $f \in \text{Lip}_{\text{loc}}(\mathbb{R}^m_+)$ and (13) holds. Accordingly, the inequality (11) must be true. Summing up, we have proved that $f \in \mathcal{C}$ if and only if (12) is true. The remaining part of Theorem 4.1 is now obvious from the very definition of the class $\mathcal{C}_a$. This concludes our proof of Theorem 4.1. □

In general, the following properties are immediate consequences of either Theorem 4.1 or the very definition of the class $\mathcal{C}_a$:

(a) $\mathcal{C}_a$ is convex.
(b) $\mathcal{C}_a \subseteq \mathcal{C}_b$ if $a \leq b$.
(c) $c \in \mathcal{C}_a$ if $c \geq 0$.
(d) $\lambda \mathcal{C}_a = \mathcal{C}_{\lambda a}$, $\lambda > 0$.
(e) $\theta \mathcal{C}_a \subseteq \mathcal{C}_a$, $0 < \theta < 1$.
(f) $\mathcal{C}_a + \mathcal{C}_b \subseteq \mathcal{C}_{a+b}$.
(g) $f, g \in \mathcal{C}_a \Rightarrow \max(f, g) \in \mathcal{C}_a$.
(h) $f, g \in \mathcal{C}_a \Rightarrow \min(f, g) \in \mathcal{C}_a$.
(i) If T is an n by n permutation matrix or an n by n diagonal matrix with positive entries, then $f(x) \in \mathcal{C}_a \Rightarrow f(Tx) \in \mathcal{C}_a$.
(j) For any $i = 1, \ldots, m$, $c_{b,i} \in \mathcal{C}_a$ if and only if $b \leq a$.
(k) For any $i = 1, \ldots, m$, $\lambda c_{a,i} \notin \mathcal{C}_a$ if $\lambda > 1$.
(l) For any $i = 1, \ldots, m$, $p_{b,i} \in \mathcal{C}_a$ if and only if $b \leq a$.
(m) For any $i = 1, \ldots, m$, $\lambda p_{a,i} \notin \mathcal{C}_a$ if $\lambda > 1$.

(n) $f \in \mathcal{C}_a$ if f is non-negative and concave.
(o) $f \in \mathcal{C}_a \Rightarrow e^{r\tau} S_\tau f \in \mathcal{C}_a$.

Here, for the sake of completeness, we indicate a proof of Property **(n)**. To begin with it is well known that a concave function f on $\mathbb{R}_+^m$ belongs to the class $\mathrm{Lip}_{\mathrm{loc}}(\mathbb{R}_+^m)$ and that the convex set $\{(x,t) \in \mathbb{R}_+^m \times \mathbb{R}; t \leq f(x)\}$ has a hyperplane of support at each point $(x, f(x))$, $x \in \mathbb{R}_+^m$ (see e.g. Hörmander [17]). Moreover, by the Rademacher theorem referred to above, there exists a set $D \subseteq \mathbb{R}_+^m$ such that f is differentiable at each point of D and such that the complement of D in $\mathbb{R}_+^m$ is a null set. Accordingly,

$$f(y) \leq f(x) + \langle \nabla f(x), y - x \rangle, \quad x \in D, \ y \in \mathbb{R}_+^m.$$

Thus, given $x \in D$, we have $\langle \nabla f(x), x \rangle \leq f(x)$ as f is non-negative. But here $\nabla f(x) \geq 0$ since the function $h(s) = f(x + sy)$, $s \geq 0$, is non-decreasing for all $x, y \in \mathbb{R}_+^m$. This proves (11) so that $f \in \mathcal{C}_a$ for every $a > 0$.

Throughout the remaining part of this paper we assume that

$$\max_{i=1,\ldots,m} \sigma_i = \sigma_m.$$

Now given $a > 0$ and a continuous function $f : \mathbb{R}_+^m \to [0, +\infty[$, set for fixed $\tau > 0$,

$$g = g_\tau = \frac{1}{\sigma_m \sqrt{\tau}} \ln(1 + f/a). \tag{14}$$

We shall say that the function f belongs to the class $\mathcal{C}_{a,m}$ if the function

$$\Phi^{-1}(\mathbb{P}[g_\tau(x_1 e^{\sigma_1 \sqrt{\tau} \langle c_1, G \rangle}, \ldots, x_m e^{\sigma_m \sqrt{\tau} \langle c_m, G \rangle}) \leq s]) - s, \quad s > 0$$

is non-decreasing for every $x \in \mathbb{R}_+^m$ and every $\tau > 0$. It is readily seen that any $f \in \mathcal{C}_{a,m}$ is a payoff function. The class $\mathcal{C}_{a,m}$ turns out to be optimal in connection with a certain isoperimetric inequality we prove below. However, before stating this result we want to prove

THEOREM 4.2. *For any $a > 0$,*

$$\mathcal{C}_a \subseteq \mathcal{C}_{a,m}.$$

PROOF. Suppose $f \in \mathcal{C}_a$ and let g be as in (14), where $\tau > 0$ is fixed. We now claim that

$$g(ze^\xi) \leq g(z) + \frac{\|\xi\|_\infty}{\sigma_m \sqrt{\tau}}$$

if $z \in \mathbb{R}_+^m$ and $\xi \in \mathbb{R}^m$. But

$$g(ze^\xi) = \frac{1}{\sigma_m \sqrt{\tau}} \ln((a + f(ze^\xi))/a)$$

and since $f \in \mathcal{C}_a$, Theorem 4.1 yields

$$g(ze^\xi) \leq \frac{1}{\sigma_m \sqrt{\tau}} \ln((a + f(z))e^{\|\xi\|_\infty}/a)$$

and the claim above follows at once.

To complete the proof of Theorem 4.2 we represent the standard Gaussian random vector G in $\mathbb{R}^n$ as the identity mapping in $\mathbb{R}^n$ and put for any fixed $x \in \mathbb{R}_+^n$ and $s > 0$,
$$A(s) = [g(x_1 e^{\sigma_1 \sqrt{\tau} \langle c_1, G \rangle}, \ldots, x_m e^{\sigma_m \sqrt{\tau} \langle c_m, G \rangle}) \leq s].$$
Then, if $\varepsilon > 0$,
$$A(s) + \bar{B}(0; \varepsilon) \subseteq A(s + \varepsilon)$$
and the isoperimetric inequality for G gives
$$\Phi^{-1}(\mathbb{P}[A(s+\varepsilon)]) \geq \Phi^{-1}(\mathbb{P}[A(s)]) + \varepsilon.$$
Since $x \in \mathbb{R}_+^m$ and $\tau > 0$ are arbitrary, $f \in \mathcal{C}_{a,m}$ and Theorem 4.2 is proved. $\square$

In what follows we shall write $\psi \in \mathcal{V}(\varphi)$ if $\psi \in \mathcal{V}_0(\varphi)$ and
$$\overline{\lim}_{s \to +\infty} s^{-p}(|\varphi(s)| + |\psi(s)|) < +\infty$$
for an appropriate $p > 0$.

THEOREM 4.3. *Suppose* $\psi \in \mathcal{V}(\varphi)$. *Then, if* $f \in \mathcal{C}_{a,m}$ *and*
$$u_{\varphi \circ f}(\tau, x) = u_{\varphi \circ c_{a,m}}(\tau, y)$$
where $x, y \in \mathbb{R}_+^m$ *and* $\tau > 0$ *are fixed,*
$$u_{\psi \circ f}(\tau, x) \leq u_{\psi \circ c_{a,m}}(\tau, y).$$

PROOF. In the proof, without loss of generality, we assume that $\varphi(0) = \psi(0) = 0$. We have
$$c_{a,m}(y e^{r\tau} M_\sigma^W(\tau)) = (y_m e^{(r - \sigma_m^2/2)\tau + \sigma_m W_m(\tau)} - a)^+$$
and hence
$$c_{a,m}(y e^{r\tau} M_\sigma^W(\tau)) = a(e^{a_m + \sigma_m W_m(\tau)} - 1)^+$$
for a suitable constant a_m. Setting $B_m = W_m(\tau)/\sqrt{\tau}$, we get
$$c_{a,m}(y e^{r\tau} M_\sigma^W(\tau)) = a(e^{\sigma_m \sqrt{\tau}(B_m - b_m)} - 1)^+$$
for a suitable constant b_m. Thus
$$c_{a,m}(y e^{r\tau} M_\sigma^W(\tau)) = a(e^{\sigma_m \sqrt{\tau}(B_m - b_m)^+} - 1).$$
Now define
$$j(s) = a(e^{\sigma_m \sqrt{\tau} s} - 1), \quad s \geq 0$$
and set $\varphi_0 = \varphi(j)$ so that
$$\varphi(c_{a,m}(y e^{r\tau} M_\sigma^W(\tau))) = \varphi_0((B_m - b_m)^+)$$
and
$$\mathbb{E}[\varphi(c_{a,m}(y e^{r\tau} M_\sigma^W(\tau)))] = \int_0^{+\infty} \mathbb{P}[(B_m - b_m)^+ > s] \, d\varphi_0(s)$$

since $\varphi_0(0) = 0$.

In the next step we introduce the function $g = g_\tau$ by the equation (14) and have $f = j(g)$ and $\varphi(f) = \varphi_0(g)$. Thus

$$\varphi(f(xe^{r\tau}M_\sigma^W(\tau))) = \varphi_0(g(xe^{r\tau}M_\sigma^W(\tau)))$$

and

$$\mathbb{E}[\varphi(f(xe^{r\tau}M_\sigma^W(\tau)))] = \int_0^{+\infty} \mathbb{P}[g(xe^{r\tau}M_\sigma^W(\tau)) > s]\,d\varphi_0(s).$$

Further, we define

$$h(s) = \mathbb{P}[g(xe^{r\tau}M_\sigma^W(\tau)) \leq s], \quad s \geq 0$$

and have

$$h(s) = \mathbb{P}[g(z_1 e^{\sigma_1\sqrt{\tau}\langle c_1, G\rangle}, \ldots, z_m e^{\sigma_m\sqrt{\tau}\langle c_m, G\rangle}) \leq s], \quad s \geq 0$$

for appropriate $z_1, \ldots, z_m \in \mathbb{R}_+$.

Now suppose $s_0 \geq 0$ and

$$h(s_0) \geq \mathbb{P}[(B_m - b_m)^+ \leq s_0]. \tag{15}$$

We then have

$$h(s_0 + \varepsilon) \geq \mathbb{P}[(B_m - b_m)^+ \leq s_0 + \varepsilon], \quad \varepsilon > 0$$

because $f \in \mathcal{C}_{a,m}$ and B_m is a $N(0;1)$-distributed random variable. To complete the proof of Theorem 4.3, first set $\psi_0 = \psi(j)$ so that

$$\mathbb{E}[\psi(c_{a,m}(ye^{r\tau}M_\sigma^W(\tau)))] = \int_0^{+\infty} \mathbb{P}[(B_m - b_m)^+ > s]\,d\psi_0(s)$$

and

$$\mathbb{E}[\psi(f(xe^{r\tau}M_\sigma^W(\tau)))] = \int_0^{+\infty} \mathbb{P}[g(xe^{r\tau}M_\sigma^W(\tau)) > s]\,d\psi_0(s)$$

since $\psi_0(0) = 0$. Moreover, let $d\psi_0 = \lambda d\varphi_0$, where the function λ is nondecreasing, and let s_* denote the infimum over all $s_0 \geq 0$ such that (15) holds. Here, by convention, the infimum over the empty set equals $+\infty$. The extreme cases $s_* = 0$ and $s_* = +\infty$ are simple and so we concentrate on the case $0 < s_* < +\infty$. Then, for any $S \in\,]s_*, +\infty[$,

$$\int_0^S \mathbb{P}[g(xe^{r\tau}M_\sigma^W(\tau)) > s]\,d\psi_0(s) - \int_0^S \mathbb{P}[(B_m - b_m)^+ > s]\,d\psi_0(s)$$

$$= \int_0^{s_*} (\mathbb{P}[g(xe^{r\tau}M_\sigma^W(\tau)) > s] - \mathbb{P}[(B_m - b_m)^+ > s])\lambda(s)\,d\varphi_0(s)$$

$$+ \int_{s_*}^S (\mathbb{P}[g(xe^{r\tau}M_\sigma^W(\tau)) > s] - \mathbb{P}[(B_m - b_m)^+ > s])\lambda(s)\,d\varphi_0(s).$$

Here the right-hand side does not exceed

$$\lambda(s_*) \int_0^{s_*} \left(\mathbb{P}[g(xe^{r\tau}M_\sigma^W(\tau)) > s] - \mathbb{P}[(B_m - b_m)^+ > s]\right) d\varphi_0(s)$$

$$+ \lambda(s_*) \int_{s_*}^S \left(\mathbb{P}[g(xe^{r\tau}M_\sigma^W(\tau)) > s] - \mathbb{P}[(B_m - b_m)^+ > s]\right) d\varphi_0(s),$$

which is equal to

$$\lambda(s_*) \int_0^S \left(\mathbb{P}[g(xe^{r\tau}M_\sigma^W(\tau)) > s] - \mathbb{P}[(B_m - b_m)^+ > s]\right) d\varphi_0(s).$$

By letting S tend to plus infinity it is immediate that

$$\int_0^{+\infty} \mathbb{P}[g(xe^{r\tau}M_\sigma^W(\tau)) > s] \, d\psi_0(s) \leq \int_0^{+\infty} \mathbb{P}[(B_m - b_m)^+ > s] \, d\psi_0(s)$$

and Theorem 4.3 follows at once. $\square$

If X is a non-negative random variable with positive expectation, we set

$$D_{\text{rel}}[X] = \frac{\sqrt{\mathbb{E}[X^2] - (\mathbb{E}[X])^2}}{\mathbb{E}[X]}.$$

Moreover, if f is a payoff function, we use the notation

$$Z(\tau, x; f) = e^{-r\tau} f(xe^{r\tau} M_\sigma^W(\tau)).$$

Note that

$$u_f(\tau, x) = \mathbb{E}[Z(\tau, x; f)]$$

by (4).

COROLLARY 4.1. *Suppose $x, y \in \mathbb{R}_+^m$. If $f \in \mathcal{C}_{a,m}$ and*

$$u_f(\tau, x) \geq u_{c_{a,m}}(\tau, y) \tag{16}$$

then

$$D_{\text{rel}}[Z(\tau, x; f)] \leq D_{\text{rel}}[Z(\tau, y; c_{a,m})].$$

PROOF. If there is equality in (16) the conclusion in Corollary 4.1 follows from Theorem 4.3. To prove the general case it therefore suffices to show that the function

$$F(y; a) = \frac{\mathbb{E}[Z^2(\tau, y, c_{a,m})]}{(\mathbb{E}[Z(\tau, y, c_{a,m})])^2}$$

is a non-increasing function of y_m. To this end, first choose $0 < b < a$ and note that $\theta c_{b,m} \in \mathcal{C}_a$ for all $0 < \theta \leq 1$. Since $c_{b,m} \geq c_{a,m}$ there is a $\theta \in]0, 1]$ such that

$$u_{\theta c_{b,m}}(\tau, y) = u_{c_{a,m}}(\tau, y).$$

Accordingly, in view of Theorem 4.3,

$$u_{(\theta c_{b,m})^2}(\tau, y) \leq u_{c_{a,m}^2}(\tau, y)$$

and hence
$$F(y; b) \leq F(y; a).$$
Now since
$$F\left(\frac{a}{b}y; a\right) = F(y; b)$$
we are done. This completes our proof of Corollary 4.1. □

EXAMPLE 4.1. The use of the Monte Carlo method in computing option prices goes back to Boyle [10]. It is especially attractive for options depending on several assets (see, for example, Barraquand [2]). To estimate the price function $u_f(\tau, x)$ in this way, let $Z_1, \ldots, Z_N$ be stochastically independent copies of $Z(\tau, x; f)$. Then the arithmetic mean
$$\bar{Z}_N = \frac{1}{N} \sum_1^N Z_j$$
is an unbiased estimator of $u_f(\tau, x)$. The variance of $\bar{Z}_N$ equals $1/N$ times the variance of $Z(\tau, x; f)$ and, assuming $u_f(\tau, x) > 0$, the Chebychev inequality yields
$$\mathbb{P}\left[\left|\frac{\bar{Z}_N - u_f(\tau, x)}{u_f(\tau, x)}\right| \geq \varepsilon\right] \leq \frac{1}{\varepsilon^2 N}(D_{\text{rel}}[Z(\tau, x; f)])^2, \; \varepsilon > 0.$$
Therefore, it is interesting to have an explicit upper bound of $D_{\text{rel}}[Z(\tau, x; f)]$.

As an example, consider the special case $f = f_{\max}$. If $f = f_{\max}$, clearly $f \in \mathcal{C}_a$ for all $a > 0$. Now, if $a > 0$, (16) is true with $y = x$ and Corollary 4.1 yields
$$D_{\text{rel}}[Z(\tau, x; f_{\max})] \leq \lim_{a \to 0^+} D_{\text{rel}}[Z(\tau, x; c_{a,m})].$$
Thus
$$D_{\text{rel}}[Z(\tau, x; f_{\max})] \leq \sqrt{e^{\sigma_m^2 \tau} - 1}.$$
A completely different approximate method for computing the value of the option on the maximum (or minimum) of several assets is treated by Boyle and Tse [12]. □

The next theorem shows that the class $\mathcal{C}_{a,m}$ in Theorem 4.3 is the best possible. Indeed, we have

THEOREM 4.4. *Let f be a payoff function in $\mathbb{R}_+^m$ and suppose $a > 0$. Furthermore, suppose*
$$u_{\psi \circ f}(\tau, x) \leq u_{\psi \circ c_{a,m}}(\tau, y)$$
for all $x, y \in \mathbb{R}_+^m$, all $\tau > 0$, and all ψ and bounded φ such that $\psi \in \mathcal{V}(\varphi)$ and
$$u_{\varphi \circ f}(\tau, x) = u_{\varphi \circ c_{a,m}}(\tau, y).$$
Then $f \in \mathcal{C}_{a,m}$.

PROOF. To begin with, given $0 < b < c$, define $\varphi_\alpha(s) = (s-\alpha)^+$, $\alpha > 0$, and $\varphi_{b,c}(s) = \varphi_b(s) - \varphi_c(s)$ so that

$$\varphi_{b,c}(s) = \min((s-b)^+, c-b).$$

First we assume that $x \in \mathbb{R}_+^m$ and $\tau > 0$ are fixed and

$$0 < \mathbb{E}[\varphi_{b,c}(f(xe^{r\tau}M_\sigma^W(\tau)))] < c-b. \tag{17}$$

Now let $\theta > 0$ be such that

$$\mathbb{E}[\varphi_{b,c}(f(xe^{r\tau}M_\sigma^W(\tau)))] = \mathbb{E}[\varphi_{b,c}(c_{a,m}(\theta xe^{r\tau}M_\sigma^W(\tau)))]. \tag{18}$$

We next choose $h > 0$ so small that $b < b+h < c$ and define $\kappa(s) = -\min(s,h)$. Then $-\varphi_{b,b+h} = \kappa \circ \varphi_{b,c}$ and thus

$$\mathbb{E}[\varphi_{b,b+h}(f(xe^{r\tau}M_\sigma^W(\tau)))] \geq \mathbb{E}[\varphi_{b,b+h}(c_{a,m}(\theta xe^{r\tau}M_\sigma^W(\tau)))].$$

In the following, if $A \subseteq \mathbb{R}$, the function χ_A defined on $\mathbb{R}$ equals one in A and zero off A. Using this notation,

$$h\chi_{]b,+\infty[}(s) \geq \varphi_{b,b+h} \tag{19}$$

and hence

$$h\mathbb{P}[f(xe^{r\tau}M_\sigma^W(\tau)) > b] \geq \mathbb{E}[\varphi_{b,b+h}(c_{a,m}(xe^{r\tau}M_\sigma^W(\tau)))].$$

Thus

$$h\mathbb{P}[f(xe^{r\tau}M_\sigma^W(\tau)) > b]$$
$$\geq \mathbb{E}[((\theta x_m e^{r\tau}M_{\sigma_m}^{W_m}(\tau) - a)^+ - b)^+] - \mathbb{E}[((\theta x_m e^{r\tau}M_{\sigma_m}^{W_m}(\tau) - a)^+ - b - h)^+].$$

From this we get

$$\mathbb{P}[f(xe^{r\tau}M_\sigma^W(\tau)) > b] \geq -\frac{d}{db}\mathbb{E}[((\theta x_m e^{r\tau}M_{\sigma_m}^{W_m}(\tau) - a)^+ - b)^+]$$

where the right-hand side equals

$$\mathbb{P}[(\theta x_m e^{r\tau}M_{\sigma_m}^{W_m}(\tau) - a)^+ - b > 0] = \mathbb{P}[\theta x_m e^{r\tau}M_{\sigma_m}^{W_m}(\tau) - a - b > 0]$$

and, accordingly,

$$\mathbb{P}[f(xe^{r\tau}M_\sigma^W(\tau)) > b] \geq \Phi\left(\frac{1}{\sigma_m\sqrt{\tau}}\ln\frac{\theta x_m e^{(r-\sigma_m^2/2)\tau}}{a+b}\right). \tag{20}$$

Now suppose $b < c-h < c$ and observe that $\varphi_{c-h,c} = \varphi_{c-h-b} \circ \varphi_{b,c}$. Remembering (18), we have

$$\mathbb{E}[\varphi_{c-h,c}(f(xe^{r\tau}M_\sigma^W(\tau)))] \leq \mathbb{E}[\varphi_{c-h,c}(c_{a,m}(\theta xe^{r\tau}M_\sigma^W(\tau)))].$$

Furthermore, since

$$h\chi_{]c,+\infty[}(s) \leq \varphi_{c-h,c}(s) \tag{21}$$

it follows that
$$h\mathbb{P}[f(xe^{r\tau}M_\sigma^W(\tau)) > c] \le \mathbb{E}[\varphi_{c-h,c}(c_{a,m}(xe^{r\tau}M_\sigma^W(\tau)))]$$

and as above we conclude that
$$\mathbb{P}[f(xe^{r\tau}M_\sigma^W(\tau)) > c] \le \Phi\left(\frac{1}{\sigma_m\sqrt{\tau}}\ln\frac{\theta x_m e^{(r-\sigma_m^2/2)\tau}}{a+c}\right). \qquad (22)$$

Comparing (20) and (22) it follows that
$$(a+b)e^{\sigma_m\sqrt{\tau}\Phi^{-1}(\mathbb{P}[f(xe^{r\tau}M_\sigma^W(\tau))>b])} \ge (a+c)e^{\sigma_m\sqrt{\tau}\Phi^{-1}(\mathbb{P}[f(xe^{r\tau}M_\sigma^W(\tau))>c])}.$$

Clearly, this inequality also holds if (17) is violated and we conclude that the function
$$\sigma_m\sqrt{\tau}\Phi^{-1}(\mathbb{P}[f(xe^{r\tau}M_\sigma^W(\tau)) \le s]) - \ln(a+s), \quad s \ge 0$$
is non-decreasing. Since this is true for all $x \in \mathbb{R}_+^m$ and all $\tau > 0$ we conclude that the function f belongs to the class $\mathcal{C}_{a,m}$, which completes our proof of Theorem 4.4. □

Suppose now that f is a general payoff function. The expectation at time t of the value of the derivative security $\mathcal{U}_f^T$ at the maturity date T equals $v_f(\tau, X(t); 0)$, where
$$v_f(\tau, x; 0) = \mathbb{E}[f(xe^{\mu\tau+\sigma W(\tau)})].$$
Here we employ the vector notation $\mu = (\mu_1, \ldots, \mu_m), \sigma = (\sigma_1, \ldots, \sigma_m)$, and $W(\tau) = (W_1(\tau), \ldots, W_m(\tau))$. Thus
$$v_f(\tau, x; 0) = e^{r\tau}u_f(\tau, xe^{(\mu-r+\sigma^2/2)\tau}).$$

Now, suppose $t < t_* \le T$ and set $\tau_* = T - t_*$. If $\tau_* > 0$, the expectation at time t of the value of $\mathcal{U}_f^T$ at time t_* equals $v_f(\tau, X(t); \tau_*)$, where
$$v_f(\tau, x; \tau_*) = \mathbb{E}[u_f(\tau_*, xe^{\mu(t_*-t)+\sigma W^0(t_*-t)})]$$
and where W^0 is a stochastically independent copy of W. Hence
$$v_f(\tau, x; \tau_*) = \mathbb{E}[e^{-r\tau_*}f(xe^{\mu(t_*-t)+\sigma W^0(t_*-t)}e^{r\tau_*}M_\sigma^W(\tau_*))].$$
Since $t_* - t = \tau - \tau_*$, we get
$$v_f(\tau, x; \tau_*) = \mathbb{E}[e^{-r\tau_*}f(xe^{(\mu+\sigma^2/2)(\tau-\tau_*)}e^{r\tau_*}M_\sigma^W(\tau))].$$
Thus
$$v_f(\tau, x; \tau_*) = e^{r(\tau-\tau_*)}u_f(\tau, xe^{(\mu-r+\sigma^2/2)(\tau-\tau_*)}).$$
Alternatively, it is simple to derive the same formula using the semi-group property of the family $(S_\tau)_{\tau>0}$.

Theorem 4.3 thus has the following consequence:

COROLLARY 4.2. *Let $\psi \in \mathcal{V}(\varphi)$. Then, if $f \in \mathcal{C}_{a,m}$ and*
$$v_{\varphi \circ f}(\tau, x; \tau_*) = v_{\varphi \circ c_{a,m}}(\tau, y; \tau_*)$$
where $x, y \in \mathbb{R}_+^m$ and $\tau \geq \tau_* \geq 0$ are fixed,
$$v_{\psi \circ f}(\tau, x; \tau_*) \leq v_{\psi \circ c_{a,m}}(\tau, y; \tau_*).$$

5. Extremal Properties of Puts

Given $a > 0$, we define
$$\mathcal{P}_a = (a - \mathcal{C})^+.$$
Stated more explicitly, a function $f \in \mathcal{P}_a$ if and only if $f \in \text{Lip}_{\text{loc}}(\mathbb{R}_+^m)$, $0 \leq f < a$ and
$$\langle x, |\nabla f(x)| \rangle + f(x) \leq a, \quad \text{a.e.}$$
In view of Theorem 4.1 we now have

THEOREM 5.1. *Suppose $a > 0$ and let $f : \mathbb{R}_+^m \to [0, a[$. Then $f \in \mathcal{P}_a$ if and only if*
$$\frac{1}{a - f(xe^\xi)} \leq \frac{e^{\|\xi\|_\infty}}{a - f(x)}, \quad x \in \mathbb{R}_+^m, \ \xi \in \mathbb{R}_+^m.$$

In general, the following properties are immediate consequences of either Theorem 5.1 or the very definition of the class $\mathcal{P}_a$:

(a) $\mathcal{P}_a$ is convex.
(b) $\mathcal{P}_a \subseteq \mathcal{P}_b$ if $a \leq b$.
(c) $c \in \mathcal{P}_a$ if $0 \leq c < a$.
(d) $\lambda \mathcal{P}_a = \mathcal{P}_{\lambda a}$, $\lambda > 0$.
(e) $\theta \mathcal{P}_a \subseteq \mathcal{P}_a$, $0 < \theta < 1$.
(f) $\mathcal{P}_a + \mathcal{P}_b \subseteq \mathcal{P}_{a+b}$.
(g) $f, g \in \mathcal{P}_a \Rightarrow \max(f, g) \in \mathcal{P}_a$.
(h) $f, g \in \mathcal{P}_a \Rightarrow \min(f, g) \in \mathcal{P}_a$.
(i) If T is an n by n permutation matrix or an n by n diagonal matrix with positive entries, then $f(x) \in \mathcal{P}_a \Rightarrow f(Tx) \in \mathcal{P}_a$.
(j) For any $i = 1, \ldots, m$, $p_{b,i} \in \mathcal{P}_a$ if and only if $b \leq a$.
(k) For any $i = 1, \ldots, m$, $\lambda p_{a,i} \notin \mathcal{P}_a$ if $\lambda > 1$.
(l) $f \in \mathcal{P}_a$ if $0 \leq f < a$ is convex.
(m) $f \in \mathcal{P}_a \Rightarrow e^{r\tau} S_\tau f \in \mathcal{P}_a$.

We are now going to introduce slightly larger classes of payoff functions than the classes $\mathcal{P}_a$, $a > 0$. To this end, let $a > 0$ be given and suppose $f : \mathbb{R}_+^m \to [0, a[$ is a continuous function and set for fixed $\tau > 0$,
$$g = g_\tau = -\frac{1}{\sigma_m \sqrt{\tau}} \ln(1 - f/a). \tag{23}$$

We shall say that the the function f belongs to the class $\mathcal{P}_{a,m}$ if the function

$$\Phi^{-1}(\mathbb{P}[g_\tau(x_1 e^{\sigma_1\sqrt{\tau}\langle c_1, G\rangle}, \ldots, x_m e^{\sigma_m\sqrt{\tau}\langle c_m, G\rangle}) \leq s]) - s, \quad s > 0$$

is non-decreasing for every $x \in \mathbb{R}_+^m$ and $\tau > 0$. Here, again, $G = (G_1, \ldots, G_n)$ denotes the standard Gaussian random vector in $\mathbb{R}^n$ with stochastically independent $N(0;1)$-distributed components.

We now set, for any $f \in \mathcal{P}_{a,m}$,

$$I_a(f) = \frac{af}{a-f}$$

and have

$$(1 - f/a)(1 + I_a(f)/a) = 1.$$

THEOREM 5.2. (a) *The map I_a is a bijection of $\mathcal{P}_{a,m}$ onto $\mathcal{C}_{a,m}$.*
(b) *The restriction map of I_a to $\mathcal{P}_a$ is a bijection of $\mathcal{P}_a$ onto $\mathcal{C}_a$.*

PROOF. Part (a) follows at once from the equations (14) and (23). Moreover Part (b) is an immediate consequence of Theorems 4.1 and 5.1. □

THEOREM 5.3. *Suppose $\psi \in \mathcal{V}(\varphi)$. Then, if $f \in \mathcal{P}_{a,m}$ and*

$$u_{\varphi \circ f}(\tau, x) = u_{\varphi \circ p_{a,m}}(\tau, y)$$

where $x, y \in \mathbb{R}_+^m$ and $\tau > 0$ are fixed,

$$u_{\psi \circ f}(\tau, x) \leq u_{\psi \circ p_{a,m}}(\tau, y).$$

PROOF. Set $f^* = I_a(f)$ and $p_{a,m}^* = I_a(p_{a,m})$. Then

$$f = \frac{af^*}{a+f^*}$$

and

$$p_{a,m} = \frac{ap_{a,m}^*}{a+p_{a,m}^*}.$$

Moreover, we define

$$\varphi^*(s) = \varphi\left(\frac{as}{a+s}\right), \quad s \geq 0.$$

Then

$$u_{\varphi \circ f}(\tau, x) = u_{\varphi^* \circ f^*}(\tau, x)$$

and

$$u_{\varphi \circ p_{a,m}}(\tau, y) = u_{\varphi^* \circ p_{a,m}^*}(\tau, y).$$

From the definition of the map I_a it follows that

$$p_{a,m}^*(v) = \left(\frac{a^2}{v_m} - a\right)^+, \quad v \in \mathbb{R}_+^m$$

and using (4) we conclude that

$$u_{\varphi^* \circ p_{a,m}^*}(\tau, y) = u_{\varphi^* \circ c_{a,m}}(\tau, z)$$

where $z \in \mathbb{R}_+^m$ and
$$z_m = \frac{a^2 e^{-2r\tau + \sigma_m^2 \tau}}{y_m}.$$
The result is now an immediate consequence of Theorem 4.3. This completes our proof of Theorem 5.3. □

THEOREM 5.4. *Suppose $a > 0$ and let $f : \mathbb{R}_+^m \to [0, a[$ be a payoff function. Furthermore, suppose*
$$u_{\psi \circ f}(\tau, x) \leq u_{\psi \circ p_{a,m}}(\tau, y)$$
for all $x, y \in \mathbb{R}_+^m$, all $\tau > 0$, and all ψ and bounded φ such that $\psi \in \mathcal{V}(\varphi)$ and
$$u_{\varphi \circ f}(\tau, x) = u_{\varphi \circ p_{a,m}}(\tau, y).$$
Then $f \in \mathcal{P}_{a,m}$.

PROOF. By exploiting the map I_a as in the proof of Theorem 5.3 the result follows at once from Theorem 4.4. □

The next result follows from Theorem 5.3 in the same way as Corollary 4.2 follows from Theorem 4.3.

COROLLARY 5.1. *Let $\psi \in \mathcal{V}(\varphi)$. Then, if $f \in \mathcal{P}_{a,m}$ and*
$$v_{\varphi \circ f}(\tau, x; \tau_*) = v_{\varphi \circ p_{a,m}}(\tau, y; \tau_*),$$
where $x, y \in \mathbb{R}_+^m$ and $\tau > \tau_* \geq 0$ are fixed,
$$v_{\psi \circ f}(\tau, x; \tau_*) \leq v_{\psi \circ p_{a,m}}(\tau, y; \tau_*).$$

References

[1] Bakry, D. and Ledoux, M. Lévy-Gromov isoperimetric inequality for an infinite dimensional diffusion generator. Invent. math. **123** (1995), 259–281.

[2] Barraquand, J. Numerical valuation of high dimensional multivariate European securities. Paris Research Laboratory, Report No. 26 (1993).

[3] Berwald, L. Verallgemeinerungen eines Mittelwertsatzes von J. Favard für positive konkave Funktionen. Acta Math. **79** (1947), 16–37.

[4] Black, F. and Scholes, M. The pricing of options and corporate liabilities. Journal of Political Economy **72** (1973), 164–175.

[5] Bobkov, S. An isoperimetric inequality on the discrete cube and an elementary proof of the isoperimetric inequality in Gauss space. Ann. Prob. **25** (1997), 206–214.

[6] Borell, Ch. Integral inequalities for generalized concave or convex functions. J. Math. Anal. Appl. **43** (1973), 419–440.

[7] Borell, Ch. The Brunn-Minkowski inequality in Gauss space. Invent. math. **30** (1975), 207–216.

[8] Borell, Ch. Gaussian Radon measures on locally convex spaces. Math. Scand. **38** (1976), 265–284.

[9] Borell, Ch. Geometric properties of some familiar diffusions in $\mathbb{R}^n$. Ann. Probability **21** (1993), 482–489.

[10] Boyle P. P. Options: A Monte Carlo approach. Journal of Financial Economics **4** (1977), 323–338.

[11] Boyle P. P. The quality option and timing options in future contracts. The Journal of Finance **44** (1989), 101–113.

[12] Boyle, P. P. and Tse, Y. K. An algorithm for computing values of options on the maximum or minimum of several assets. Journal of Finance and Quantitative Analysis **25** (1990), 215–227.

[13] Davidovič, Ju. S., Korenbljum, B. I., Hacet, T. A property of logarithmically concave functions. Soviet Math. Dokl. **10** (1969), 477–480.

[14] Duffie, D. Dynamic Asset Pricing Theory. Princeton University Press, Princeton (1992).

[15] Federer, H. Geometric Measure Theory. Springer, Berlin and Heidelberg (1969).

[16] Harrison, J. M. and Pliska S. R. Martingales and stochastic integrals in the theory of continuous trading. Stochastic Processes and their Applications **11** (1981), 215–216.

[17] Hörmander, L. Notions of Convexity. Birkhäuser, Boston, Basel, and Berlin (1994).

[18] Johnson, H. Options on the maximum or minimum of several assets. Journal of Financial and Quantitative Analysis **22** (1987), 277–283.

[19] Margrabe, W. The value of an option to exchange one asset for another. Journal of Finance **33** (1978), 177–186.

[20] Merton, R. The theory of rational option pricing. Bell Journal of Economics and Management Science **4** (1973), 141–183.

[21] Prékopa, A. On logarithmic concave measures and functions. Acta Sci. Math. (Szeged) **34** (1973), 335–343.

[22] Stulz, R. Options on the minimum or the maximum of two risky assets: Analysis and Applications. Journal of Financial Economics **10** (1982), 161–185.

[23] Sudakov, V. N. and Tsyrelson, B. S. Extremal properties of half-spaces for spherically invariant measures. J. Sov. Math. **9** (1978), 9–18. Translated from Zap. Nauchn. Sem. Leninggrad. Otdel. Mat. Inst. Steklov **41** (1974), 14–24.

[24] Tong, Y. L. The Multivariate Normal Distribution. Springer, Berlin and New York (1990).

CHRISTER BORELL
CHALMERS UNIVERSITY OF TECHNOLOGY
DEPARTMENT OF MATHEMATICS
CTH&GU
S-412 96 GÖTEBORG
SWEDEN
   borell@math.chalmers.se

Random Points in Isotropic Convex Sets

JEAN BOURGAIN

ABSTRACT. Let K be a symmetric convex body of volume 1 whose inertia tensor is isotropic, i.e., for some constant L we have $\int_K \langle x, y \rangle^2 \, dx = L^2 |y|^2$ for all y. It is shown that if m is about $n(\log n)^3$ then with high probability, this tensor can be approximately realised by an average over m independent random points chosen in K,
$$\frac{1}{m} \sum_{i=1}^{m} \langle x_i, y \rangle^2.$$

Our aim is to prove the following fact:

PROPOSITION. *Let $K \subset \mathbb{R}^n$ be a convex centrally symmetric body of volume 1, in isotropic position, i.e.,*

$$\int_K \langle x, e_i \rangle \langle x, e_j \rangle \, dx = L^2 \delta_{ij} \quad \text{where } L = L_K (\gtrsim 1). \tag{1}$$

Fix $\delta > 0$ and choose m random points $x_1, \ldots, x_m \in K$, where

$$m > C(\delta) n (\log n)^3. \tag{2}$$

Then, with probability $> 1 - \delta$,

$$(1-\delta) L^2 < \frac{1}{m} \sum_{i=1}^{m} |\langle x_i, y \rangle|^2 < (1+\delta) L^2 \tag{3}$$

for all $y \in S^{n-1} = [|y| = 1]$.

We first use the following probabilistic estimate:

LEMMA 1. *Let $f_1, \ldots, f_m$ be independent copies of a random variable f satisfying*

$$\int f^2 = 1, \tag{4}$$

$$\|f\|_{\psi_1} < C \quad (\text{where } \psi_1(t) = e^t), \tag{5}$$

$$\|f\|_\infty < B. \tag{6}$$

(Here and in the sequel we use c and C to denote positive constants, not necessarily the same each time.) Let $\varepsilon > 0$ and assume $B > 1/\varepsilon$ say. Then

$$\operatorname{mes}\left[(1-\varepsilon)m < \sum_{i=1}^m f_i^2 < (1+\varepsilon)m\right] > 1 - e^{-c\frac{\varepsilon}{B}m}. \tag{7}$$

PROOF (standard). For real λ (to be specified),

$$\int e^{\lambda(\sum_{i=1}^m (f_i^2 - 1))} = \left(\int e^{\lambda(f^2-1)}\right)^m. \tag{8}$$

By (4)

$$\int e^{\lambda(f^2-1)} = 1 + \sum_{j\geq 2} \frac{1}{j!}\lambda^j \int (f^2-1)^j. \tag{9}$$

From (5) and (6),

$$\int (1+|f|)^{2j} < \min\left((Cj)^{2j}, (1+B)^j(Cj)^j\right). \tag{10}$$

for each j. Hence, substituting (10) in (9),

$$\int e^{\lambda(f^2-1)} < 1 + \sum_{j\geq 2}(C\lambda)^j(j \wedge B)^j < 1 + C\lambda^2 \tag{11}$$

provided

$$\lambda < \frac{c}{B}. \tag{12}$$

for an appropriate c. Thus $(8) < (1+C\lambda^2)^m < e^{C\lambda^2 m}$ and from this fact and Tchebychev's inequality

$$\operatorname{mes}\left[\left|\frac{1}{m}\sum_{i=1}^m (f_i^2 - 1)\right| > \varepsilon\right] < e^{-\lambda m\varepsilon}e^{C\lambda^2 m} < e^{-c\frac{\varepsilon}{B}m} \tag{13}$$

for appropriate λ satisfying (12) (and since $1/\varepsilon < B$). □

Recall the important fact (following from the Brunn–Minkowski inequality) that, for K convex with $\operatorname{Vol} K = 1$, there is equivalence

$$\|\langle y, x\rangle\|_{L^{\psi_1}(K, dx)} \sim \|\langle y, x\rangle\|_{L^2(K, dx)} \tag{14}$$

(with an absolute constant). Hence, in our situation

$$\|\langle y, x\rangle\|_{L^{\psi_1}(K, dx)} < CL \quad \text{if } |y| = \|y\|_2 \leq 1. \tag{15}$$

It follows that

$$\operatorname{mes}\left[x \in K \,\big|\, |x| > \lambda L\sqrt{n}\right] < e^{-C\lambda} \quad \text{for } \lambda > 1. \tag{16}$$

The next estimate may be refined significantly in terms of an estimate on the ℓ^2-operator norm (see remark at the end) but for our purposes the following cruder form is sufficient.

LEMMA 2. *Let K be as above and $x_1, \ldots, x_m$ random points in K. Then, with probability $> 1 - \delta$,*

$$\left| \sum_{i \in E} x_i \right| < C(\delta) L \log n \bigl(|E|^{1/2} n^{1/2} + |E| \bigr) \tag{17}$$

holds for all subsets $E \subset \{1, \ldots, m\}$.

PROOF. Write

$$\left| \sum_{i \in E} x_i \right|^2 = \sum_{i \in E} |x_i|^2 + 2 \sum_{\substack{i \neq j \\ i,j \in E}} \langle x_i, x_j \rangle. \tag{18}$$

From (15), we may clearly assume

$$|x_i| < C L \log n \sqrt{n} \quad \text{for all} \quad i = 1, \ldots, m.$$

Hence the first term of (18) may be assumed bounded by $C L^2 (\log n)^2 n |E|$.

To estimate the second term of (18), we use a standard decoupling trick. We can find subsets E_1, E_2 of E satisfying $E_1 \cap E_2 = \emptyset$, $|E_1| \geq |E_2|$, and

$$\sum_{i \neq j, i,j \in E} \langle x_i, x_j \rangle \leq 4 \sum_{i \in E_1} \left| \left\langle x_i, \sum_{j \in E_2} x_j \right\rangle \right|. \tag{19}$$

Hence we are reduced to bounding expressions of the form (19).

Rewrite

$$\sum_{i \in E_1} \left| \left\langle x_i, \sum_{j \in E_2} x_j \right\rangle \right| = \left| \sum_{j \in E_2} x_j \right| \sum_{i \in E_1} |\langle x_i, y_{E_2}(x) \rangle| \tag{20}$$

where

$$y_{E_2}(x) = \frac{\sum_{j \in E_2} x_j}{\left| \sum_{j \in E_2} x_j \right|} \; ; \quad \text{thus } |y_{E_2}| = 1. \tag{21}$$

Observe that the system $(x_i)_{i \in E_1}$ is independent of y_{E_2}, since $E_1 \cap E_2 = \emptyset$. Fix size scales $|E_1| \sim m_1$, $|E_2| \sim m_2$, $m \geq m_1 \geq m_2 \geq 1$.

Thus for fixed $m_1 > m_2$, (E_1, E_2) run over at most $m^{C m_1}$ pairs of subsets of $\{1, \ldots, m\}$. For given y, $|y| = 1$, (15) easily implies that

$$\int e^{\frac{C}{L} \sum_{i \in E_1} |\langle x_i, y \rangle|} \prod_{i \in E_1} dx_i < 2^{|E_1|}; \tag{22}$$

hence, for $\mu > C$,

$$\operatorname{mes} \left[(x_i)_{1 \leq i \leq m} \in K^m \; \Big| \; \sum_{i \in E_1} |\langle x_i, y_{E_2}(x) \rangle| > \mu L |E_1| \right] < e^{-c \mu |E_1|}. \tag{23}$$

Consequently, from (20) and the preceding, we may write

$$\sum_{i \in E_1} \left|\left\langle x_i, \sum_{j \in E_2} x_j \right\rangle\right| < \left|\sum_{j \in E_2} x_j\right| \mu L |E_1| \qquad (24)$$

for all $|E_1| \sim m_1$, $|E_2| \sim m_2$, $E_1 \cap E_2 = \emptyset$ provided

$$m^{Cm_1} e^{-\mu m_1} < 2^{-m_1}; \quad \text{thus } \mu \sim \log m \sim \log n. \qquad (25)$$

Thus, letting $\mu \sim \log n$, (24) may be assumed valid for all $E_1, E_2 \subset \{1, \ldots, m\}$ with $E_1 \cap E_2 = \emptyset$.

Substituting (19), (24) in (18) thus yields, for all $E \subset \{1, \ldots, m\}$,

$$\left|\sum_{i \in E} x_i\right|^2 \leq CL^2 (\log n)^2 \, n \, |E| + CL(\log n)|E| \max_{E_2 \subset E} \left|\sum_{j \in E_2} x_j\right| \qquad (26)$$

and (17) immediately follows. $\square$

PROPOSITION. *Fix $\delta > 0$ and choose random points $x_1, \ldots, x_m \in K$, with $m > C(\delta)n(\log n)^3$. Then with probability $> 1 - \delta$*

$$(1-\delta)L^2 < \frac{1}{m} \sum_{i=1}^{m} |\langle x_i, y \rangle|^2 < (1+\delta)L^2 \quad \text{for all } y \in S^{n-1}. \qquad (27)$$

PROOF. Restrict y to a $\frac{\delta}{10}$-dense set $\mathcal{F}_\delta$ in the unit sphere S^{n-1}, $\#\mathcal{F}_\delta < \left(\frac{C}{\delta}\right)^n$. Fix $y \in \mathcal{F}$ and define

$$f = f^y(x) = \begin{cases} \frac{1}{L}|\langle x, y \rangle| & \text{if } |\langle x, y \rangle| < C_1 (\log n) L, \\ 0 & \text{otherwise} \end{cases} \qquad (28)$$

(with C_1 to be specified).

Thus

$$1 - \int f^2 = \frac{1}{L^2} \int_{K \cap |\langle x,y \rangle| > C_1 (\log n) L} |\langle x, y \rangle|^2 dx < e^{-c_1 \log n}. \qquad (29)$$

Applying Lemma 1 with $B = C_1 \log n$, $\varepsilon = \frac{\delta}{10}$, it follows that for a random choice $x_1, \ldots, x_m$ of points in K, with probability $> 1 - e^{-c(\varepsilon/\log n)m}$,

$$\int f^2 - \varepsilon < \frac{1}{m} \sum_{i=1}^{m} f^y(x_i)^2 < \left(\int f^2 + \varepsilon\right); \qquad (30)$$

hence, by (28) and (29),

$$\left|1 - \frac{1}{L^2 m} \sum \{\langle x_i, y \rangle^2 \mid |\langle x_i, y \rangle| < C_1 L \log n\}\right| < \varepsilon + \left(1 - \int f^2\right) < 2\varepsilon. \qquad (31)$$

Letting

$$\left(\tfrac{C}{\delta}\right)^n e^{-c(\varepsilon/\log n)m} \ll 1, \quad \text{i.e., } m \gtrsim \frac{1}{\varepsilon} \log \frac{1}{\delta} (\log n) n, \qquad (32)$$

we may then assume (31) for all $y \in \mathcal{F}_\delta$.

On the other hand, from Lemma 2, a random choice $\{x_i \mid i = 1, \ldots, m\}$ of m points in K will also with probability $> 1 - \delta$ satisfy (17) for all $E \subset \{1, \ldots, m\}$. This permits to estimate $\#E_\beta$, where for given y satisfying $|y| = 1$,

$$E_\beta = E_\beta(y) = \{i = 1, \ldots, m \mid |\langle x_i, y \rangle| > \beta\}, \quad \beta > C_1 (\log n) L. \tag{33}$$

Indeed, it follows from (17) that

$$\tfrac{1}{2}\beta |E_\beta| < CL \log n \left(|E_\beta|^{1/2} n^{1/2} + |E_\beta| \right) \tag{34}$$

hence

$$|E_\beta| < C \frac{L^2 (\log n)^2 n}{\beta^2} \tag{35}$$

from the choice of β. Consequently

$$\frac{1}{L^2 m} \sum \{\langle x_i, y\rangle^2 \mid |\langle x_i, y\rangle| \geq C_1 L \log n\} < \frac{1}{L^2 m} \sum_{\substack{n > \beta > C_1 L \log n \\ \beta \text{ dyadic}}} \beta^2 |E_\beta|$$

$$< C(\delta) (\log n)^3 \frac{n}{m} < \frac{\delta}{10} \tag{36}$$

by the choice of m.

Finally, combining (36) and (31), it follows that for all $y \in F_\delta$

$$\left| 1 - \frac{1}{L^2 m} \sum_{i=1}^{m} \langle x_i, y \rangle^2 \right| < 2\varepsilon + \frac{\delta}{10} < \frac{\delta}{3} \tag{37}$$

and therefore also (27). □

REMARK. By refining a bit the method of proof of Lemma 2, one may obtain the following result: Let $x_1, \ldots, x_n$ be a choice of n independent vectors in $\mathbb{R}^n$ according to a probability measure μ on $\mathbb{R}^n$ satisfying

$$\|\langle x, y \rangle\|_{L^{\psi_1}(\mu(dx))} < \frac{1}{\sqrt{n}} \quad \text{for all } y \in S^{n-1}. \tag{38}$$

Then, with probability $> 1 - \delta$, one gets for the matrix $(x_1, \ldots, x_n)$ the bound

$$\|(x_1, \ldots, x_n)\|_{B(\ell_n^2)} < C(\delta) \left(\int \left(\max_{1 \leq i \leq n} |x_i| \right) d\mu + 1 \right). \tag{39}$$

This is the same estimate as one would get assuming an L^{ψ_2}-bound

$$\|\langle x, y \rangle\|_{L^{\psi_2}(\mu(dx))} < \frac{1}{\sqrt{n}} \quad \text{for } y \in S^{n-1} \tag{40}$$

instead of (38).

JEAN BOURGAIN
SCHOOL OF MATHEMATICS
OLDEN LANE
INSTITUTE FOR ADVANCED STUDY
PRINCETON, NJ 08540
UNITED STATES OF AMERICA
   bourgain@math.ias.edu

Threshold Intervals under Group Symmetries

JEAN BOURGAIN AND GIL KALAI

ABSTRACT. This article contains a brief description of new results on threshold phenomena for monotone properties of random systems. These results sharpen recent estimates of Talagrand, Russo and Margulis. In particular, for isomorphism invariant properties of random graphs, we get a threshold whose length is only of order $1/(\log n)^{2-\varepsilon}$, instead of previous estimates of the order $1/\log n$. The new ingredients are delicate inequalities in the spirit of harmonic analysis on the Cantor group.

A subset A of $\{0,1\}^n$ is called monotone if the conditions $x \in A$, $x' \in \{0,1\}^n$ and $x_i \leq x'_i$ for $i = 1, \ldots, n$ imply $x' \in A$. For $0 \leq p \leq 1$, define μ_p the product measure on $\{0,1\}^n$ with weights $1-p$ at 0 and p at 1. Thus

$$\mu_p(\{x\}) = (1-p)^{n-j} p^j \quad \text{where} \quad j = \#\{i = 1, \ldots, n \mid x_i = 1\}. \tag{1}$$

If A is monotone, then $\mu_p(A)$ is clearly an increasing function of p. Considering A as a "property", one observes in many cases a threshold phenomenon, in the sense that $\mu_p(A)$ jumps from near 0 to near 1 in a short interval when $n \to \infty$. Well known examples of these phase transitions appear for instance in the theory of random graphs. A general understanding of such threshold effects has been pursued by various authors (see for instance Margulis [M] and Russo [R]). It turns out that this phenomenon occurs as soon as A depends little on each individual coordinate (Russo's zero-one law). A precise statement was given by Talagrand [T] in the form of the following inequality.

Define for $i = 1, \ldots, n$

$$A_i = \{x \in \{0,1\}^n \mid x \in A, \, U_i x \notin A\} \tag{2}$$

where $U_i(x)$ is obtained by replacement of the i-th coordinate x_i by $1 - x_i$ and leaving the other coordinates unchanged. The number $\mu_p(A_i)$ is the *influence* of the i-th coordinate (with respect to μ_p). Let

$$\gamma = \sup_{i=1,\ldots,n} \mu_p(A_i). \tag{3}$$

Then
$$\frac{d\mu_p(A)}{dp} \geq c \frac{\log(1/\gamma)}{p(1-p)\log(2/p(1-p))} \mu_p(A)\left(1 - \mu_p(A)\right), \tag{4}$$
where $c > 0$ is some constant.

A simple relation due to Margulis and Russo is
$$\frac{d\mu_p}{dp} = 2/p \sum_{i=1}^{n} \mu_p(A_i). \tag{5}$$

As the right side of (5) represents the sum of the influences it follows that a small threshold interval corresponds to a large sum of influences. In [T], (4) is deduced from an inequality of the form
$$\mu_p(A)\left(1 - \mu_p(A)\right) \leq C(p) \sum_{i=1}^{n} \frac{\mu_p(A_i)}{\log(1/\mu_p(A_i))}. \tag{6}$$

The proof of this last inequality relies on the paper by Kahn, Kalai and Linial [KKL], where it is shown that always
$$\sup_{1 \leq i \leq n} \mu_{1/2}(A_i) \geq c \frac{\log n}{n}. \tag{7}$$

Friedgut and Kalai [FK] used an extension of (7) given in [BKKKL] to show that for properties which are invariant under the action of a transitive permutation group the threshold interval is $O(1/\log n)$ and proposed some conjectures on the dependence of the threshold interval on the group.

Our aim here is to obtain a refinement and strengthening of the preceding in the context of "G-invariant" properties. Let f be a 0, 1-valued function on $\{0,1\}^n$ and G a subgroup of the permutation group on n elements $\underline{n} = \{1, 2, \ldots, n\}$. Say that f is G-invariant provided
$$f(x_1, \ldots, x_n) = f(x_{\pi(1)}, \ldots, x_{\pi(n)}) \text{ for all } x \in \{0,1\}^n, \pi \in G.$$

Given G, define for $1 \leq t \leq n$
$$\phi(t) = \phi_G(t) = \min_{S \subset \underline{n}, |S| = t} \log(\#\{\pi(S) \mid \pi \in G\})$$
and for all $\tau > 0$
$$a_\tau(G) = \sup\{\phi(t) \mid \phi(t) > t^{1+\tau}\}.$$
Observe that since $\phi(t) \leq \log\binom{n}{t}$, necessarily $a_\tau(G) \lesssim (\log n)^{1/\tau}$.

THEOREM 1. *Assume G transitive and A a monotone G-invariant property. Then for all $\tau > 0$*
$$\frac{d\mu_p(A)}{dp} > c_\tau a_\tau(G) \mu_p(A)\left(1 - \mu_p(A)\right),$$
provided $p(1-p)$ stays away from zero in a weak sense, say
$$\log(p(1-p))^{-1} \lesssim \log \log n.$$

It follows that in particular the threshold interval is at most
$$C_\tau a_\tau(G)^{-1} \text{ for all } \tau > 0.$$
Previous results as mentioned above only yield estimates of the form $(\log n)^{-1}$ and the main point of this work is to provide a method going beyond this. For crossing the $(\log n)^{-1}$ bar we need a complicated harmonic analysis argument. This may be useful in related combinatorial problems.

Theorem 1 is deduced from (5) and the following fact, independent of monotonicity assumptions.

THEOREM 2. *Assume that A is G-invariant and (12) holds. Then for all $\tau > 0$*
$$\sum \mu_p(A_i) > c_\tau a_\tau(G)\mu_p(A)(1 - \mu_p(A)).$$

For primitive permutation groups Theorem 1 and the excellent knowledge of primitive permutation groups [C, KL] (based on the classification theorem for finite simple groups) imply a close to complete description of the possible threshold interval of a G-invariant property, depending on the structure of G. (Recall that a permutation group $G \subset S_n$ is primitive if it is impossible to partition $\underline{n}$ to blocks $B_1, \ldots B_t$, $t > 1$ so that every element in G permute the blocks among themselves.) It turns out that there are some gaps in the possible behaviors of the largest threshold intervals. This interval is proportional to $n^{-1/2}$ for S_n and A_n but at least $\log^{-2} n$ for any other group. The worst threshold interval can be proportional to $\log^{-c} n$ for c belonging to arbitrary small intervals around the following values: $2, \frac{3}{2}, \frac{4}{3}, \frac{5}{4}, \ldots$, or for c which tends to zero as a function of n in an arbitrary way. This (and more) is summarized in the next theorem. First we need a few definitions. For a permutation group $G \subset S_n$ let
$$T_G(\varepsilon) = \sup\{q - p : \mu_p(A) = \varepsilon, \mu_q(A) = 1 - \varepsilon\},$$
where the supremum is taken over all monotone subsets of $\{0,1\}^n$ which are invariant under G. A composition factor of group G is a quotient group H/H' where H is a normal subgroup of G and H' is a normal subgroup of H. A section of G is a quotient H/H' where H is an arbitrary subgroup of G and H' is a normal subgroup of H.

THEOREM 3. *Let $G \subset S_n$ be a primitive permutation group.*

1. *If $G = S_n$ or $G = A_n$ then $T_G(\varepsilon) = \log(1/\varepsilon)/n^{1/2}$.*
2. *If $G \ne S_n, A_n$, $T_G(\varepsilon) \geq c_1 \log(1/\varepsilon)/\log^2 n$.*
3. *For every integer $r > 0$ and real numbers $\delta > 0$ and $\varepsilon > 0$, if $T_G(\varepsilon) \leq c_2 \log(1/\varepsilon)/(\log n)^{(1+1/(r+1))}$ then already*
$$T_G(\varepsilon) \leq c_3(\delta) \log(1/\varepsilon)/(\log n)^{(1+1/r-\delta)}.$$
4. *If G does not involve as composition factors alternating groups of high order then $T_G(\varepsilon) \geq \log(1/\varepsilon)/\log n \log \log n$.*

5. Let $n = \binom{m}{r}$ and G is S_m acting on r-subsets of $[m]$. Then for every $\delta > 0$
$$(\log(1/\varepsilon)/\log^{(1+1/(r-1))} n) \leq T_G(\varepsilon) \leq c(\delta)(\log(1/\varepsilon)/\log^{(1+1/(r-1)-\delta)} n)$$

6. For $G = \text{PSL}(m,q)$ acting on the projective space over F_q, for fixed q,
$$T_G(\varepsilon) = O(\log(1/\varepsilon)/\log n \log \log n)$$

7. For every function $w(n)$ such that $\log w(n)/\log \log n \to 0$ there are primitive group $G_n \subset S_n$ such that $T_{G_n}(\varepsilon)$ behaves like $\log(1/\varepsilon)/\log n \cdot w(n)$.

8. For every $w(n) > 1$ such that $w(n) = O(\log \log n)$ there are primitive group $G_n \subset S_n$ which do not involve alternating groups of high order as composition factors such that $T_{G_n}(\varepsilon)$ behaves like $\log(1/\varepsilon)/(\log n \cdot w(n))$.

9. If G does not involve as sections alternating groups of high order then $T_G(\varepsilon) \geq O(\log(1/\varepsilon)/\log n)$.

The preceding yields a particularly satisfying result on the size of the maximal threshold for monotone graph properties. In the particular case of monotone graph properties on N vertices, we get $n = \binom{N}{2}$ and G is induced by permuting the vertices. One gets essentially

$$\phi(t) \sim \log\left(\frac{N}{\sqrt{t}}\right)$$

in this situation and the conclusion of Theorem 1 is that any threshold interval is at most $C_\tau (\log N)^{-2+\tau}$, with $\tau > 0$. This is essentially the sharp result, since, fixing $M \sim \log N$, the property for a graph on N vertices to contain a clique of size M yields a threshold interval $\sim (\log N)^{-2}$.

More details and the proofs appear in [BK].

References

[B] B. Bollobás, *Random graphs*, Academic Press, London, 1985.

[BKKKL] J. Bourgain, J. Kahn, G. Kalai, Y. Katznelson and N. Linial, "The influence of variables in product spaces", *Israel J. Math.* **77** (1992), 55–64.

[BK] J. Bourgain and G. Kalai, "Influences of variables and threshold intervals under group symmetries," *Geometric and Functional Analysis* **7**:3 (1997), 438–461.

[C] P. Cameron, "Finite permutation groups and finite simple groups", *Bull. London Math. Soc.* **13** (1981), 1–22.

[FK] E. Friedgut and G. Kalai, "Every monotone graph property has a sharp threshold", *Proc. Amer. Math. Soc.* **124** (1996), 2993–3002.

[KKL] J. Kahn, G. Kalai, and N. Linial, "The influence of variables on Boolean functions", Proc. 29th IEEE FOCS **58-80**, IEEE, NY (1988).

[KL] P. Kleidman and M. Liebeck, *The subgroup structure of the finite classical groups*, Cambridge University Press, Cambridge and New York, 1990.

[M] G. Margulis, "Probabilistic characteristic of graphs with large connectivity", *Problemy Peredači Informacii* **10**:2 (1974) (in Russian); translation in *Problems Info. Transmission*, 1977.

[R] L. Russo, "An approximative zero-one law", *Z. Wahrsch. Verw. Gebiete* **61** (1982), 129–139.

[T] M. Talagrand, "On Russo's approximative zero-one law", *Annals of Prob.* **22**:3 (1994), 1576–1587.

JEAN BOURGAIN
SCHOOL OF MATHEMATICS
OLDEN LANE
INSTITUTE FOR ADVANCED STUDY
PRINCETON, NJ 08540
UNITED STATES OF AMERICA
  bourgain@math.ias.edu

GIL KALAI
INSTITUTE OF MATHEMATICS
HEBREW UNIVERSITY OF JERUSALEM
JERUSALEM
ISRAEL
  kalai@math.huji.ac.il

On a Generalization of the Busemann–Petty Problem

JEAN BOURGAIN AND GAOYONG ZHANG

ABSTRACT. The generalized Busemann–Petty problem asks: If K and L are origin-symmetric convex bodies in $\mathbb{R}^n$, and the volume of $K \cap H$ is smaller than the volume of $L \cap H$ for every i-dimensional subspace H, $1 < i < n$, does it follow that the volume of K is smaller than the volume of L? The hyperplane case $i = n-1$ is known as the Busemann–Petty problem. It has a negative answer when $n > 4$, and has a positive answer when $n = 3, 4$. This paper gives a negative answer to the generalized Busemann–Petty problem for $3 < i < n$ in the stronger sense that the integer i is not fixed. For the 2-dimensional case $i = 2$, it is proved that the problem has a positive answer when L is a ball and K is close to L.

1. Introduction

Denote by $\mathrm{vol}_i(\cdot)$ the i-dimensional Lebesgue measure, and denote by $G_{i,n}$ the Grassmann manifold of i-dimensional subspaces of $\mathbb{R}^n$. The *generalized Busemann–Petty problem* asks:

GBP. *If K and L are origin-symmetric convex bodies in $\mathbb{R}^n$, is there the implication*

$$\mathrm{vol}_i(K \cap \xi) \leq \mathrm{vol}_i(L \cap \xi), \quad \forall \xi \in G_{i,n} \implies \mathrm{vol}_n(K) \leq \mathrm{vol}_n(L)? \quad (1.1)$$

The case of $i = 1$ is trivially true. The hyperplane case $i = n - 1$ is well-known as the *Busemann–Petty problem* (see [BP] and [Bu]). Many authors contributed to the solution of the Busemann–Petty problem (see [Ba] [Bo] [G1] [Gia] [Gie] [GR] [Ha] [Lu] [LR] [Pa] [Z1]). The problem has a negative answer when $n > 4$ (see [G1], [Pa] and [Z2]), and it has a positive answer when $n = 3, 4$ (see [G2] and [Z4]). The notion of *intersection body*, introduced by Lutwak [Lu], plays an

---

1991 *Mathematics Subject Classification.* Primary: 52A20; secondary: 52A40, 53C65.

Key words and phrases. Convex body, geometric inequality, cross section, Radon transform.

Research supported in part by NSF Grant DMS–9304580.

important role in the solution of the Busemann–Petty problem. It relates to the positivity of the inverse spherical Radon transform.

Because of the special feature of the answer to the Busemann–Petty problem, it is interesting to consider the generalized Busemann–Petty problem. What are the dimensions of cross sections and ambient spaces so that the generalized Busemann–Petty problem has a positive or negative answer? By introducing the notion of i-*intersection body* and using techniques in functional analysis and Radon transforms on Grassmannians, it is proved in [Z3] that the answer to the generalized Busemann–Petty problem is equivalent to the existence of origin-symmetric convex bodies which are not i-intersection bodies. When $3 < i < n$, we give a negative answer to the problem. The argument shows that cylinders are not i-intersection bodies if $3 < i < n$. We also give a partial answer to the case of 2-dimensional sections. We remark that one of the results in [Z3] that no polytope is an i-intersection body is not correct.

It is shown in [Z3] that the generalized Busemann–Petty problem has a positive answer if K is an i-intersection body, in particular, if K is a ball in $\mathbb{R}^n$. However, when L is a ball, the generalized Busemann–Petty problem may still have a negative answer. For instance, Keith Ball observed that one can construct counterexamples by using the techniques in [Ba] and letting $K = $ the unit cube, $L = $ a ball of appropriate radius when n and i are sufficiently large. We prove that, when L is a ball and K is sufficiently close to L, the generalized Busemann–Petty problem of 2-dimensional sections has a positive answer. The result is contained in the following theorem.

THEOREM 1.1. *Let K be a centered convex body and let B_n be the standard unit ball in $\mathbb{R}^n$. There exists $\delta_0 > 0$ which only depends on the dimension so that if $\mathrm{dist}(K, B_n) < \delta_0$ then*

$$\mathrm{vol}_2(K \cap \xi) \leq \mathrm{vol}_2(B_n \cap \xi), \quad \forall \xi \in G_{2,n} \implies \mathrm{vol}_n(K) \leq \mathrm{vol}_n(B_n).$$

Let ω_n be the volume of B_n. By the homogeneity of the inequalities in the last implication, we obtain the following corollary.

COROLLARY 1.2. *Let K be a centered convex body in $\mathbb{R}^n$. There exists $\delta_0 > 0$ which only depends on the dimension so that if the distance of K to a ball is less than δ_0, then*

$$\mathrm{vol}_n(K)^{\frac{2}{n}} \leq \frac{\omega_n^{\frac{2}{n}}}{\pi} \max_{\xi \in G_{2,n}} \mathrm{vol}_2(K \cap \xi). \tag{1.2}$$

Inequality (1.2) is proved for any centered convex bodies in $\mathbb{R}^3$ in [G3]. It might be still true for any centered convex bodies in all dimensions as well.

Note that, for the generalized Busemann–Petty problem, the dimension i of sections in the implication (1.1) is fixed. It is natural to ask what will happen if the dimension i of sections is not fixed but takes different values. We would like to thank V.D. Milman who brought our attention to this question. Our answer is contained in the following theorem.

THEOREM 1.3. *There exist centered convex bodies of revolution K and L so that, for all $3 < i < n$,*
$$\mathrm{vol}_i(K \cap \xi) < \mathrm{vol}_i(L \cap \xi), \quad \forall \xi \in G_{i,n},$$
but
$$\mathrm{vol}_n(K) > \mathrm{vol}_n(L).$$

This result is best possible in the class of convex bodies of revolution. It is proved that, *if K is a centered convex body of revolution, then*
$$\mathrm{vol}_i(K \cap \xi) \le \mathrm{vol}_i(L \cap \xi), \quad \forall \xi \in G_{i,n}, \quad \Longrightarrow \quad \mathrm{vol}_n(K) \le \mathrm{vol}_n(L),$$
when $i = 2$, or 3. See [G1], [Z2] and [Z3].

The proofs of Theorems 1.1 and 1.3 use the tools of Radon transforms on Grassmannians. We give definitions and basic facts of the Radon transforms for later use.

Let $C_e(S^{n-1})$ be the space of continuous even functions on the unit sphere S^{n-1}, and denote by $C(G_{i,n})$ the space of continuous functions on $G_{i,n}$. The *Radon transform*, for $2 \le i \le n-1$,
$$\mathrm{R}_i : C_e(S^{n-1}) \longrightarrow C(G_{i,n})$$
is defined by
$$(\mathrm{R}_i f)(\xi) = \frac{1}{i\omega_i} \int_{u \in S^{n-1} \cap \xi} f(u)\, du, \quad \xi \in G_{i,n},\ f \in C_e(S^{n-1}),$$
where ω_i and du are the volume and the surface area element of the i-dimensional unit ball, respectively.

Let ρ_K be the *radial function* of a centered convex body K in $\mathbb{R}^n$ given by
$$\rho_K(u) = \max\{\lambda \ge 0 : \lambda u \in K\}, \quad u \in S^{n-1}.$$

The Radon transform R_i is closely connected with the central sections of centered bodies by the following formula
$$(\mathrm{R}_i \rho_K^i)(\xi) = \frac{1}{\omega_i} \mathrm{vol}_i(K \cap \xi), \quad \xi \in G_{i,n}. \tag{1.3}$$

The *dual transform* R_i^t of R_i is the map $C(G_{i,n}) \to C_e(S^{n-1})$ given by
$$(\mathrm{R}_i^t g)(u) = \int_{u \in \xi \in G_{i,n}} g(\xi)\, d\xi, \quad u \in S^{n-1},\ g \in C(G_{i,n}).$$

We have the following duality [He, pp. 144 and 161]:
$$\langle \mathrm{R}_i f, g \rangle = \langle f, \mathrm{R}_i^t g \rangle, \quad f \in C_e(S^{n-1}),\ g \in C(G_{i,n}), \tag{1.4}$$
where $\langle \cdot, \cdot \rangle$ is the usual inner product of functions in homogeneous spaces.

2. Two-Dimensional Sections

In this section we give the proof of Theorem 1.1. One technical part of the proof is to approximate arbitrary convex bodies by smooth convex bodies quantitatively. We use convolutions on the rotation group SO(n) of $\mathbb{R}^n$.

Let G be a compact Lie group. Let $C(G)$ be the space of continuous functions on G with the uniform topology. For $f, g \in C(G)$, the *convolution* $f * g \in C(G)$ of f and g is defined by

$$(f * g)(u) = \int_{v \in G} f(uv^{-1})g(v)\, dv = \int_{v \in G} f(v)g(v^{-1}u)\, dv,$$

where dv is the invariant probability measure of G.

Associated with a convex body K is its *support function* h_K defined on S^{n-1} by

$$h_K(u) = \max\{\langle u, x \rangle : x \in K\}, \quad u \in S^{n-1},$$

where $\langle u, x \rangle$ is the usual inner product of u and x in $\mathbb{R}^n$. The *polar body* K^* of K is defined by

$$K^* = \{x \in \mathbb{R}^n : \langle x, y \rangle \leq 1 \text{ for all } y \in K\}.$$

Its support function is given by

$$h_{K^*}(u) = \rho_K^{-1}(u), \quad u \in S^{n-1}.$$

If h_K is the support function of K, and f is a positive function on SO(n), then $f * h_K$ is the support function of another convex body. Moreover, the convolution preserves the symmetry of the convex body. For a proof of this fact and more details on convex bodies and convolutions, see [GZ].

LEMMA 2.1. *Let G be a compact Lie group of dimension m. If f is Lipschitz continuous on G, then there exists $\delta_0 > 0$ which depends only on G and the Lipschitz constant of f, so that for any $\delta < \delta_0$ there exists C^∞ positive function ϕ_δ satisfying*

$$|\phi_\delta * f - f| < \delta, \quad \|\phi_\delta * f\|_{C^2} < \delta^{-m-3}.$$

PROOF. Let B_δ be the geodesic ball of radius δ at the unit of G. Let ϕ be a C^∞ nonnegative function which is strictly positive inside $B_{\delta/2}$ but is zero outside B_δ. Let $\exp : T_e G \to G$ be the exponential map. Condiser the C^∞ function

$$\phi_\delta(x) = a_\delta^{-1} \phi\big(\exp(\delta^{-1} \exp^{-1}(x))\big),$$

where $a_\delta = \int_G \phi\big(\exp(\delta^{-1} \exp^{-1}(x))\big)\, dx$. When δ is small,

$$a_\delta \sim c\delta^m, \tag{2.1}$$

for some constant c.

Since f is Lipschitz continuous, we have

$$|\phi_\delta * f(x) - f(x)| = \left|\int_G \phi_\delta(y)f(yx)\,dy - \int_G \phi_\delta(y)f(x)\,dy\right|$$
$$\leq \int_G \phi_\delta(y)|f(yx) - f(x)|\,dy \leq c_1\delta. \qquad (2.2)$$

From the following equalities,

$$\phi_\delta * f(x) = \int_G \phi_\delta(xy^{-1})f(y)\,dy = a_\delta^{-1}\int_G \phi\big(\exp(\delta^{-1}\exp^{-1}(xy^{-1}))\big)f(y)\,dy,$$

the second order derivatives of $\phi_\delta * f$ yield a factor δ^{-2}. Therefore, when δ is small (depending on the upper bound of f), (2.1) gives

$$\|\phi_\delta * f\|_{C^2} \leq c_2 \delta^{-m-2}. \qquad (2.3)$$

Note that c_1 and c_2 only depend on G and the Lipschitz constant of f. From (2.2) and (2.3), the required inequalities follow immediately. $\square$

LEMMA 2.2. *Let K be a centered convex body in $\mathbb{R}^n$. There exists $\delta_0 > 0$ which only depends on the dimension and the diameters of K and its polar body, so that for any $\delta < \delta_0$ there exists a centered convex body K_δ with C^∞ radial function ρ_{K_δ} so that*

$$|\rho_{K_\delta} - \rho_K| < \delta, \quad \|\rho_{K_\delta}\|_{C^2} < \delta^{-n^2}. \qquad (2.4)$$

PROOF. Consider the support function h_{K^*} of the polar body K^* of K. Since the sphere $S^{n-1} = \mathrm{SO}(n)/\mathrm{SO}(n-1)$ is a homogeneous space, the support function h_{K^*} on S^{n-1} can be viewed as a function on $\mathrm{SO}(n)$. From Lemma 2.1, for any $\delta < \delta_0$ there exists C^∞ function ϕ_δ so that

$$|\phi_\delta * h_{K^*} - h_{K^*}| < \delta, \quad \|\phi_\delta * h_{K^*}\|_{C^2} < \delta^{-m-3},$$

where $m = \dim \mathrm{SO}(n) = \frac{1}{2}(n^2 - n)$. The number δ_0 depends on the dimension and the Lipschitz constant of h_{K^*}. Since the Lipschitz constant of h_{K^*} depends only on the dimension and the diameter of K^*, the number δ_0 only depends on the dimension and the diameter of K^*.

Define a centered convex body K_δ by

$$\rho_{K_\delta}^{-1} = h_{K_\delta^*} = \phi_\delta * h_{K^*}.$$

Therefore, ρ_{K_δ} is C^∞ and satisfies the inequalities

$$|\rho_{K_\delta}^{-1} - \rho_K^{-1}| < \delta, \quad \|\rho_{K_\delta}^{-1}\|_{C^2} < \delta^{-m-3}.$$

By using the fact that the function ρ_{K_δ} and its first order derivative are bounded by a constant depending on the diameter of K, we conclude (2.4) from the last two inequalities. $\square$

We will use the symbol $\lesssim$ which means that the expression on the left-hand side of the symbol is less than the expression on the right-hand side by a constant factor depending only on the dimension.

LEMMA 2.3. *Let F be a uniform Lipschitz function on S^{n-1}. If, for $0 < \delta < 1$,*

$$\|R_2 F\|_2 < \delta, \tag{2.5}$$

then

$$\|F\|_\infty \lesssim \delta^{\frac{1}{n+3}}. \tag{2.6}$$

PROOF. Consider the spherical harmonic expansion of F, $F = \sum Y_k$. Then

$$\|R_2 F\|_2 \sim \left(\sum k^{-1} \|Y_k\|_2^2 \right)^{\frac{1}{2}}, \tag{2.7}$$

where $\sim$ means that the quantities on both sides of it are bounded by each other with constant factors depending only on the dimension. See [St].

For $0 < r < 1$, let

$$P_r F = \sum r^k Y_k.$$

Since the Lipschitz constant of F is uniformly bounded on S^{n-1}, we have

$$\|P_r F - F\|_\infty \lesssim (1-r)^{\frac{1}{2}}. \tag{2.8}$$

See [BL].

From (2.7) and (2.5), we obtain

$$\|P_r F\|_\infty < \sum r^k \|Y_k\|_\infty \lesssim \sum r^k k^{\frac{n-1}{2}} \|Y_k\|_2$$
$$\lesssim \left(\sum r^k k^{\frac{n-1}{2}} k^{\frac{1}{2}} \right) \|R_2 F\|_2 \lesssim (1-r)^{-\frac{n}{2}-1} \delta. \tag{2.9}$$

Thus by (2.8), (2.9) and choosing $r = 1 - \delta^{\frac{2}{n+3}}$, we have

$$\|F\|_\infty \lesssim (1-r)^{\frac{1}{2}} + (1-r)^{-\frac{n}{2}-1} \delta \leq \delta^{\frac{1}{n+3}}.$$

This completes the proof. □

PROOF OF THEOREM 1.1. Consider the ratio

$$\bar{V}_n(K) = \frac{\text{vol}_n(K)}{\text{vol}_n(B_n)}.$$

Choose the invariant probability measures on the sphere S^{n-1} and on the Grassmannian $G_{2,n}$. From (1.4), we have

$$\bar{V}_n(K) = \int \rho_K^n = \int [R_2(R_2^t R_2)^{-1} \rho_K^{n-2}](R_2 \rho_K^2). \tag{2.10}$$

The implication in Theorem 1.1 becomes

$$\bar{V}_2(K \cap \xi) \leq 1, \quad \xi \in G_{2,n} \implies \bar{V}_n(K) \leq 1. \tag{2.11}$$

Assume that
$$\begin{cases} \bar{V}_n(K) > 1, \\ \bar{V}_2(K \cap \xi) \leq 1, \quad \xi \in G_{2,n}, \end{cases} \quad (2.12)$$
and
$$\text{dist}(K, B_n) = \delta_0 > 0. \quad (2.13)$$

From Lemma 2.2, there exists K_1 such that
$$|\rho_{K_1} - \rho_K| < \delta, \quad \|\rho_{K_1}\|_{C^2} < \delta^{-n^2}, \quad (2.14)$$
where $\delta = \delta_0^N$, and N is a constant to be chosen which only depends on the dimension.

Apply (2.10) to K_1. Then by (2.12) and (2.14) we have
$$\bar{V}_n(K) + o(\delta) = \int [\mathbf{R}_2(\mathbf{R}_2^t\mathbf{R}_2)^{-1}\rho_{K_1}^{n-2}](\mathbf{R}_2\rho_{K_1}^2), \quad (2.15)$$
and
$$\int \mathbf{R}_2(\mathbf{R}_2^t\mathbf{R}_2)^{-1}\rho_{K_1}^{n-2} = \int \rho_{K_1}^{n-2} \leq \left(\int \rho_{K_1}^n\right)^{\frac{n-2}{n}}$$
$$= \bar{V}_n(K_1)^{\frac{n-2}{n}} + o(\delta) \leq \bar{V}_n(K) + o(\delta). \quad (2.16)$$

From (2.15) and (2.16), we obtain
$$o(\delta) \leq \int [\mathbf{R}_2(\mathbf{R}_2^t\mathbf{R}_2)^{-1}\rho_{K_1}^{n-2}](\mathbf{R}_2\rho_{K_1}^2 - 1). \quad (2.17)$$

The assumption (2.13) gives
$$|\rho_K - 1| \leq \delta_0, \quad (2.18)$$
$$|\rho_{K_1} - 1| \leq 2\delta_0. \quad (2.19)$$

Hence
$$\|\nabla \rho_{K_1}\| \lesssim \delta_0^{\frac{1}{2}}. \quad (2.20)$$

A proof of the last inequality can be found in [Bo]. From (2.20), one has
$$\left\|\nabla\left(\rho_{K_1}^{n-2} - \int \rho_{K_1}^{n-2}\right)\right\| \lesssim \delta_0^{\frac{1}{2}}.$$

If E_{2-n} extends a function on S^{n-1} to a homogeneous function of degree $2-n$ in $\mathbb{R}^n$, and S restricts a function in $\mathbb{R}^n$ to S^{n-1}, then,
$$\left\|S(-\Delta_{\mathbb{R}^n})^{\frac{1}{2}}E_{2-n}\left(\rho_{K_1}^{n-2} - \int \rho_{K_1}^{n-2}\right)\right\|_{BMO} \lesssim \delta_0^{\frac{1}{2}}. \quad (2.21)$$

Use here the fact that L^∞-control on the tangential derivative yields BMO-control on the normal derivative.

By (2.14), the inequality (2.21) implies
$$\left\|S(-\Delta_{\mathbb{R}^n})^{\frac{1}{2}}E_{2-n}\left(\rho_{K_1}^{n-2} - \int \rho_{K_1}^{n-2}\right)\right\|_\infty \lesssim \delta_0^{\frac{1}{2}}\log(\delta^{-n^2}/\delta_0^{\frac{1}{2}}) < \delta_0^{\frac{1}{3}}.$$

The last inequality used the fact that the constant N only depends on the dimension. Since $S(-\Delta_{\mathbb{R}^n})^{\frac{1}{2}} E_{2-n} = (R_2^t R_2)^{-1}$ (see [St]), it follows that

$$\|(R_2^t R_2)^{-1}(\rho_{K_1}^{n-2} - \int \rho_{K_1}^{n-2})\|_\infty < \delta_0^{\frac{1}{3}}. \tag{2.22}$$

From (2.18) and (2.22), we have

$$\|(R_2^t R_2)^{-1}\rho_{K_1}^{n-2} - 1\|_\infty = \|(R_2^t R_2)^{-1}\rho_{K_1}^{n-2} - \int \rho_{K_1}^{n-2}\|_\infty + o(\delta_0) \lesssim \delta_0^{\frac{1}{3}}. \tag{2.23}$$

It follows in particular from (2.23) that

$$\frac{1}{2} < R_2(R_2^t R_2)^{-1}\rho_{K_1}^{n-2} < 2. \tag{2.24}$$

Assumption (2.12) means that

$$R_2 \rho_K^2 \leq 1. \tag{2.25}$$

It follows from (2.17), (2.24) and (2.25) that

$$\int [R_2(R_2^t R_2)^{-1}\rho_{K_1}^{n-2}](1 - R_2\rho_K^2) \leq o(\delta), \tag{2.26}$$

$$\int |1 - R_2\rho_K^2| \leq o(\delta). \tag{2.27}$$

By Lemma 2.3, this yields for small δ_0

$$\delta_0 \sim \|1 - \rho_K^2\|_\infty < \delta_0^{\frac{N}{n+3}}, \tag{2.28}$$

Hence, for N large enough, a contradiction follows. Therefore, the assumption (2.12) is impossible, and the implication (2.11) is true. This completes the proof. □

3. High-Dimensional Sections

In this section we give a proof for Theorem 1.3. The following lemma gives the the Radon transforms of functions which are $SO(n-1)$ invariant.

LEMMA 3.1. *Let g be a continuous function on S^{n-1} which is $SO(n-1)$ invariant. Then*

$$R_i g(u) = \frac{c_1}{\cos \phi} \int_\phi^{\frac{\pi}{2}} g(v) \left(1 - \frac{\cos^2 \psi}{\cos^2 \phi}\right)^{\frac{i-3}{2}} \sin \psi \, d\psi \tag{3.1}$$

$$\int_{S^{n-1}} g \, dv = c_2 \int_0^{\frac{\pi}{2}} g(v) \sin^{n-2} \psi \, d\psi, \tag{3.2}$$

where ϕ and ψ are the angles of the unit vectors u and v with the x_n-axis, respectively.

PROOF. The proof of formula (3.1) is similar to that of Lemma 2.1 in [Z1]. See also Lemma 8 in [Z2]. (3.2) follows from the spherical coordinates. □

LEMMA 3.2. *There exist a convex body K and a C^∞ function g so that*
$$R_i(\rho_K^{i-4}g) < 0, \quad \langle \rho_K^{n-4}, g \rangle > 0, \tag{3.3}$$
for all integers $3 < i < n$.

PROOF. Consider a convex body of revolution K. From (3.1) and (3.2), the inequalities in (3.3) become

$$\int_\phi^{\frac{\pi}{2}} \rho_K(\psi)^{i-4} g(\psi) \left(1 - \frac{\cos^2 \psi}{\cos^2 \phi}\right)^{\frac{i-3}{2}} \sin \psi \, d\psi < 0, \tag{3.4}$$

$$\int_0^{\frac{\pi}{2}} \rho_K^{n-4}(\psi) g(\psi) \sin^{n-2} \psi \, d\psi > 0. \tag{3.5}$$

We need to choose K and g so that the last two inequalities are satisfied.

Let K be a cylinder. Then
$$\rho_K(\psi) = \frac{1}{\sin \psi}, \quad \psi_1 \leq \psi \leq \frac{\pi}{2},$$
for some $\psi_1 > 0$. Choose g so that $g(\psi) = 0$ when $0 \leq \psi \leq \psi_1$. Then (3.4) and (3.5) can be written as

$$\int_\phi^{\frac{\pi}{2}} g(\psi) \left(1 - \frac{\cos^2 \psi}{\cos^2 \phi}\right)^{\frac{i-3}{2}} \sin^{5-i} \psi \, d\psi < 0, \tag{3.6}$$

$$\int_{\psi_1}^{\frac{\pi}{2}} g(\psi) \sin^2 \psi \, d\psi > 0, \quad \psi_1 \leq \phi \leq \frac{\pi}{2}. \tag{3.7}$$

Let $\cos \psi = t$. Then (3.6) and (3.7) become

$$\int_0^x g(\psi(t)) \left(1 - \frac{t^2}{x^2}\right)^{\frac{i-3}{2}} (1-t^2)^{\frac{4-i}{2}} \, dt < 0 \tag{3.8}$$

and

$$\int_0^{x_1} g(\psi(t))(1-t^2)^{\frac{1}{2}} \, dt > 0, \quad 0 \leq x \leq x_1, \tag{3.9}$$

where $x_1 = \cos \psi_1$.

Let $f(t) = g(\psi(t))(1-t^2)^{\frac{1}{2}}$. We write (3.8) and (3.9) as

$$\int_0^x f(t) \left(\frac{x^2 - t^2}{1-t^2}\right)^{\frac{i-3}{2}} dt < 0 \tag{3.10}$$

and

$$\int_0^{x_1} f(t) \, dt > 0, \quad 0 \leq x \leq x_1. \tag{3.11}$$

Let

$$g(t, x) = \left(\frac{x^2 - t^2}{1-t^2}\right)^{\frac{i-3}{2}}.$$

When $i > 3$, $g(t,x)$ is strictly decreasing for $t \in [0, x]$. Choose $f(t)$ such that

$$f(t) < 0 \quad \text{if } 0 < t < x_0,$$
$$f(t) > 0 \quad \text{if } x_0 < t < x_1$$
$$f(t) = 0 \quad \text{if } x_1 < t < 1,$$

and

$$\int_0^{x_1} f(t)\,dt = 0.$$

It follows that

$$\int_0^x f(t)g(t,x)\,dt = \int_0^{x_0} f(t)g(t,x)\,dt + \int_{x_0}^x f(t)g(t,x)\,dt$$
$$< \int_0^{x_0} f(t)g(x_0,x)\,dt + \int_{x_0}^x f(t)g(x_0,x)\,dt$$
$$= g(x_0,x) \int_0^x f(t)\,dt \leq 0.$$

By a small perturbation of f, there is

$$\int_0^x f(t)g(t,x)\,dt < 0$$

and

$$\int_0^{x_1} f(t)\,dt > 0, \quad 0 < x < x_1.$$

Therefore, one can choose f so that (3.10) and (3.11) are true. This proves the lemma. □

PROOF OF THEOREM 1.3. By the Lemma 3.2, there is a C^∞ convex body K of positive curvature and a C^∞ function g on S^{n-1} so that for all $3 < i < n$,

$$R_i\left(\rho_K^{i-4} g\right) < 0, \quad \langle \rho_K^{n-4}, g \rangle > 0.$$

Define a centered convex body of revolution K_ε by

$$\rho_{K_\varepsilon}^4 = \rho_K^4 + \varepsilon g,$$

for $\varepsilon > 0$ small. We have

$$V(K_\varepsilon) - V(K) = \frac{n}{4} \langle \rho_K^{n-4}, g \rangle \varepsilon + o(\varepsilon),$$

$$\mathrm{vol}_i(K_\varepsilon \cap \xi) - \mathrm{vol}_i(K \cap \xi) = \frac{i}{4} R_i(\rho_K^{i-4} g)(\xi) \varepsilon + o(\varepsilon).$$

Therefore, when ε is small enough, we have $V(K_\varepsilon) > V(K)$ and

$$\mathrm{vol}_i(K_\varepsilon \cap \xi) < \mathrm{vol}_i(K \cap \xi), \quad \forall \xi \in G_{i,n}, \ 3 < i < n. \quad \square$$

References

[Ba] K. Ball, "Some remarks on the geometry of convex sets", pp. 224–231 in *Geometric aspects of Functional Analysis* 1986-1987, edited by J. Lindenstrauss and V. D. Milman, Lecture Notes in Math. **1317**, Springer, Berlin, 1988.

[Bo] J. Bourgain, "On the Busemann–Petty problem for perturbations of the ball", *Geom. Funct. Anal.* **1** (1991), 1–13.

[BL] J. Bourgain and J. Lindenstrauss, "Projection bodies ", pp. 250–270 in *Geometric aspects of Functional Analysis* 1986-1987, edited by J. Lindenstrauss and V. D. Milman, Lecture Notes in Math. **1317**, Springer, Berlin, 1988.

[Bu] H. Busemann, "Volumes and areas of cross sections", *Amer. Math. Monthly* **67** (1960), 248–250 and 671.

[BP] H. Busemann and C. M. Petty, "Problems on convex bodies", *Math. Scand.* **4** (1956), 88–94.

[G1] R. J. Gardner, "Intersection bodies and the Busemann–Petty problem", *Trans. Amer. Math. Soc.* **342** (1994), 435–445.

[G2] R. J. Gardner, "A positive answer to the Busemann–Petty problem in three dimensions", *Annals of Math.* **140** (1994), 435–447.

[G3] R. J. Gardner, *Geometric tomography*, Cambridge University Press, New York, 1995.

[Gia] A. Giannopoulos, "A note on a problem of H. Busemann and C. M. Petty concerning sections of symmetric convex bodies", *Mathematika* **37** (1990), 239–244.

[Gie] M. Giertz, "A note on a problem of Busemann", *Math. Scand.* **25** (1969), 145–148.

[GR] E. Grinberg and I. Rivin, "Infinitesimal aspects of the Busemann and Petty problem", *Bul. London Math. Soc.* **22** (1990), 478–484.

[GZ] E. Grinberg and G. Zhang, "Convolutions, transforms and convex bodies", to appear in *Proc. London Math. Soc.*.

[Ha] H. Hadwiger, "Radialpotenzintegrale zentralsymmetrischer Rotationsk örper und Ungleichheitsaussagen Busemannscher Art", *Math. Scand.* **23** (1968), 193–200.

[He] S. Helgason, *Groups and geometric analysis*, Academic Press, 1984.

[LR] D. G. Larman and C. A. Rogers, "The existence of a centrally symmetric convex body with central cross-sections that are unexpectedly small", *Mathematika* **22** (1975), 164–175.

[Lu] E. Lutwak, "Intersection bodies and dual mixed volumes", *Adv. Math.* **71** (1988), 232–261.

[Pa] M. Papadimitrakis, "On the Busemann–Petty problem about convex, centrally symmetric bodies in $\mathbb{R}^n$", *Mathematika* (1994).

[Sc] R. Schneider, "Convex bodies: The Brunn–Minkowski theory", Cambridge University Press, 1993.

[St] R. Strichartz, "L^p estimates for Radon transforms in Euclidean and non-Euclidean spaces", *Duke Math. J.* **48** (1981), 699–727.

[Z1] G. Zhang, "Centered bodies and dual mixed volumes", *Trans. Amer. Math. Soc.* **345** (1994), 777–801.

[Z2] G. Zhang, "Intersection bodies and Busemann–Petty inequalities in $\mathbb{R}^4$", *Annals of Math.* **140** (1994), 331–346.

[Z3] G. Zhang, "Sections of convex bodies", *Amer. Jour. Math.* **118** (1996), 319–340.

[Z4] G. Zhang, "A positive answer to the Busemann–Petty problem in four dimensions", preprint.

JEAN BOURGAIN
SCHOOL OF MATHEMATICS
OLDEN LANE
INSTITUTE FOR ADVANCED STUDY
PRINCETON, NJ 08540
UNITED STATES OF AMERICA
  bourgain@math.ias.edu

GAOYONG ZHANG
DEPARTMENT OF MATHEMATICS
POLYTECHNIC UNIVERSITY
6 METROTECH CENTER
BROOKLYN, NY 11201
UNITED STATES OF AMERICA
  gzhang@math.poly.edu

Isotropic Constants of Schatten Class Spaces

SEAN DAR

ABSTRACT. We study the value of the isotropic constant of the unit ball in the Schatten class spaces C_p^n. We prove that, for $2 \leq p \leq \infty$, this value is bounded by a fixed constant, whereas for $1 \leq p < 2$ it is bounded by $c\,(\log n)^{1/p-1/2}$, where c is a fixed constant.

The isotropic constant of a convex symmetric body $K \subset \mathbb{R}^d$ is a highly important quantity. One of its equivalent definitions is

$$L_K = \min_{T \in SL(d)} \frac{1}{\sqrt{d}\,|K|^{1/d}} \sqrt{\frac{1}{|K|} \int_K \|Tu\|_2^2 \, du},$$

where $|K|$ stands for the volume of K and $\|\cdot\|_2$ is the usual Euclidean norm. See [MP] for other formulations and a full discussion.

Estimation of L_K is one of the central problems on the border between local theory and convexity. Bourgain [Bo] has shown that $L_K \lesssim d^{1/4} \log d$ for any $K \subset \mathbb{R}^d$, where by $A \lesssim B$ we mean that $A \leq cB$ for some universal constant c. He also raised the problem of universal boundedness of the isotropic constant independent of the dimension.

Till now there was no improvement of Bourgain's result in the general case, but boundedness of the isotropic constant was established for some families of bodies (see [MP,Ba,J1,J2], for example), covering the unit balls of most of the classical spaces.

The case of the Schatten class spaces C_p^n, however, was left out.

In [D] we showed that $L_{B_{C_1^n}} \lesssim \sqrt{\log n}$. Here we extend the result to all values of p using a simpler argument.

First we shall recall the definiton of the Schatten class spaces. If u is an $n \times n$ matrix, u^*u is a positive definite symmetric matrix, i.e., orthogonally diagonalizable with nonnegative eigenvalues $\eta_1 \geq \eta_2 \geq \ldots \geq \eta_n \geq 0$. Set $\lambda_i(u) = \sqrt{\eta_i}$. The numbers $\lambda_1(u) \geq \lambda_2(u) \geq \cdots \geq \lambda_n(u) \geq 0$ are called the characteristic values of u.

The Schatten class space C_p^n is the n^2-dimensional space of all $n \times n$ real matrices equipped with the norm $\|u\|_{C_p^n} = (\sum_{i=1}^n \lambda_i(u)^p)^{1/p}$.

We'll denote by K_p the unit ball of C_p^n. It is well known (see [TJ], for example) that $|K_p|^{1/n^2} \sim 1/n^{(1/2)+(1/p)}$.

PROPOSITION. *For $2 \leq p \leq \infty$, L_{K_p} is bounded by a universal constant.*
For $1 \leq p < 2$, we have $L_{K_p} \lesssim (\log n)^{(1/p)-(1/2)}$.

PROOF. For $2 \leq p \leq \infty$, we have

$$K_p \subset n^{(1/2)-(1/p)} K_2 \sim \sqrt{n^2} |K_p|^{1/n^2} B_{l_2^{n^2}}.$$

Hence

$$L_{K_p} \leq \frac{1}{\sqrt{n^2}|K_p|^{1/n^2}} \sqrt{\frac{1}{|K_p|}\int_{K_p} \|u\|_2^2 \, du} \lesssim \sqrt{\frac{1}{|K_p|}\int_{K_p}\|u\|_{K_p}^2 \, du} \leq 1.$$

Our treatment of the case $1 \leq p < 2$ will rely on the following fact:

For any $t \geq 1$, $\left| K_p \setminus \left(\frac{2pt \log n}{n} \right)^{1/p - 1/2} B_{l_2^{n^2}} \right| \lesssim \frac{1}{n^{2(t-2)n}} |K_p|.$ \quad (*)

Before proving this, let's see that it implies the desired result. Indeed,

$$\frac{1}{|K_p|}\int_{K_p}\|u\|_2^2 \, du = \int_0^\infty \frac{|K_p \setminus rB_{l_2^{n^2}}|}{|K_p|} 2r \, dr$$

$$= \left(\frac{2p\log n}{n}\right)^{2(1/p-1/2)} 2\left(\frac{1}{p}-\frac{1}{2}\right)$$

$$\times \int_0^\infty \frac{|K_p \setminus (\frac{2pt\log n}{n})^{1/p-1/2} B_{l_2^{n^2}}|}{|K_p|} t^{(2/p)-2} \, dt$$

$$\leq \left(\frac{2p\log n}{n}\right)^{2(1/p-1/2)} 2\left(\frac{1}{p}-\frac{1}{2}\right)\left(2 + \int_2^\infty \frac{t^{(2/p)-2} \, dt}{n^{(t-2)n}}\right)$$

$$\sim \left(\frac{\log n}{n}\right)^{2(1/p-1/2)} \sim n^2 |K_p|^{2/n^2} (\log n)^{2(\frac{1}{p}-\frac{1}{2})}.$$

Therefore, $L_{K_p} \lesssim (\log n)^{1/p-1/2}$.

It remains to prove (*):

Any $u \in K_p$ can be decomposed as $w + (u-w)$ in such a way that w is a rank-one operator with norm $\lambda_1(u)$ not greater than 1 and $u - w$ has characteristic values $\lambda_2(u), \ldots, \lambda_n(u), 0$.

If $u \in K_p \setminus \left(\frac{2pt\log n}{n}\right)^{1/p-1/2} B_{l_2^{n^2}}$, then

$$\lambda_1(u) \geq \frac{\|u\|_2^{2/2-p}}{\|u\|_p^{p/2-p}} \geq \left(\frac{2pt\log n}{n}\right)^{1/p},$$

so we'll have

$$\|u - w\|_p = (\|u\|_p^p - \lambda_1(u)^p)^{1/p} \leq \left(1 - \frac{2pt \log n}{n}\right)^{1/p} \leq 1 - \frac{2t \log n}{n}$$

for this rank-one operator $w \in K_p$.

Now take a $\frac{\log n}{n}$-net $\mathcal{N}$ for $B_{l_2^n}$. This can be done with

$$|\mathcal{N}| \leq \left(1 + \frac{2}{\frac{\log n}{n}}\right)^n = \left(1 + \frac{2n}{\log n}\right)^n.$$

Then $\mathcal{N} \otimes \mathcal{N}$ is a $\frac{2\log n}{n}$-net for the rank-one operators with norm at most 1, so we have

$$K_p \setminus \left(\frac{2pt \log n}{n}\right)^{1/p - 1/2} B_{l_2^{n^2}} \subset \bigcup_{w \in \mathcal{N} \otimes \mathcal{N}} \left(w + \left(1 - \frac{2(t-1)\log n}{n}\right) K_p\right).$$

Hence

$$\left|K_p \setminus \left(\frac{2pt \log n}{n}\right)^{1/p - 1/2} B_{l_2^{n^2}}\right| \leq |\mathcal{N} \otimes \mathcal{N}| \left(1 - \frac{2(t-1)\log n}{n}\right)^{n^2} |K_p|$$

$$\leq \left(1 + \frac{2n}{\log n}\right)^{2n} \frac{1}{n^{2(t-1)n}} |K_p|$$

$$\lesssim \frac{1}{n^{2(t-2)n}} |K_p|,$$

and we are done. □

Added in proof. Boundedness of the isotropic constant of the Schatten classes was very recently established using more complicated methods by H. Konig, M. Meyer and A. Pajor.

References

[Ba] Ball, K. Isometric problems in l_p and the sections of convex sets, Ph.D. Dissertation, Trinity College, Cambridge, 1986.

[Bo] Bourgain, J. On the ditribution of polynomials on high dimensional convex sets, pp. 127-137 in GAFA Seminar 1989–90, Lecture Notes in Math. 1469, Springer, 1991,

[D] Dar, S. Remarks on Bourgain's Problem on Slicing of Convex Bodies, in Operator Theory: Advances and Applications 77, 1995.

[J1] Junge, M. Hyperplane conjecture for subspaces of l_p, preprint, 1991.

[J2] Junge, M. Proportional subspaces of spaces with unconditional basis have good volume properties, in Operator Theory: Advances and Applications, 77, 1995.

[MP] Milman, V. D. and Pajor, A. Isotropic position and inertia ellipsoids and zonoids of the unit ball of a normed n-dimensional space, pp. 69–109 in Geometrical aspects of functional analysis, Israel Seminar, 1987–88, Lect. Notes in Math. 1376, Springer, 1989.

[TJ] Tomczak-Jaegermann, N. Banach–Mazur Distance and Finite-dimensional Operator Ideal, Pitman Monographs 38, Longman, London, 1989.

SEAN DAR
DEPARTMENT OF MATHEMATICS
TEL AVIV UNIVERSITY
RAYMOND AND BEVERLY SACKLER FACULTY OF EXACT SCIENCES
RAMAT AVIV, TEL AVIV 69978
ISRAEL
    sean@math.tau.ac.il

On the Stability of the Volume Radius

EFIM D. GLUSKIN

ABSTRACT. The volume radius of a given n-dimensional body is the radius of a euclidean ball having the same volume as this body. We prove that the volume radius of a given convex symmetric n-dimensional body with diameter at most $\sqrt{n}$ is almost equal to the volume radius of a body obtained by the intersection of this body with n other bodies whose polars are bounded by 1 mean width.

In the last decade, interest in the problem of bounds for volumes of convex bodies was renewed mainly because of its applications to Banach Space Geometry and related topics. At the end of the 80's sharp bounds for volume radius of convex polytopes with given distance between antipodal faces were found independently by several authors: Carl and Pajor [1], Bourgain, Lindenstrauss and Milman [2], Gluskin [3]. Closely related results were obtained by Vaaler [4], Dilworth and Szarek [5] and Bárány and Furedi [6]. See also Ball and Pajor [7] where, following Kashin's conjecture, the problem was considered as a limiting case of a series of Vaaler-type results. Moreover in [3] it was observed that the volume radius of a unit cube has a certain stability property with respect to cutting the cube by a sequence of bands (see Proposition 1 below for the exact formulation). Some of Kashin's ideas enabled us to use this property for an alternative proof of Spencer's theorem [8] on a lacunary analogue of the Rudin–Shapiro polynomials (see [3]). Later Kashin [9] used the same approach for finite dimensional analogues of Menshov's correction theorem.

Here we continue to study this property. We show that it holds not only for cubes but also for a wide class of bodies. It is then observed that the condition on the width of the bands which appeared in [3] is very close to Talagrand's [10] description of bounded Gaussian processes. This observation permits us to restate the result in an invariant form which is more convenient for application. Moreover the result of [3] is extended to the intersection of a given body with a sequence of cylinders. This is made possible by a suitable generalization of the Khatri–Sidak theorem [11, 12].

Supported in part by the United States – Israel Binational Science Foundation.

We denote by $|x|$ the standard Euclidean norm of $x \in \mathbb{R}^n$. The unit ball of $\mathbb{R}^n$ is denoted by D_n, where

$$D_n = \{x \in \mathbb{R}^n : |x| \leq 1\}.$$

We say that a convex body $V \subset \mathbb{R}^n$ is absolutely convex if $V = -V$. For such a body and $x \in \mathbb{R}^n$ one defines $\|x\|_V$ by setting

$$\|x\|_V = \inf\{\lambda \in \mathbb{R}_+ : x \in \lambda V\}.$$

A linear operator $S : \mathbb{R}^n \to \mathbb{R}^k$ is called a partial isometry if $|Sx| = |x|$ for any x orthogonal to $\ker S$. For such an S and positive r we denote $W(S,r) = \{x \in \mathbb{R}^n : Sx \in rD_k\}$. Note that for $k=1$ a partial isometry is just some norm one linear functional and the corresponding set $W(S,r)$ is a band of width $2r$. As usual, vol or vol_n is the standard Lebesgue measure on $\mathbb{R}^n$. The canonical Gaussian measure on $\mathbb{R}^n$ is denoted by γ_n; it is a probability measure with density $(2\pi)^{-n/2} \exp(-|x|^2/2)$.

As usual, we denote by χ_V the indicator function of the set V, such that $\chi_V(x) = 1$ if $x \in V$ and $\chi_V(x) = 0$ if $x \notin V$.

THEOREM. *For any $\varepsilon > 0$ there exists a positive constant $C = C(\varepsilon) < \infty$ such that the following assertion holds. For a given n, let $K \subset \mathbb{R}^n$ be an absolutely convex body such that $K \subset \sqrt{n} D_n$. Then, for any n absolutely convex bodies $V_1, \ldots, V_n \subset \mathbb{R}^n$ satisfying*

$$\int_{\mathbb{R}^n} \|x\|_{V_i} \, d\gamma_n(x) \leq 1 \quad \text{for } i = 1, 2, \ldots, n,$$

the following inequality holds:

$$(1-\varepsilon) \leq \left(\frac{\text{vol}\big(K \cap C(V_1 \cap V_2 \cap \cdots \cap V_n)\big)}{\text{vol } K} \right)^{1/n} \leq 1.$$

Talagrand [10, Theorem 2] proved that for any n and any absolutely convex body $V \subset \mathbb{R}^n$ satisfying

$$\int \|x\|_V \, d\gamma_n(x) \leq 1$$

there exists a sequence of norm one linear functionals $f_i \in (\mathbb{R}^n)^*$ possessing the property that

$$\bigcap_{i=1}^{\infty} W\left(f_i, \sqrt{\log(2+i)}\right) \subset C_T V,$$

where $C_T < \infty$ is a universal constant. By Talagrand's result, the theorem is equivalent to the following proposition:

PROPOSITION 1. *For a given n let $K \subset \mathbb{R}^n$ be an absolutely convex body satisfying $K \subset \sqrt{n}D_n$. Then, for any sequence of norm one linear functionals $f_i \in (\mathbb{R}^n)^*$, $i = 1, 2, \ldots$, with some constant $C = C(\varepsilon)$ depending on ε only, we have*

$$(1-\varepsilon) \leq \left(\frac{\mathrm{vol}(K \cap (\bigcap_{i=1}^{\infty} W(f_i, r_i)))}{\mathrm{vol}\,K}\right)^{1/n} \leq 1,$$

where $r_i \geq C\sqrt{\log(2 + i/n)}$.

PROOF. It is clear that, for any body $V \subset \lambda D_n$,

$$1 \leq \left(\frac{1}{\sqrt{2\pi}}\right)^n \frac{\mathrm{vol}\,V}{\gamma_n(V)} \leq e^{\lambda^2/2}.$$

In particular, for K as in the Proposition, with $C_1 = C_1(\varepsilon)$,

$$C_1\sqrt{2\pi}\left(\gamma_n\left(\frac{1}{C_1}K\right)\right)^{1/n} \geq \sqrt{1-\varepsilon}\,(\mathrm{vol}\,K)^{1/n}.$$

On the other hand,

$$\left(\mathrm{vol}\left(K \cap \left(\bigcap_{i=1}^{\infty} W(f_i, r_i)\right)\right)\right)^{1//n} \geq C_1\sqrt{2\pi}\gamma_n\left(\frac{1}{C_1}K \cap \frac{1}{C_1}\left(\bigcap_{i=1}^{\infty} W(f_i, r_i)\right)\right).$$

By the Khatri–Sidak theorem,

$$\gamma_n\left(\frac{1}{C_1}K \cap \frac{1}{C_1}\left(\bigcap_{i=1}^{\infty} W(f_i, r_i)\right)\right) \geq \gamma_n\left(\frac{1}{C_1}K\right) \prod_{i=1}^{\infty} \gamma_n(W(f_i, r_i/C_1)).$$

The elementary bound $\gamma_n(W(f,r)) = \gamma_1((-r,r)) \geq 1 - e^{-r^2/2}$ implies that

$$\left(\prod_{i=1}^{\infty} \gamma_n(W(f_i, r_i/C_1))\right)^{1/n} \geq \sqrt{1-\varepsilon} \tag{1}$$

for $r_i = C\sqrt{\log(2+i/n)}$ with $C = C(\varepsilon)$, and the proposition follows. □

REMARK 1. The particular case of the cube instead of the general convex body K was considered in [3]. In the first version of the paper the proof of Proposition 1 followed the same scheme as that of [3]. A significant simplification of the proof was found by the referee. I wish to express my gratitude to him for his suggestion to publish his proof here.

REMARK 2. We say that a body V satisfies the positive correlation property (PCP) if for any absolutely convex body W and for any positive constant λ one has $\gamma_n(\lambda V \cap W) \geq \gamma_n(\lambda V)\gamma_n(W)$. The proof of Proposition 1 shows the following fact:

For any $\varepsilon > 0$ there exists a constant $C > 0$ such that for any absolutely convex body $K \subset \sqrt{n}D_n$ and for any sequence of bodies V_j satisfying the PCP one has
$$1 - \varepsilon \leq \left(\frac{\mathrm{vol}(K \cap (\bigcap_{j=1}^{\infty} V_j))}{\mathrm{vol}\, K}\right)^{1/n} \leq 1,$$
providing $\left(\prod_{j=1}^{\infty} \gamma_n(V_j/C)\right) \geq (1 - \varepsilon/2)^n$.

The Gaussian Correlation Conjecture states that any absolutely convex body satisfies the PCP. In these terms the Khatri–Sidak theorem states that any band satisfies the PCP. A slight modification of its proof leads to this result:

LEMMA. *Let $V \subset \mathbb{R}^n$ be an absolutely convex body and $S: \mathbb{R}^n \to \mathbb{R}^k$ be a partial isometry of rank $S = k$. Then, for any positive r, we have*
$$\gamma_n(V \cap W(S, r)) \geq \gamma_n(V)\gamma_n(W(S, r)) = \gamma_n(V)\gamma_k(rD_k).$$

In other words, any cylinder satisfies the PCP. Certainly one can use the lemma to obtain analogues of Proposition 1 for the case of the intersection with a sequence of cylinders. We omit the precise statement, which is rather complicated. For the reader's convenience, we outline a proof of the lemma.

Let us fix some partial isometry S from $\mathbb{R}^n$ onto $\mathbb{R}^k$. It is well known that its conjugate S^* is a partial isometry from $\mathbb{R}^k$ to $\mathbb{R}^n$, which is right inverse to S, and $Q = \mathrm{Id}_{\mathbb{R}^n} - S^*S$ is an orthogonal projection on $\ker S$. For $\alpha \in [0, \pi/2]$ let the operator $T(\alpha): \mathbb{R}^{n+k} \to \mathbb{R}^{n+k}$ be defined by
$$T(\alpha)(x, y) = (Qx + \cos\alpha\, S^*Sx - \sin\alpha\, S^*y,\; \sin\alpha\, Sx + \cos\alpha\, y)$$
for $(x, y) \in \mathbb{R}^n \times \mathbb{R}^k = \mathbb{R}^{n+k}$. For a given absolutely convex $V_1 \subset \mathbb{R}^n$ and $V_2 \subset \mathbb{R}^k$ we denote
$$\widetilde{V}_1(\alpha) = T(\alpha)(V_1 \times \mathbb{R}^k) \quad \text{and} \quad \widetilde{V}_2(\alpha) = T(\alpha)(\mathbb{R}^n \times V_2).$$
Consider the function
$$f_{V_1, V_2}(\alpha) = \mathrm{vol}\{\widetilde{V}_1(\alpha) \cap \widetilde{V}_2(0) \cap D_{n+k}\},$$
or equivalently
$$f_{V_1, V_2}(\alpha) = \int_{\widetilde{V}_2(0) \cap D_{n+k}} \chi_{\widetilde{V}_1(0)}(T^{-1}(\alpha)z)\, dz.$$

It is easy to see that f_{V_1, V_2} is absolutely continuous and one has the following equality for a.e. $\alpha \in [0, \pi/2]$:
$$\frac{d}{d\alpha} f_{V_1, V_2}(\alpha) = \int_{\partial(\widetilde{V}_2(0) \cap D_{n+k})} \chi_{\widetilde{V}_1(0)}(T^{-1}(\alpha)z) \langle \widetilde{n}(z), Jz\rangle\, d\widetilde{\mu}(z),$$
where $\widetilde{\mu}$ is the Lebesgue measure on $\partial(\widetilde{V}_2(0) \cap D_{n+k})$, $\widetilde{n}$ is outer normal to $\partial(\widetilde{V}_2(0) \cap D_{n+k})$ and operator J is given by $J = T(\alpha)\frac{d}{d\alpha}T^{-1}(\alpha)$. Straightforward

computation shows that $J(x,y) = (S^*y, -Sx)$ for $(x,y) \in \mathbb{R}^n \times \mathbb{R}^k$. Since $\langle z, Jz \rangle = 0$ for any $z \in \mathbb{R}^{n+k}$ one can rewrite the previous equality as

$$\frac{d}{d\alpha} f_{V_1,V_2} = -\int_{\partial V_2} d\mu(y) \int_{W_y} \langle n(y), Sx \rangle \, dx, \tag{2}$$

where μ is the Lebesgue measure on ∂V_2, $n(y)$ is outer normal to ∂V_2 at point y and

$$W_y = \{x \in \sqrt{(1-|y|^2)_+} \, D_n : Qx + \cos \alpha \, S^*Sx \in V_1 - \sin \alpha \, S^*y\} .$$

PROPOSITION 2. *Let $p : \mathbb{R}_+^1 \to \mathbb{R}_+^1$ be some nonincreasing positive function and ν be an absolutely continuous measure on $\mathbb{R}^{n+k}$ with density $p(|x|)$. Then, for any absolutely convex body $V_1 \subset \mathbb{R}^n$ and any positive r the function*

$$h(\alpha) = \nu\{\widetilde{V}_1(\alpha) \cap r(\mathbb{R}^n \times D_k)\}$$

is nondecreasing on $[0, \pi/2]$.

SKETCH OF THE PROOF. It is clear that without loss of generality, we can consider the case $p = \chi_{[0,1]}$ only. In this case $h(\alpha) = f_{V,rD_k}(\alpha)$. By Lemma 2 from [3] one has

$$\int_{W_y} \langle S^*y, x \rangle \, dx \leq 0 . \tag{3}$$

Taking into account that $n(y) = r^{-1}y$ for $y \in \partial(rD_k) = rS^{k-1}$ we get Proposition 2 from (2) and (3). □

The inequality $h(0) \leq h(\pi/2)$ with $p(t) = (2\pi)^{-n/2} e^{-t^2/2}$ proves the lemma. □

NOTE. When this paper was almost complete, we learned from S. Bobkov that Schechtman, Schlumprecht and Zinn [13] had found a very short proof of the Khatri–Sidak theorem, based on the Borel–Prékopa–Leindler characterization of log-concave measures. A simple modification of the method of [13] leads to an alternative proof of the lemma.

REMARK 3. It is fairly simple to see that the conditions of Proposition 1 on r_i are optimal for $i \gg n$; see [3, 14]. For $i < n$ this is not so. In fact the proof of Proposition 1 shows that it holds under better conditions on r_i, $i < n$ than those stated. To see this one has to take into account that we only need to estimate the product in (1) and to use for example the inequality $\gamma_1((-r,r)) \geq 2(2\pi)^{-1/2} r e^{-r^2/2}$.

The same remark holds in the case of the cylinders' intersection.

Acknowledgements

Part of this work was done while the author was visiting at the Institute for Advanced Study, Princeton and MSRI, Berkeley. I wish to thank both Institutions for their hospitality.

I also wish to express my gratitude to Professors V. Milman, G. Schechtman, B. Shiff and N. Tomczak–Jaegerman for their interest in this work, to S. Bobkov for bringing my attention to the paper [13] and to the referee for his valuable remarks.

References

[1] Carl, B., Pajor, A., Gelfand numbers of operators with values in a Hilbert space, Invent. Math. 94 (1988), 479–504.

[2] Bourgain, J., Lindenstrauss, J., Milman, V. D., Approximation of zonoids by zonotopes, Acta Math. 162 (1989), 73–141.

[3] Gluskin, E. D., "Extremal properties of orthogonal parallelepipeds and their applications to the geometry of Banach space", Math., USSR Sbornik 64 (1989), 85–96.

[4] Vaaler, J. D., A geometric inequality with applications to linear forms, Pacific J. Math. 83 (1979), 543–553.

[5] Dilworth, S., Szarek, S., The cotype constant and an almost Euclidean decomposition for finite-dimensional normed spaces, Israel J. Math. 52 (1985), 82–96.

[6] Bárány, Furedi, Z., Computing the volume is difficult, Discrete Comput. Geom. 2 (1987), 319–326.

[7] Ball, K., Pajor, A., Convex bodies with few faces, Proc. AMS 110, N1 (1990), 225–231.

[8] Spencer, J., Six standard deviations suffice, Trans. Amer. Math. Soc. 289 (1985), N2, 679–706.

[9] Kashin, B. S., Analogue of Men'shov's theorem, "on correction" for discrete orthonormal systems, Mathematical Notes (translation of Mathematicheskie Zametki), 46 (1989), N6, 934–939.

[10] Talagrand, M., Regularity of Gaussian processes, Acta Math. 159 (1987), 99–149.

[11] Khatri, C. G., On certain inequalities for normal distributions and their applications to simultaneous confidence bounds, Ann. Math. Statist. 38 (1967), 1853–1867.

[12] Sidak, Z., Rectangular confidence regions for the means of multivariate normal distributions, J. Amer. Statist. Assoc. 62 (1967), 626–633.

[13] Schechtman, G., Schlumprecht, Th., Zinn, J., On the Gaussian measure of the intersection of symmetric, convex sets, Annals of Probability (to appear).

[14] Gluskin, E. D., Deviation of Gaussian vector from a subspace of ℓ_∞^n and random subspaces of ℓ_∞^N, Leningrad Math. J., 1, 1990, N5, 1165–1175.

EFIM D. GLUSKIN
SCHOOL OF MATHEMATICAL SCIENCES
SACKLER FACULTY OF EXACT SCIENCES
TEL AVIV UNIVERSITY
RAMAT-AVIV, 69978, ISRAEL

Polytope Approximations of the Unit Ball of ℓ_p^n

W. TIMOTHY GOWERS

ABSTRACT. A simple and explicit method is given for approximating the unit ball of ℓ_p^n by polytopes. The method leads to a natural generalization of ℓ_p-spaces with good duality and interpolation properties.

1. Introduction

The classical spaces ℓ_p and L_p are the best known and in many ways most fundamental examples of Banach spaces. In view of their interesting properties it is natural to ask whether the role of the function t^p in these spaces can be played by other more general functions. This question was answered by Orlicz, who defined a certain class of functions, now known as Orlicz functions, and associated with each one a sequence space and a function space, now called an Orlicz sequence space and Orlicz function space. The Orlicz spaces are generally regarded as the correct and most natural spaces to associate with given Orlicz functions.

One of the aims of this paper is to cast doubt on that view, at least in its isometric interpretation. We shall do this by discussing a different generalization which arises geometrically and has two desirable isometric properties lacked by Orlicz spaces. First, the dual of one of our spaces is isometric to another such space. Second, complex interpolation between two of our spaces yields a third in a natural way. Irritatingly, we have not managed to establish whether our new spaces are *isomorphic* to Orlicz spaces, in which case they are a useful renorming of them, or whether they are completely different. Our route to the new generalization starts with an unusual (perhaps even eccentric) problem which will be described below, and which relates more to the polytope approximations of the title.

The results of this paper originated with the following line of thought. A common way of constructing a finite-dimensional normed space which lacks symmetry properties, invented by Gluskin [G], is to take a small number of antipodal pairs of points at random from the unit sphere of ℓ_2^n and to take their convex hull as the unit ball of the space. Alternatively, and dually, one can take the convex polytope defined by the hyperplanes tangent to the sphere at the given points. However, such spaces have properties which a general space cannot be expected to have. For example, geometrically their unit balls are far from being typical centrally symmetric polytopes, having in the first case very few vertices and in the second case very few faces. A related fact is that the convex hull of a small number of points gives a space with very small cotype constants. Is there some way of constructing a convex body which is in any useful sense completely generic?

We shall return to this question at the end of the paper, but it is not our main concern. Instead, we shall give an (incomplete) investigation of what happens when one mixes the process of taking the convex hull of a few points with the dual process. One is led naturally away from generic spaces and towards very special ones. At the end of the paper we make several suggestions for how the investigation might be continued.

2. A Polytope Approximation of the Unit Ball of ℓ_2^n

Recall that a basis $e_1, \ldots, e_n$ of an n-dimensional normed space is said to be 1-*symmetric* if, for every choice of scalars $a_1, \ldots, a_n$, every choice of signs $\varepsilon_1, \ldots, \varepsilon_n$ with $\varepsilon_i = \pm 1$, and every permutation π of $\{1, 2, \ldots, n\}$, we have the equality

$$\left\| \sum_{i=1}^n \varepsilon_i a_i e_{\pi(i)} \right\| = \left\| \sum_{i=1}^n a_i e_i \right\|.$$

For $n \geq 1$ let $\mathcal{F}_n$ stand for the set of normed spaces of the form $(\mathbb{R}^n, \|\cdot\|)$ for which the standard basis $e_1, \ldots, e_n$ is normalized and 1-symmetric. We also define two operations S and T which map $\bigcup_{n=1}^\infty \mathcal{F}_n$ to $\bigcup_{n=1}^\infty \mathcal{F}_{n+1}$ as follows. Given $X \in \mathcal{F}_n$, we set $T(X)$ to be the unique normed space in $\mathcal{F}_{n+1}$ such that for any $\boldsymbol{a} = \sum_1^{n+1} a_i e_i \in \mathbb{R}^{n+1}$ with $a_1 \geq \cdots \geq a_{n+1} \geq 0$, $\|\boldsymbol{a}\|_{T(X)} = \|\sum_1^n a_i e_i\|_X$. The norm on $T(X)$ is as small as it can be given that it is 1-symmetric and that its restriction to any n coordinates yields the norm on X. The operation S is defined in the opposite way: the norm on $S(X)$ is as big as it can be under the same conditions. Thus $S(X) = (T(X^*))^*$. Alternatively, given the unit ball of X, embed it into $\mathbb{R}^{n+1}$ in the $n+1$ natural ways (up to symmetry) given by ignoring each of the coordinates in turn. The convex hull of these $n+1$ copies is the unit ball of $S(X)$. Note that the unit ball of $T(X)$ has a geometric description as well. The Cartesian product of one of the $n+1$ images of the unit ball of X with the direction not used is a cylinder. The unit ball of $T(X)$ is the intersection of these $n+1$ cylinders.

Now, if we start with the single space in $\mathcal{F}_1$, the normalized 1-dimensional space, which we shall call $\mathbb{R}$, and apply S $n-1$ times, we clearly end up with ℓ_1^n. Similarly, if we apply T $n-1$ times we obtain ℓ_∞^n. What happens if we alternate

S with T? The answer is the main result of this section and the motivation for the rest of the paper.

THEOREM 1. *Let $n \geqslant 2$ and let X be a space in $\mathcal{F}_n$ obtained from the space $\mathbb{R}$ by applying the operations S and T alternately. Then $d(X, \ell_2^n) = \sqrt{2}$.*

The proof of this theorem is based on two lemmas: the first is trivial and the second also easy.

LEMMA 2. *If $X, Y \in \mathcal{F}_n$ and $\|\boldsymbol{a}\|_X \leqslant \|\boldsymbol{a}\|_Y$ for every $\boldsymbol{a} \in \mathbb{R}^n$, then $\|\boldsymbol{a}\|_{SX} \leqslant \|\boldsymbol{a}\|_{SY}$ and $\|\boldsymbol{a}\|_{TX} \leqslant \|\boldsymbol{a}\|_{TY}$ for every $\boldsymbol{a} \in \mathbb{R}^{n+1}$.*

LEMMA 3. *Suppose $X = \ell_2^n$. Then $\|\boldsymbol{a}\|_{\ell_2^{n+2}} \leqslant \|\boldsymbol{a}\|_{TSX}$ for every $\boldsymbol{a} \in \mathbb{R}^{n+2}$.*

Before proving Lemma 3, let us see why the two lemmas are sufficient to prove the main result. In the argument that follows and for the rest of the paper it will be convenient, when $X, Y \in \mathcal{F}_n$, to use the abbreviation $c_1 X \leqslant c_2 Y$ for the statement that $c_1\|\boldsymbol{a}\|_X \leqslant c_2\|\boldsymbol{a}\|_Y$ for any vector $\boldsymbol{a} \in \mathbb{R}^n$.

PROOF OF THEOREM 1. First consider the space $(TS)^k(\mathbb{R})$ for some $k \in \mathbb{N}$. It is clear from the two lemmas that $(TS)^k(\mathbb{R}) \geqslant \ell_2^{2k+1}$. Moreover, since $T(\mathbb{R}) \geqslant 2^{-1/2} \ell_2^2$, we also have that $(TS)^k T(\mathbb{R}) \geqslant 2^{-1/2} \ell_2^{2k+2}$, and hence that $S(TS)^{k-1} T(\mathbb{R}) = (ST)^k(\mathbb{R}) \geqslant 2^{-1/2} \ell_2^{2k+1}$. However, $(TS)^k(\mathbb{R})$ and $(ST)^k(\mathbb{R})$ are dual to each other, so we have the relations

$$2^{-1/2} \ell_2^{2k+1} \leqslant (ST)^k(\mathbb{R}) \leqslant \ell_2^{2k+1} \leqslant (TS)^k(\mathbb{R}) \leqslant 2^{1/2} \ell_2^{2k+1}.$$

It follows immediately also that

$$2^{-1/2} \ell_2^{2k} \leqslant T(ST)^{k-1}(\mathbb{R}) \leqslant \ell_2^{2k} \leqslant S(TS)^{k-1}(\mathbb{R}) \leqslant 2^{1/2} \ell_2^{2k}.$$

This establishes that $d(X, \ell_2^n) \leqslant \sqrt{2}$, conditional on the truth of Lemma 3. It is well known that the distance from a space in $\mathcal{F}_n$ to ℓ_2^n is attained by the identity map. Considering the norms in X and ℓ_2^n of $\boldsymbol{e}_1$ and $\boldsymbol{e}_1 + \boldsymbol{e}_2$, we obtain the reverse inequality. $\square$

PROOF OF LEMMA 3. Throughout this proof, all sequences will be assumed to be positive and decreasing. Writing $X = \ell_2^n$ we have, for any $\boldsymbol{a} \in \mathbb{R}^{n+1}$ and $\boldsymbol{b} \in \mathbb{R}^{n+2}$,

$$\|\boldsymbol{a}\|_{SX} = \max\left\{ \sum_1^{n+1} f_i a_i : \sum_1^n f_i^2 \leqslant 1 \right\}$$

and

$$\|\boldsymbol{b}\|_{TSX} = \max\left\{ \sum_1^{n+1} f_i b_i : \sum_1^n f_i^2 \leqslant 1 \right\}.$$

Hence it is enough to show that, for any $a_1 \geqslant \cdots \geqslant a_{n+2} \geqslant 0$ with $\sum_1^{n+2} a_i^2 = 1$, we can find $f_1 \geqslant \cdots \geqslant f_{n+1} \geqslant 0$ such that $\sum_1^n f_i^2 \leqslant 1$ and $\sum_1^{n+1} f_i a_i \geqslant 1$. We do this by setting $f_i = \lambda a_i$ for $1 \leq i \leq n$, where $\lambda = (1 - a_{n+1}^2 - a_{n+2}^2)^{-1/2}$, and setting $f_{n+1} = f_n$. Then certainly $\sum_1^n f_i^2 = 1$, and moreover

$$\sum_1^{n+1} f_i a_i = \lambda \sum_1^n a_i^2 + \lambda a_n a_{n+1} = \frac{1 - a_{n+1}^2 - a_{n+2}^2 + a_n a_{n+1}}{(1 - a_{n+1}^2 - a_{n+2}^2)^{1/2}}$$

$$\geq \frac{1 - a_{n+2}^2}{(1 - 2a_{n+2}^2)^{1/2}} \geq 1. \qquad \square$$

The proof of Theorem 1 can be generalized very easily to approximate ℓ_p^n in a similar way when p or its conjugate is an integer. However, with a little more work, one can approximate ℓ_p^n for an arbitrary p. We shall do this in Section 4. First, we show how to calculate the norm of a vector in a space of the form $U_1(U_2(\ldots U_{n-1}(\mathbb{R})\ldots))$ where each U_i is either S or T. We shall call such a space an ST-space.

3. An Algorithm for Calculating the Norm

Let $X \in \mathcal{F}_n$ and let $\boldsymbol{a} = \sum_1^{n+1} a_i \boldsymbol{e}_i$ be a vector in $\mathbb{R}^{n+1}$ such that $a_1 \geq \cdots \geq a_{n+1} \geq 0$. Then by definition $\|\boldsymbol{a}\|_{TX} = \|\sum_1^n a_i \boldsymbol{e}_i\|_X$. We shall now show how to calculate $\|\boldsymbol{a}\|_{SX}$. Given two vectors $\boldsymbol{a}$ and $\boldsymbol{b}$ in $\mathbb{R}^n$, write $\boldsymbol{a}^* = (a_1^*, \ldots, a_n^*)$ and $\boldsymbol{b}^* = (b_1^*, \ldots, b_n^*)$ for their (non-negative) decreasing rearrangements. Write $\boldsymbol{a} \prec \boldsymbol{b}$ if $\sum_{i=1}^k a_i^* \leq \sum_{i=1}^k b_i^*$ for every $k \leq n$. It is well known that $\boldsymbol{a} \prec \boldsymbol{b}$ if and only if $\boldsymbol{a}$ lies in the convex hull of all vectors that can be obtained from $\boldsymbol{b}$ by permuting the coordinates and changing some of their signs. Therefore, if $\|\cdot\|$ is a 1-symmetric norm and $\boldsymbol{a} \prec \boldsymbol{b}$, we know that $\|\boldsymbol{a}\| \leq \|\boldsymbol{b}\|$.

Given $X \in \mathcal{F}_n$ and $\boldsymbol{a} \in \mathbb{R}^{n+1}$, we define a function $\|\cdot\|'$ on $\mathbb{R}^{n+1}$ by

$$\|\boldsymbol{a}\|' = \min\{\|\boldsymbol{b}\|_X : \boldsymbol{b} \in X, \boldsymbol{a} \prec \boldsymbol{b}\},$$

where $\boldsymbol{a} \prec \boldsymbol{b}$ means of course that $\boldsymbol{a} \prec \boldsymbol{b}'$, where $\boldsymbol{b}'$ is the image of $\boldsymbol{b}$ in $\mathbb{R}^{n+1}$ under the obvious inclusion. (Note that the existence of the above minimum follows easily from compactness.)

LEMMA 4. *Let $X \in \mathcal{F}_n$ and $\boldsymbol{a} \in \mathbb{R}^{n+1}$. Then $\|\boldsymbol{a}\|_{SX} = \|\boldsymbol{a}\|'$.*

PROOF. Given two vectors $\boldsymbol{a}_1$ and $\boldsymbol{a}_2$ in $\mathbb{R}^{n+1}$, let $\boldsymbol{b}_i \in \mathbb{R}^n$ be such that $\boldsymbol{a}_i \prec \boldsymbol{b}_i$ and $\|\boldsymbol{b}_i\|_X$ is minimal, for $i = 1, 2$. Then we certainly have $\boldsymbol{a}_1 + \boldsymbol{a}_2 \prec \boldsymbol{b}_1^* + \boldsymbol{b}_2^*$, and thus, since X is a 1-symmetric space,

$$\|\boldsymbol{a}_1 + \boldsymbol{a}_2\|' \leq \|\boldsymbol{b}_1^* + \boldsymbol{b}_2^*\|_X \leq \|\boldsymbol{b}_1^*\|_X + \|\boldsymbol{b}_2^*\|_X = \|\boldsymbol{b}_1\|_X + \|\boldsymbol{b}_2\|_X = \|\boldsymbol{a}_1\|' + \|\boldsymbol{a}_2\|'.$$

It follows that $\|\cdot\|'$ is a norm. By the discussion above concerning the relation $\prec$, the unit ball of the space $(\mathbb{R}^{n+1}, \|\cdot\|')$ is contained in that of SX. Since it also contains the $n+1$ natural images of $B(X)$ in $\mathbb{R}^{n+1}$, we have that $\|\cdot\|' = \|\cdot\|_{SX}$, as stated. $\qquad \square$

In the next lemma we identify a vector at which the minimum of the set

$$\{\|\boldsymbol{b}\|_X : \boldsymbol{b} \in X, \boldsymbol{a} \prec \boldsymbol{b}\}$$

is attained. We write $x \vee y$ for $\max\{x, y\}$.

LEMMA 5. *Let $X \in \mathcal{F}_n$, let $\boldsymbol{a} = \sum_1^{n+1} a_i \boldsymbol{e}_i$ be a vector in $\mathbb{R}^{n+1}$ for which $a_1 \geqslant \cdots \geqslant a_{n+1} > 0$ and let $\gamma \geqslant 0$ be the unique number for which*

$$\sum_1^n (a_i \vee \gamma) = \sum_1^{n+1} a_i.$$

Then $\|\boldsymbol{a}\|_{SX} = \left\|\sum_1^n (a_i \vee \gamma) \boldsymbol{e}_i\right\|_X$.

PROOF. We insist that $a_{n+1} > 0$ for convenience: it is obvious how to calculate $\|\boldsymbol{a}\|_{SX}$ if $a_{n+1} = 0$. This gives us the uniqueness of γ. Set $\boldsymbol{a}' = \sum_1^n (a_i \vee \gamma) \boldsymbol{e}_i$. We need to show that if $\boldsymbol{b} \in \mathbb{R}^n$ and $\boldsymbol{a} \prec \boldsymbol{b}$ then $\boldsymbol{a}' \prec \boldsymbol{b}$. It is clear that $\boldsymbol{a} \prec \boldsymbol{a}'$, and by our earlier remarks and the symmetry of X, we will also have $\|\boldsymbol{a}'\|_X \leqslant \|\boldsymbol{b}\|_X$. We may clearly suppose that $b_i = b_i^*$ for every i and that $\sum_1^n b_i = \sum_1^{n+1} a_i$. Since $\boldsymbol{a} \prec \boldsymbol{b}$ we then have, for every $1 \leqslant k \leqslant n$, that $\sum_k^n b_i \leqslant \sum_k^{n+1} a_i$. Let k be maximal such that $a_k > \gamma$. Clearly $k < n$, and $\gamma = (n-k)^{-1} \sum_{k+1}^{n+1} a_i$. For $l \leqslant k$ it is obvious that $\sum_1^l (a_i \vee \gamma) \leqslant \sum_1^l b_i$. When $l > k$ we have

$$\sum_1^l (a_i \vee \gamma) = \sum_1^{n+1} a_i - (n-l)\gamma = \sum_1^{n+1} a_i - \frac{n-l}{n-k} \sum_{k+1}^{n+1} a_i$$

$$\leqslant \sum_1^{n+1} a_i - \frac{n-l}{n-k} \sum_{k+1}^n b_i \leqslant \sum_1^{n+1} a_i - \sum_{l+1}^n b_i = \sum_1^l b_i.$$

This proves the lemma. □

To conclude this section we shall show how to calculate the norm of the vector $(3, 3, 3, 2, 2, 2, 1, 1, 1)$ in the spaces $(ST)^4(\mathbb{R})$ and $(TS)^4(\mathbb{R})$ by repeated application of Lemma 5 and the definition of the operation T. At each stage, we replace the vector we have by one of length one less, while preserving its norm. At a "T" stage, we simply remove the last coordinate. At an "S" stage, we apply Lemma 5. The process is summarized in the table at the top of the next page.

Notice that $\|(3, 3, 3, 2, 2, 2, 1, 1, 1)\|_2 = \sqrt{42}$, and, as must be the case by the proof of Theorem 1, $\sqrt{21} \leqslant 5\frac{1}{4} \leqslant \sqrt{42} \leqslant 8 \leqslant \sqrt{84}$.

4. Approximating ℓ_p^n by an ST-Space

In order to show that ℓ_p^n can be approximated by an ST-space, it turns out to be convenient and natural to generalize the notion of ST-space to function spaces (although these do not appear in the final statement). We shall prove an inequality which is a little more sophisticated than Lemma 3. However, our main difficulty is notational rather than conceptual. In the next section, we shall

	3	3	3	2	2	2	1	1	1
$\xrightarrow{S}$	3	3	3	2	2	2	$1\frac{1}{2}$	$1\frac{1}{2}$
$\xrightarrow{T}$	3	3	3	2	2	2	$1\frac{1}{2}$
$\xrightarrow{S}$	3	3	3	$2\frac{1}{2}$	$2\frac{1}{2}$	$2\frac{1}{2}$
$\xrightarrow{T}$	3	3	3	$2\frac{1}{2}$	$2\frac{1}{2}$
$\xrightarrow{S}$	$3\frac{1}{2}$	$3\frac{1}{2}$	$3\frac{1}{2}$	$3\frac{1}{2}$
$\xrightarrow{T}$	$3\frac{1}{2}$	$3\frac{1}{2}$	$3\frac{1}{2}$
$\xrightarrow{S}$	$5\frac{1}{4}$	$5\frac{1}{4}$
$\xrightarrow{T}$	$5\frac{1}{4}$

	3	3	3	2	2	2	1	1	1
$\xrightarrow{T}$	3	3	3	2	2	2	1	1
$\xrightarrow{S}$	3	3	3	2	2	2	2
$\xrightarrow{T}$	3	3	3	2	2	2
$\xrightarrow{S}$	3	3	3	3	3
$\xrightarrow{T}$	3	3	3	3
$\xrightarrow{S}$	4	4	4
$\xrightarrow{T}$	4	4
$\xrightarrow{S}$	8

explain a different and in some ways more satisfactory approach, which shows again that ℓ_p^n can be approximated by an ST-space, while avoiding the use of a technical lemma from this section (Lemma 7 below).

Given a real number $t > 0$, let $F[0,t]$ denote the vector space of step functions from the closed interval $[0,t]$ to $\mathbb{R}$, and let $\mathcal{G}_t$ denote the set of normed spaces $(F[0,t], \|\cdot\|)$ such that $\|f\| = \|f^*\|$ for any $f \in F[0,t]$, where f^* is the decreasing rearrangement of f. Note that we do not ask for these normed spaces to be complete. If $s > t$ and $f \in F[0,s]$ we shall write f_t for the restriction of the function f to the interval $[0,t]$. For $n \in \mathbb{N}$ let I_n denote the linear map from $\mathbb{R}^n$ to $F[0,n]$ determined by $e_j \mapsto \chi_{[j-1,j)}$. We shall prove a result about norms on $F[0,t]$ but our interest will eventually be in subspaces of the form $I_n(\mathbb{R}^n)$.

We now define operations which are similar to S and T, but which map $\mathcal{G}_{t_1}$ to $\mathcal{G}_{t_2}$, where $t_1 < t_2$. Given $\alpha > 1$, define an operation $T_\alpha : \bigcup_{t>0} \mathcal{G}_t \longrightarrow \bigcup_{t>0} \mathcal{G}_{\alpha t}$ as follows. If $X \in \mathcal{G}_t$ and $f \in F[0,\alpha t]$ then $\|f\|_{T_\alpha(X)} = \|(f^*)_t\|_X$. Thus, the norm on the space $T_\alpha(X)$ is as small as it can be given that $T_\alpha(X)$ is in the set $\mathcal{G}_{\alpha t}$ and that the norm on $T_\alpha(X)$ coincides with that on X for functions supported on the interval $[0,t]$. As in the discrete case, S_α is defined in the opposite way: the norm on $S_\alpha(X)$ is as large as it can be under the same conditions. The following lemma we state without proof, since the analogy with Lemma 5 is very close.

LEMMA 6. *Let $t > 0$, $\alpha > 1$ and $X \in \mathcal{G}_t$. Let $f \in F[0,\alpha t]$ be a step function satisfying $f = f^*$ and $f(s) > 0$ for some $s > t$. Then if $\gamma > 0$ is the unique number for which*

$$\int_0^t (f(t) \vee \gamma)\, dt = \int_0^{\alpha t} f(t)\, dt,$$

we have $\|f\|_{S_\alpha(X)} = \|(f \vee \gamma)_t\|_X$.

We now come to the main lemma of this section. As in the first section, if X and Y are spaces in $\mathcal{G}_t$ and c_1 and c_2 are positive constants, then we shall write $c_1 X \leqslant c_2 Y$ if $c_1 \|f\|_X \leqslant c_2 \|f\|_Y$ for every $f \in F[0,t]$. For $1 \leqslant p \leqslant \infty$ and $t > 0$

let L_p^t denote the space $(F[0,t], \|\cdot\|_p)$, where $\|\cdot\|_p$ is the usual norm on $L_p[0,t]$ restricted to $F[0,t]$.

LEMMA 7. *Let $1 < p < \infty$, $t > 0$, $\alpha > 1$ and $\beta = \alpha^{p-1}$. Then*
$$S_\alpha(T_\beta(L_p^t)) \leqslant L_p^{\alpha\beta t} \leqslant T_\beta(S_\alpha(L_p^t)).$$

PROOF. We shall prove the right-hand inequality only. The left-hand one can be proved by a similar argument, or else by using duality. Suppose then that the right-hand inequality does not hold. In that case we can find $N \in \mathbb{N}$ and sequences $0 = x_0 < x_1 < \cdots < x_N = \alpha\beta t$ and $\lambda_1 \geqslant \lambda_2 \geqslant \cdots \geqslant \lambda_N \geqslant 0$ such that, setting $A_i = [x_{i-1}, x_i)$ and $f = \sum_1^N \lambda_i \chi_{A_i}$, we have
$$\int_0^{\alpha\beta t} f(x)^p \, dx > \int_0^t (f(x) \vee \gamma)^p \, dx$$
where
$$\int_0^t (f(x) \vee \gamma) \, dt = \int_0^{\alpha t} f(x) \, dx.$$

Let i_0 be the minimal index i for which $\lambda_i \leqslant \gamma$ and set $s = x_{i_0-1}$. Thus $f(x) \leqslant \gamma$ if and only if $x \geqslant s$. Without loss of generality, there exists i_1 such that $x_{i_1} = \alpha t$. We have
$$\gamma(t - s) = \int_s^{\alpha t} f(x) \, dx,$$
and, by assumption,
$$\int_0^{\alpha\beta t} f(x)^p \, dx - \int_0^t (f(x) \vee \gamma)^p \, dx$$
$$= \int_s^{\alpha\beta t} f(x)^p \, dx - (t-s)^{-p} \left(\int_s^{\alpha t} f(x) \, dx \right)^p (t-s) > 0.$$

Write A for $(t-s)^{p-1} \int_s^{\alpha\beta t} f(x)^p \, dx$ and B for $\left(\int_s^{\alpha t} f(x) \, dx \right)^p$. Our hypothesis now reads that $A > B$. If we fix the sequence $x_0, \ldots, x_N$, we may assume, by a simple compactness argument, that $\lambda_1 \geqslant \cdots \geqslant \lambda_N \geqslant 0$ have been chosen so as to maximize the ratio A/B. Pick $i_0 \leqslant i < i_1$. We find
$$\frac{\partial}{\partial \lambda_i} \frac{A}{B} = \frac{1}{B^2} \left(B(t-s)^{p-1}(x_i - x_{i-1}) p \lambda_i^{p-1} - Ap \left(\int_s^{\alpha t} f(x) \, dx \right)^{p-1} (x_i - x_{i-1}) \right)$$

But since $\lambda_i \leqslant \gamma$, we have also $\left(\int_s^{\alpha t} f(x) \, dx \right)^{p-1} \geqslant (t-s)^{p-1} \lambda_i^{p-1}$. Since $A > B$ we obtain that $\frac{\partial}{\partial \lambda_i} \frac{A}{B} < 0$. A simple calculation shows that if we decrease λ_i very slightly, then the value of i_0 does not change. If the change to λ_i is small enough, and the resulting sequence is still a decreasing one, then it follows that A/B was not maximized by the original sequence. Hence, we obtain that $\lambda_{i_0} = \lambda_{i_0+1} = \cdots = \lambda_{i_1}$. It is obvious also that $\lambda_{i_1} = \cdots = \lambda_N$. Without loss of generality they all take the value 1.

It remains to show that
$$(\alpha\beta t - s)(t-s)^{p-1} \leqslant (\alpha t - s)^p$$
By applying the weighted arithmetic-geometric-mean inequality twice and using the fact that $\beta = \alpha^{p-1}$, we have, writing q for the conjugate index of p,

$$(\alpha\beta t - s)^{1/p}(t-s)^{1/q} = \alpha t \left(1 - \frac{s}{\alpha\beta t}\right)^{1/p}\left(1 - \frac{s}{t}\right)^{1/q}$$
$$\leqslant \alpha t \left(1 - \frac{s}{p\alpha\beta t} - \frac{s}{qt}\right) \leqslant \alpha t \left(1 - \frac{s}{\alpha t}\right) = \alpha t - s.$$

The result follows on raising both sides to the power p. $\square$

The next lemma is similar to Lemma 7, but rather easier.

LEMMA 8. *Let $t > 0$, $X \in \mathcal{G}_t$ and $\alpha, \beta > 1$. Then $S_\alpha T_\beta X \leqslant T_\beta S_\alpha X$.*

PROOF. Let $f \in F[0, \alpha\beta t]$. Without loss of generality $f = f^*$. Then $\|f\|_{S_\alpha T_\beta X} = \|(f \vee \gamma)_t\|_X$, where γ satisfies

$$\int_0^{\beta t} (f(x) \vee \gamma)\,dx = \int_0^{\alpha\beta t} f(x)\,dx$$

and $\|f\|_{T_\beta S_\alpha X} = \|(f \vee \gamma')_t\|_X$, where γ' satisfies

$$\int_0^t (f(x) \vee \gamma')\,dx = \int_0^{\alpha t} f(x)\,dx.$$

It is therefore enough to show that $\int_0^t (f(x) \vee \gamma)\,dx \leqslant \int_0^{\alpha t} f(x)\,dx$, and thus that $\gamma' \geqslant \gamma$. Let $s = \inf\{x : f(x) \leqslant \gamma\}$. If $s \geqslant t$, the result is trivial. Otherwise, we wish to show that $(t-s)\gamma \leqslant \int_s^{\alpha t} f(x)\,dx$, given that $\gamma = (\beta t - s)^{-1} \int_s^{\alpha\beta t} f(x)\,dx$. But since f is a positive decreasing function and $\alpha, \beta > 1$, we have

$$\frac{\int_s^{\alpha\beta t} f(x)\,dx}{\int_s^{\alpha t} f(x)\,dx} \leqslant \frac{\alpha\beta t - s}{\alpha t - s} \leqslant \frac{\beta t - s}{t - s},$$

which proves the lemma. $\square$

In fact, it is not hard to see that equality in the last line can occur only for functions supported on the interval $[0, t]$, or functions whose modulus is constant (except on a set of measure zero) on the interval $[0, \alpha\beta t]$.

Lemmas 7 and 8 contain the essence of the proof that we can approximate ℓ_p^n by an ST-space. The remaining arguments are easy, but this may not be immediately apparent. The reader is strongly advised to consider, for any space $X \in \mathcal{G}_t$ under discussion, a logarithmic graph of the function $\lambda(t) = \|\chi_{[0,t]}\|_X$, i.e., a graph of $\log \lambda$ against $\log t$. Note that the slope of this graph, whenever X is an ST-space, is either zero or one, and when $X = L_p^t$ it is $1/p$. The terminology that follows is needed in order to formalize arguments which from such a graph are simple.

Define an ST-sequence to be a sequence of the form $U_1, \ldots, U_k$, where each U_i is S_α or T_α for some $\alpha = \alpha_i$. Given $0 < u < t$ we shall say that a function $\lambda : [u,t] \longrightarrow \mathbb{R}$ is an ST-function if it is piecewise linear and its right derivative at any x is either $\lambda(x)/x$ or zero. Given a space $X \in \mathcal{G}_u$, we shall say that a space Y is an ST-extension of X if $Y = U_k(\ldots(U_1(X))\ldots)$ for some ST-sequence $U_1, \ldots, U_k$. Given a space $X \in \mathcal{G}_u$, there is an obvious one-to-one correspondence between ST-sequences, ST-extensions of X and ST-functions taking the value $\|\chi_{[0,u]}\|_X$ at u. (Given an ST-extension Y of X, the associated ST-function is the function $\lambda : x \mapsto \|\chi_{[0,x]}\|_Y$.) When we refer to any correspondence between ST-functions, ST-extensions and ST-sequences, we shall always mean this one. Finally, we shall call a function $\lambda : (0,t] \to \mathbb{R}$ an ST-function if its restriction to $[u,t]$ is an ST-function for every $u > 0$. In other words, the conditions are as above except that there may be infinitely many changes of slope near zero. We associate with such a function a space as follows. Suppose the changes of slope of λ occur at points $\ldots, x_{-2}, x_{-1}, x_0, x_1, \ldots, x_n$ with $x_n < t$ and $x_{-i} \to 0$. Given a step function f defined on the interval $[0,t]$, we may replace it by a step function supported on the interval $[0, x_n]$ with the same norm, either by projecting f onto this interval (at a T stage) or by using Lemma 6 as a definition (at an S stage). We can then repeat this process, obtaining a function supported on $[0, x_{n-1}]$ and so on. After finitely many stages one obtains a function $\omega \chi_{[0,s]}$ for some $\omega, s \geqslant 0$. The norm of f is then defined to be $\omega \lambda(s)$. The resulting space we will call the ST-space associated with λ, denoted L^t_λ. (Note that our notation is not at all standard: L_ϕ usually stands for the Orlicz space associated with the Orlicz function ϕ. However, the notation is so convenient here that we have adopted it.)

During the rest of this section we will make a number of uses of the simple fact that if $X \leqslant Y$ then $T_\alpha X \leqslant T_\alpha Y$ and $S_\alpha X \leqslant S_\alpha Y$ (the function-space analogue of Lemma 2).

LEMMA 9. *Let $X \in \mathcal{G}_u$, let $Y_1, Y_2 \in \mathcal{G}_t$ be two ST-extensions of X and let λ_1 and λ_2 be the associated ST-functions. Then if $\lambda_1(x) \leqslant \lambda_2(x)$ for every $x \in [u,t]$, it follows that $Y_1 \leqslant Y_2$.*

PROOF. Let $w_1, \ldots, w_N$ be the set of values, in increasing order, taken by either λ_1 or λ_2 when they are differentiable with derivative zero, or taken by λ_2 when it is maximal. Let s be maximal such that λ_1 and λ_2 are equal on the interval $[u,s]$. Then $\lambda_1(s) = \lambda_2(s) = w_i$, say. If $s < t$, then our aim is to replace λ_1 by a larger function λ_1', still dominated by λ_2, but now equal to it until they both take the value w_{i+1}. Write Z for the ST-extension of X corresponding to the restrictions of λ_1 and λ_2 to the interval $[u,s]$, and $U_1, \ldots, U_l$ for the corresponding ST-sequence. Note that the next terms in the sequences of Y_1 and Y_2 must be of the form T_α and S_β, by the maximality of s. We must now consider various cases.

First, if $i + 1 = N$ and neither λ_1 nor λ_2 has zero right derivative when they take the value w_N, then w_N is the maximum of λ_2 and the ST-sequences of

Y_1 and Y_2 must be $U_1, \ldots, U_l, T_\alpha$ and $U_1, \ldots, U_l, S_\alpha$ respectively, (where α can be shown to be w_{i+1}/w_i). In this case, let $\lambda'_1 = \lambda_2$, and observe that the only modification we have made to the ST-sequence corresponding to λ_1 is to replace the final T_α by S_α.

Otherwise, we know that at least one of λ_1 and λ_2 has zero derivative when it takes the value w_{i+1}. Suppose first that λ_1 has. Then the ST-sequence of the space Y_1 begins $U_1, \ldots, U_l, T_{\alpha_1}, S_{\alpha_2}, T_{\alpha_3}$, with $\alpha_1, \alpha_2, \alpha_3 > 1$ (and again one can show that $\alpha_2 = w_{i+1}/w_i$). We modify this sequence by exchanging T_{α_1} with S_{α_2} and define λ'_1 to be the correspondingly modified ST-function. It is not hard to show that λ'_1 is still dominated by λ_2 and that λ'_1 and λ_2 are equal until they both take the value w_{i+1}.

If on the other hand λ_2 has zero derivative when it takes the value w_{i+1}, then the ST-sequence of Y_2 begins $U_1, \ldots, U_l, S_{\beta_1}, T_{\beta_2}$, with $\beta_1, \beta_2 > 1$. We now have three further sub-cases. If the ST-sequence of Y_1 is $U_1, \ldots, U_l, T_{\alpha_1}$, then $\alpha_1 \geqslant \beta_1\beta_2$, so we can replace this ST-sequence by the equivalent sequence $U_1, \ldots, U_l, T_{\beta_1}, T_\gamma$ where $\beta_1\gamma = \alpha_1$. Now modify the sequence by changing T_{β_1} into S_{β_1} and let λ'_1 be the corresponding function. Then again λ'_1 is dominated by λ_2 but equal to it until they both equal w_{i+1}.

If the ST-sequence of Y_1 begins $U_1, \ldots, U_l, T_{\alpha_1}, S_{\alpha_2}$ and $\alpha_2 \geqslant \beta_1$, then we can replace it with the equivalent sequence $U_1, \ldots, U_l, T_{\alpha_1}, S_{\beta_1}, S_\gamma$, where $\gamma = \alpha_2/\beta_1$. We modify this sequence by exchanging T_{α_1} and S_{β_1} and let λ'_1 be the corresponding function. Once again, it has the required property.

Finally, if the ST-sequence of Y_1 begins $U_1, \ldots, U_l, T_{\alpha_1}, S_{\alpha_2}$ and $\alpha_2 < \beta_1$, then the hypothesis of this case implies that $U_1, \ldots, U_l, T_{\alpha_1}, S_{\alpha_2}$ is the whole sequence. In this case, we modify the sequence in two steps. First we exchange the T_{α_1} and the S_{α_2}. Then we replace the T_{α_1} with $S_{\gamma_1}, T_{\gamma_2}$ where $\gamma_1 = \beta_1/\alpha_2$ and $\gamma_2 = \alpha_1\alpha_2/\beta_1$.

In each of the cases above, λ'_1 is obtained from λ_1 either by changing a T_α in its ST-sequence to an S_α, or by changing the order of a consecutive pair S_α, T_β so that after the change the S_α operation is performed first, (or in the final case doing both). By the definitions of the operations S_α and T_β, the simple fact mentioned just before this lemma, and Lemma 8, each such modification increases the corresponding norm. Hence, by induction on i, one can transform λ_1 into λ_2 by applying a finite sequence of changes to the corresponding ST-sequence, each of which increases the corresponding norm. It follows that $Y_1 \leqslant Y_2$ as stated. □

COROLLARY 10. *Let $X_1, X_2 \in \mathcal{G}_u$ and let λ_1 and λ_2 be two ST-functions on the interval $[u,t]$ with $\lambda_i(u) = \|\chi_{[0,u]}\|_{X_i}$ for $i = 1, 2$. Let Y_1 and Y_2 be the corresponding ST-extensions of X_1 and X_2. Suppose there is a constant $c > 0$ such that $X_1 \leqslant cX_2$, $\lambda_1(u) = c\lambda_2(u)$ and $\lambda_1(x) \leqslant c\lambda_2(x)$ for any $x \in [u,t]$. Then*

$$Y_1 \leqslant cY_2.$$

PROOF. Without loss of generality $c = 1$. By Lemma 9 and the analogue of Lemma 2 we see that $Y_1 \leqslant Z \leqslant Y_2$, where Z is the λ_2-extension of X_1. □

COROLLARY 11. *Let λ and μ be two ST-functions defined on the interval $(0,t]$ with $\lambda(s) \leqslant c\mu(s)$ for every s. Then $L_\lambda^t \leqslant cL_\mu^t$.*

PROOF. Let f be a decreasing step function with the first step ending at u. The norm of f in an ST-space does not depend on the behaviour of the ST function below u. Therefore for values less than u we may replace λ and μ by any other ST-functions. Choose functions $\lambda_1 \leqslant c\mu_1$ such that for some $u_1 \leqslant u$ we have that $\lambda_1(u_1) = c\mu_1(u_1)$ and λ_1 and μ_1 are linear below u_1. We may then apply Corollary 10 (where X_1 and X_2 will be multiples of $L_1^{u_1}$). □

LEMMA 12. *Let $s, t > 0$, let $1 < p < \infty$, let q be the conjugate index of p and let $\alpha > 1$. Then $S_\alpha(\alpha^{-1/q} L_p^s) \leqslant L_p^{\alpha s}$ and $L_p^{\beta t} \leqslant T_\beta(\beta^{1/p} L_p^t)$.*

PROOF. The second inequality states that, for any positive decreasing function f on the interval $[0, \beta t]$, we have the inequality

$$\int_0^{\beta t} (f(x))^p \, dx \leqslant \beta \int_0^t (f(x))^p \, dx,$$

which is obvious. To prove the first, observe that by the above inequality, with $t = \alpha^{-p/q} s$ and $\beta = \alpha^{p/q}$, we have

$$\alpha^{-1/q} L_p^s \leqslant T_\beta(L_p^t),$$

so, by the extension of Lemma 2, we have

$$S_\alpha(\alpha^{-1/q} L_p^s) \leqslant S_\alpha T_\beta(L_p^t).$$

Now $\beta = \alpha^{p/q} = \alpha^{p-1}$, so, by Lemma 7,

$$S_\alpha T_\beta(L_p^t) \leqslant L_p^{\alpha \beta t} = L_p^{\alpha s}.$$
□

We are now ready for the main theorem of this section.

THEOREM 13. *Let $0 < u < t$, let $c_1 \leqslant c \leqslant c_2$ and let λ be an ST-function on the interval $[u,t]$ such that $\lambda(u) = cu^{1/p}$ and $c_1 x^{1/p} \leqslant \lambda(x) \leqslant c_2 x^{1/p}$ for any $x \in [u,t]$. Let Y be the ST-extension of cL_p^u corresponding to the function λ. Then $c_1 L_p^t \leqslant Y \leqslant c_2 L_p^t$.*

PROOF. We shall prove only the left-hand inequality: the other is similar. Without loss of generality $c_1 = 1$. Define Z to be the space $S_{c^q} L_p^{c^{-q}u}$, the norm of which, by Lemma 12, is dominated by that of cL_p^u. Notice that $\|\chi_{[0,u]}\|_Z = cu^{1/p}$.

We now define an ST-sequence $S_{\alpha_1}, T_{\beta_1}, S_{\alpha_2}, T_{\beta_2}, \ldots, S_{\alpha_k}, T_{\beta_k}$, letting μ be the function (which will be defined on the interval $[u,t]$) corresponding to the subsequence $T_{\beta_1}, S_{\alpha_2}, T_{\beta_2}, \ldots, S_{\alpha_k}, T_{\beta_k}$. Set $\alpha_1 = c^q$ and $\beta_1 = \alpha_1^{p-1} = c^p$. In general, once we have defined $\alpha_1, \ldots, \alpha_{i-1}$ and $\beta_1, \ldots, \beta_{i-1}$, we let α_i be maximal such that the resulting part of the function μ is dominated by $\min\{\lambda, t^{1/p}\}$, and then set $\beta_i = \alpha_i^{p-1}$. This construction guarantees that, whenever $\mu'(a) = 0$, either there is some b such that $\lambda'(b) = 0$ and $\lambda(b) = \mu(a)$, or $\mu(a) = t^{1/p}$. (Informally, at each stage the logarithmic graph of μ rises with slope one until it hits a horizontal part of the logarithmic graph of λ, at which point it becomes horizontal until it rejoins the line $y = x/p$.) Since λ has only finitely many changes of slope, the graph of μ reaches a height of $t^{1/p}$ after only finitely many changes of direction, and after one further step (corresponding to T_{β_k}) the process stops.

Let X be the ST-extension of Z corresponding to the function μ. Then Lemma 7 implies (after an easy induction) that $L_p^t \leqslant X$. Since $Z \leqslant cL_p^u$ and $\mu \leqslant \lambda$ with $\mu(u) = \lambda(u)$, Corollary 10 implies that $X \leqslant Y$, completing the proof of the theorem. □

COROLLARY 14. *Let λ be an ST-function defined on the interval $(0, t]$ such that $c_1 s^{1/p} \leqslant \lambda(s) \leqslant c_2 s^{1/p}$ for every s. Then $c_1 L_p^t \leqslant L_\lambda^t \leqslant c_2 L_p^t$.*

PROOF. Just as in the proof of Corollary 11, we can deduce this from Theorem 13 by considering step functions first. □

It remains to show that we can approximate ℓ_p^n by an ST-space in the sense of Section 2, whatever the value of p. Theorem 13 tells us that all we need to worry about is the norm of vectors of the form $\sum_1^k e_i$.

THEOREM 15. *Let $1 \leqslant p \leqslant \infty$. Then for any $n \in \mathbb{N}$, there exists an ST-space X such that $d(X, \ell_p^n) < 3/2$.*

PROOF. By duality, we may assume that $1 \leqslant p \leqslant 2$. Suppose $X \in \mathcal{F}_n$, $X' \in \mathcal{G}_n$ and the embedding I_n is an isometry from X to its image in X'. It is not hard to see that I_{n+1} is an isometric embedding from $T(X)$ into $T_\alpha(X')$, and also from $S(X)$ into $S_\alpha(X')$, where $\alpha = 1 + 1/n$. By Theorem 13, then, it is enough to find an ST-sequence $U_1, U_2, \ldots, U_{n-1}$ such that each U_i is either $S_{1+1/i}$ or $T_{1+1/i}$ and the corresponding ST-function λ satisfies $x^{1/p} \leqslant \lambda(x) < (3/2)x^{1/p}$. Clearly we only need to check this inequality when x is an integer. Suppose we have chosen $U_1, \ldots, U_{k-1}$. Then if $\lambda(k) \geqslant (k+1)^{1/p}$, set $U_k = T_{1+1/k}$. Otherwise, let it be $S_{1+1/k}$. Then for each k, $\lambda(k)$ is either $\lambda(k-1)$ or it is less than $k^{1/p} \cdot k/(k-1)$. In the first case we are done, by induction. In the second, we are done, unless $k = 2$. But in this case, $\lambda(k) = 2 < (3/2) \cdot 2^{1/p}$. □

Note that Theorem 13 implies easily that if $k \in \mathbb{N}$ is even and $X \in \mathcal{F}_{2^k}$ is the space $T^{2^{k-1}}(S^{2^{k-2}}(\ldots(S^4(T^2(S(\mathbb{R}))))\ldots))$, then $d(X, \ell_2^{2^k}) = \sqrt{2}$. Indeed, X embeds isometrically into the space $T_2(S_2(\ldots(T_2(S_2(L_2^1)))\ldots)) \in \mathcal{G}_{2^k}$, and if λ is the ST-function associated with this space, then $\lambda(2^l) = 2^{l/2}$ if l is even, and $2^{(l+1)/2}$ if l is odd. It follows that $\sqrt{x} \leqslant \lambda(x) \leqslant \sqrt{2x}$ for every x, which, by Theorem 13, is enough to prove the above estimate. (As in Section 2, the lower bound on the distance follows from the fact that the distance is attained by the identity map.) This can also be proved directly from Lemma 7, using a duality argument similar to that of Theorem 1. In general, to approximate ℓ_p^n to within an absolute constant by an ST-space, one needs only about $\log n$ changes of direction of the corresponding ST-function λ, or $\log n$ terms in the corresponding ST-sequence. Our arguments show also that many other ST-sequences would have worked just as well in Theorem 1.

If one wishes for a better polytope approximation of ℓ_p^n, still with an easy geometrical description, then one method is to consider appropriate subspaces. For example, if $2m = nk$, $X = S(TS)^{m-1}(\mathbb{R})$ and Y is the n-dimensional subspace of X generated by the block basis $u_i = \sum_{j=(i-1)k+1}^{ik} e_j$, where $i = 1, 2, \ldots, n$, then Theorem 13 and a straightforward calculation show that $d(Y, \ell_2^n) \leqslant 1 + C/k$ for some absolute constant C.

5. Limits of ST-Spaces

We say that a function λ is a *growth function* if it is a strictly positive uniform limit of ST-functions. It is not hard to show that $\lambda : [u, t] \to \mathbb{R}$ is a growth function if and only if $\lambda(u) > 0$ and $\lambda(x) \leqslant \lambda(y) \leqslant (y/x)\lambda(x)$ whenever $u \leqslant x \leqslant y \leqslant t$, and also if and only if there is a space $X \in \mathcal{G}_t$ such that $\lambda(x) = \|\chi_{[0,x]}\|_X$ for every $u \leqslant x \leqslant t$. (A similar statement holds also for growth functions defined on the interval $(0, t]$.) It is an easy consequence of Lemma 10 that, given $X \in \mathcal{G}_u$ and a growth function λ such that $\lambda(u) = \|\chi_{[0,u]}\|_X$, the following normed space $Y \in \mathcal{G}_t$ is well defined. Pick any sequence $\lambda_1, \lambda_2, \ldots$ of ST-functions tending uniformly to λ with $\lambda_n(u) = \lambda(u)$ and let

$$\|f\|_Y = \lim_{n \to \infty} \|f\|_{Y_n}$$

where Y_n is the ST-extension of X corresponding to λ_n. We shall call the space Y the λ-*extension* of X. Given a uniform limit $\lambda : (0, t] \to \mathbb{R}$ of ST-functions λ_n, we can define a norm $\|\cdot\|_\lambda$ by again taking $\|f\|_\lambda$ to be the limit of the norms $\|f\|_{\lambda_n}$. This is the most natural definition, in the context of ST-spaces, of a norm associated with the growth function λ. We shall denote the completion of $(F[0, t], \|\cdot\|_\lambda)$ by L_λ^t. Later, we shall show that the space L_λ^t is in some ways more natural than the Orlicz space with the same growth function.

The main task of this section is to show how to calculate the norm of a function f in a λ-extension (and consequently in a space L_λ^t). This we do by giving a continuous analogue of the algorithm defined in Section 3. Thus, we are generalizing our results so far from growth functions with logarithmic gradient 0 or 1 to growth functions with logarithmic gradient in the interval $[0, 1]$. This generalization is of some interest for its own sake, but the immediate benefit is a second proof of Theorem 15 which avoids the use of Lemma 7.

For convenience, we shall assume that our growth function λ is differentiable with non-zero derivative, and we shall show how to calculate $\|f\|_\lambda$ when f is continuously differentiable with negative derivative and both f and f' are bounded away from 0 and ∞ (this includes right and left derivatives at 0 and 1). A function f with these properties we shall call *standard*. Using rearrangement-invariance and straightforward limiting arguments, one can deal with more general f and λ.

Given a standard function f, we would like, just as in Section 3, to replace f by a function of smaller support but with the same norm. Given $y \leqslant t$ large enough it will turn out that there is a unique $x \leqslant y$ such that if we define the function g_y by

$$g_y(s) = \begin{cases} f(s) & \text{if } 0 \leqslant s \leqslant x, \\ f(x) & \text{if } x \leqslant s \leqslant y, \\ 0 & \text{if } s > y, \end{cases}$$

then $\|g_y\|_\lambda = \|f\|_\lambda$. It is clear that x decreases as y decreases. (The differentiability of f applies provided x is greater than 0, which is the reason for the condition that y should be large enough.) Of much more interest is the fact that the dependence between x and $y = y(x)$ is given by the differential equation

$$y \frac{\lambda'(y)}{\lambda(y)} \frac{dy}{dx} = -\frac{f'(x)}{f(x)}(y-x), \tag{1}$$

which can be rewritten as

$$y\, d(\log \lambda(y)) = -(y-x)\, d(\log f(x)).$$

The main task of this section is to derive equation (1). However, let us first see why it gives a new proof of Theorem 15. Note first that when $x = 0$ (and y is maximal), the function g_y is simply $f(0)$ times the characteristic function of $[0, y]$, so that $\|f\|_\lambda = \|g_y\|_\lambda = f(0)\lambda(y(0))$. Hence, the differential equation gives us a means of calculating the norm $\|\cdot\|_\lambda$. Next, observe that if $\lambda(y) = y^{1/p}$, then the solution of equation (1) is

$$y = f(x)^{-p} \left(C - \int_0^x f(s)^p\, ds \right) + x$$

for some constant C. Since $y(t) = t$, we obtain that $C = \int_0^t f(s)^p\, ds$. Hence, $y(0) = f(0)^{-p} \int_0^t f(s)^p\, ds$. Thus $f(0)(y(0))^{1/p} = \|f\|_p$. But Corollary 10 (which did not use Lemma 7) can obviously be generalized to the same result for limits of ST-functions and the corresponding spaces. Since the function $t^{1/p}$ is such a function and it gives rise to the space L_p^t, we obtain Theorem 13 and hence Theorem 15.

To obtain equation (1), let $\varepsilon > 0$ and $0 < u < t$ and let $\alpha_1, \beta_1, \ldots, \alpha_N, \beta_N$ be a sequence of real numbers greater than 1 with the following properties.

(i) $\lambda(\alpha_1\beta_1 \ldots \alpha_k\beta_k u) = \beta_1 \ldots \beta_k \lambda(u)$ for $k = 1, 2, \ldots, N$.
(ii) $\alpha_k\beta_k \leqslant 1 + \varepsilon$ for $k = 1, 2, \ldots, N$.
(iii) $\alpha_1\beta_1 \ldots \alpha_N\beta_N = t/u$.

Let $X \in \mathcal{G}_u$ and let Y be the ST-extension $S_{\beta_N}T_{\alpha_N} \ldots S_{\beta_1}T_{\alpha_1}(X)$ and notice that the growth function μ of Y has the following properties.

(a) $\mu(\alpha_1\beta_1\ldots\alpha_k\beta_k) = \lambda(\alpha_1\beta_1\ldots\alpha_k\beta_k)$ for $k = 1, 2, \ldots, N$.
(b) $(1+\varepsilon)^{-1}\lambda(s) \leqslant \mu(s) \leqslant \lambda(s)$ for $u \leqslant s \leqslant t$.

For each k, define $y_k = \alpha_1\beta_1\ldots\alpha_k\beta_k u$. Fix $k < N$, set $y = y_k$ and $\delta y = y_{k+1} - y_k$, and suppose that $x = x_k$ has been defined. Define a function g by the formula

$$g(s) = \begin{cases} f(s) & \text{if } 0 \leqslant s \leqslant x, \\ f(x) & \text{if } x \leqslant s \leqslant y, \\ 0 & \text{if } s > y, \end{cases}$$

and let δx be the unique number such that $\|h\|_\mu = \|g\|_\mu$, where

$$h(s) = \begin{cases} f(s) & \text{if } 0 \leqslant s \leqslant x + \delta x, \\ f(x + \delta x) & \text{if } x + \delta x \leqslant s \leqslant y + \delta y, \\ 0 & \text{if } s > y + \delta y. \end{cases}$$

Then define x_{k+1} to be $x + \delta x$. We shall now obtain an approximate equation relating δx and δy.

Setting $\alpha = \alpha_{k+1}$ and $\beta = \beta_{k+1}$, we know from the definition of the operation T_α that $\|g\|_\mu = \|g_1\|_\mu$, where

$$g_1(s) = \begin{cases} f(s) & \text{if } 0 \leqslant s \leqslant x, \\ f(x) & \text{if } x \leqslant s \leqslant \alpha y, \\ 0 & \text{if } s > \alpha y. \end{cases}$$

By Lemma 6 (which in this context could almost be regarded as the definition of S_β) we know also that $\int_0^t g_1(s)\,ds = \int_0^t h(s)\,ds$. That is,

$$\int_0^x f(s)\,ds + (\alpha y - x)f(x) = \int_0^{x+\delta x} f(s)\,ds + (\alpha\beta y - x - \delta x)f(x + \delta x),$$

which implies that

$$\int_x^{x+\delta x} (f(s) - f(x))\,ds + (x + \delta x)(f(x) - f(x + \delta x)) = \alpha y f(x) - \alpha\beta y f(x + \delta x).$$

Bearing in mind that $\delta x = O(\delta y)$ and that $(\alpha\beta - 1)y = \delta y$, which also implies that $\alpha - 1$ and $\beta - 1$ are $O(\delta y)$, we can simplify the above to

$$\frac{\beta - 1}{\alpha\beta - 1}\delta y + o(\delta y) = -\frac{f'(x)}{f(x)}(y - x)\delta x. \qquad (2)$$

Finally, notice that

$$\lambda'(y) = \frac{\lambda(\alpha\beta y) - \lambda(y)}{(\alpha\beta - 1)y} + o(1) = \frac{\beta - 1}{\alpha\beta - 1}\frac{\lambda(y)}{y} + o(1),$$

so that, substituting into (2), we have the estimate

$$y\frac{\lambda'(y)}{\lambda(y)}\delta y + o(\delta y) = -\frac{f'(x)}{f(x)}(y - x)\delta x. \qquad (3)$$

Letting ε, and hence δy, tend to zero, we see that in the limit as μ tends to λ, we do obtain equation (1) as claimed. Having obtained the equation for a λ-extension, it is easy to see that it is valid for the space L_λ^t as well.

We end this section with a small remark. It is easy to prove that the dual of L_λ^t is L_μ^t where $\mu(s) = s/\lambda(s)$, first when λ is an ST-function, and then, on taking limits, for an arbitrary growth function. (This statement is not quite accurate since for example the dual of L_∞^t is not L_1^t. What we mean is that the norm of a measurable function in the dual of L_λ^t is its L_μ^t-norm.) The proof comes straight from the definition of the operations S_α and T_β. Similarly, it is trivial that these spaces are normed spaces. Therefore we have an argument which makes the inequalities of Hölder and Minkowski in some sense "obvious" and "geometrical". Unfortunately, working out the details is more complicated than the usual proofs of those inequalities!

6. Two Results about ST-Spaces

We shall be concerned with two natural questions in this section. First, what is the relationship, if anything, between ST-spaces and Orlicz spaces? Second, what is the result of interpolating between two ST-spaces?

For the first question, suppose X is an Orlicz function space restricted to the interval $[0, t]$ with the norm given by

$$\|f\|_X = \inf\left\{\mu > 0 : \int_0^t \phi(|f(s)|/\mu)\, ds \leqslant 1\right\}$$

where ϕ is an Orlicz function. For this space we have $\|\chi_{[0,s]}\|_X = (\phi^{-1}(s^{-1}))^{-1}$. For $0 < u < t$ let X_u be the restriction to $[0, u]$ of X. If for $u \leqslant s \leqslant t$ we set $\lambda(s) = (\phi^{-1}(s^{-1}))^{-1}$, it is natural to ask whether the (completion of the) λ-extension of X_u is X? We shall show that it is, isometrically, if and only if, for some p, $\phi(s) = s^p$ and therefore $\lambda(s) = s^{1/p}$. In other words, ST-spaces and Orlicz spaces intersect only in the L_p-spaces.

Our proof of this is slightly indirect. Suppose ϕ is an Orlicz function, X is the corresponding Orlicz space and X_u and λ are given as above. Suppose moreover that the identity is an isometry from X to the λ-extension of X_u. Let f be any standard function. For $y < t$ sufficiently close to t we replace f by a function g_y with the properties we had in the last section. That is, for some $0 \leqslant x \leqslant y$,

$$g_y(s) = \begin{cases} f(s) & \text{if } 0 \leqslant s \leqslant x, \\ f(x) & \text{if } x \leqslant s \leqslant y, \\ 0 & \text{if } s > y, \end{cases}$$

and $\|g_y\|_\lambda = \|f\|_\lambda$. It is easy to see that, for each y, g_y is unique. Suppose that $\|f\|_X = 1$. Using the definition of the norm in X and the isometry assumption one readily obtains that

$$\int_0^x \phi(f(s))\,ds + (y-x)\phi(f(x)) = \int_0^t \phi(f(s))\,ds \tag{4}$$

If we differentiate this equation with respect to x and rearrange, we obtain that

$$\frac{\phi(f(x))}{\phi'(f(x))f(x)}\frac{dy}{dx} = -\frac{f'(x)}{f(x)}(y-x).$$

On the other hand, if we substitute $\lambda(y) = (\phi^{-1}(y^{-1}))^{-1}$ into the differential equation (1), we obtain the equation

$$\frac{1}{y\phi^{-1}(y^{-1})\phi'(\phi^{-1}(y^{-1}))}\frac{dy}{dx} = -\frac{f'(x)}{f(x)}(y-x).$$

It follows from the uniqueness of g_y that $\phi(f(x))/\phi'(f(x))f(x)$ does not depend on $f(x)$. In other words, the function $t \mapsto \phi(t)/t\phi'(t)$ is a constant function. Solving this equation gives $\phi(t) = Ct^p$ for some constants C and p. We have proved the next theorem.

THEOREM 16. *Let $X \in \mathcal{G}_t$ be the restriction of an Orlicz function space on $[0,t]$ to the interval $[0,t]$ and let Y be the completion of the λ-extension of X_u, where λ and X_u are as defined above. Then X is isometric to Y under the identity map if and only if, for some constants $C > 0$ and $1 \leqslant p \leqslant \infty$, $\|f\|_X = \|f\|_Y = C\|f\|_p$.*

It would be much more interesting to find out when ST-spaces are *isomorphic* to Orlicz spaces. There seems to be a reasonable chance that they always are, in which case they could be regarded as the "correct" renorming of Orlicz spaces.

The next result shows that ST-spaces interpolate in the way one would expect. Since we prove an isometric result, we must use complex interpolation and therefore complex scalars. One can either define the norm of any vector to be the norm of its modulus, or follow the original approach making obvious modifications. These give the same result. Also, we shall make use of the fact (related to the remark at the end of the previous section) that, if $X \in \mathcal{G}_u$, and λ is a growth function, then the dual of the λ-extension of X is the μ-extension of X^*, where $\mu(s) = s/\lambda(s)$. (Again, this statement should be interpreted somewhat loosely.) For the basic facts and notation to do with interpolation, see [BL].

THEOREM 17. *Let $X \in \mathcal{G}_u$, let λ and μ be growth functions and let X_λ and X_μ be the λ- and μ-extensions of X respectively. Given $0 < \theta < 1$, let $\nu = \lambda^\theta \mu^{1-\theta}$ and let X_ν be the ν-extension of X. Then*

$$(X_\lambda, X_\mu)_{[\theta]} = X_\nu.$$

PROOF. The proof of this is very similar to the standard proof that L_p-spaces interpolate in the way one would expect. Let f be a standard function with

$\|f\|_{X_\nu} = 1$ and let $y = y(x)$ be defined by the differential equation

$$y \frac{\nu(y)}{\nu(y)} \frac{dy}{dx} = -\frac{f'(x)}{f(x)}(y-x)$$

and the initial condition $y(t) = t$. Let $D \subset \mathbb{C}$ be the set $\{z \in \mathbb{C} : 0 \leqslant \operatorname{Re}(z) \leqslant 1\}$. Fixing y, we may now define, for every $z \in D$, a function g_z by the differential equation

$$y\left((1-z)\frac{\lambda'(y)}{\lambda(y)} + z\frac{\mu'(y)}{\mu(y)}\right)\frac{dy}{dx} = -\frac{g'_z(x)}{g_z(x)}(y-x)$$

and the initial condition

$$g_z(0) = \lambda(y(0))^{-(1-z)}\mu(y(0))^{-z}.$$

It is easy to check that

$$g_z(x) = g_0(x)^{1-z} g_1(x)^z. \tag{3}$$

Now set $\tilde{f}(z, x) = \exp(\varepsilon z^2 - \varepsilon \theta^2) g_z(x)$. Thus $\tilde{f} : D \times [0, t] \to \mathbb{C}$. We have the properties of $\tilde{f}$ necessary to estimate $\|f\|_{(X_\lambda, X_\mu)_{[\theta]}}$. First, for each x it is clear that $\tilde{f}(z, x)$ is analytic in z on the interior of D. Second, $\|\tilde{f}(ir, \cdot)\|_{X_\lambda}$ and $\|\tilde{f}(1+ir, \cdot)\|_{X_\mu}$ both tend to zero as $|r|$ tends to infinity. Moreover, we have $\|\tilde{f}(ir, \cdot)\|_{X_\lambda} \leqslant 1$ and $\|\tilde{f}(1+ir, \cdot)\|_{X_\mu} \leqslant \exp(\varepsilon)$ for every $r \in \mathbb{R}$. It follows that $\|f\|_{(X_\lambda, X_\mu)_{[\theta]}} \leqslant \exp(\varepsilon)$. Since $\varepsilon > 0$ was arbitrary, we have shown that $\|f\|_{(X_\lambda, X_\mu)_{[\theta]}} \leqslant \|f\|_{X_\nu}$ for any function f.

Conversely, suppose that $\|f\|_{(X_\lambda, X_\mu)_{[\theta]}} = 1$. This tells us that for each $\varepsilon > 0$ there exists a function $\tilde{f}$ with the above properties. We also know that

$$\|f\|_{X_\nu} = \sup\left\{|\langle f, h\rangle| : h \text{ standard}, \|h\|_{X_{\nu_1}} = 1\right\}$$

where $\nu_1(s) = s/\nu(s)$.

Given a standard function h with $\|h\|_{\nu_1} \leqslant 1$, let $\tilde{h} : D \times [0, t] \to \mathbb{C}$ be given by the method used to construct $\tilde{f}$ from f, replacing all the spaces in that construction by their duals. Then set

$$F(z) = \int_0^t \tilde{f}(z, x) \tilde{h}(z, x)\, dx$$

for every $z \in D$. Then F is analytic on the interior of D, continuous on D, and $F(ir) \leqslant 1$ and $F(1+ir) \leqslant \exp(2\varepsilon)$ for every $r \in \mathbb{R}$. By the Hadamard three-line theorem (see [BL]) we obtain that

$$|\langle f, h\rangle| \leqslant |F(\theta)| \leqslant \exp(2\varepsilon).$$

Since $\varepsilon > 0$ was arbitrary, we have $\|f\|_{X_\nu} \leqslant 1$. $\square$

It is not hard to deduce from Theorem 17 that, with the same notation, we have also $(L_\lambda^t, L_\mu^t)_{[\theta]} = L_\nu^t$.

7. Suggestions for Further Research

There are many questions one can ask arising from the results of the previous sections. Some of this section is extremely speculative.

(A) The most urgent question is whether ST-spaces are renormings of Orlicz spaces or something quite different. This question has been mentioned earlier in the paper so we shall not say much more about it. Whatever the answer, one can ask whether ST-spaces always contain some ℓ_p almost isometrically. We have an argument, whose details are yet to be checked, that they do; it uses the similar result of Lindenstrauss and Tzafriri for Orlicz spaces, for which see [LT].

(B) It is likely that the map taking a growth function λ to the rearrangement-invariant function space $L_\lambda(\mathbb{R})$ with that growth function can be characterized amongst all such maps. Here are some properties that might be involved:

(i) $t^{1/p} \mapsto L_p(\mathbb{R})$.
(ii) $t/\lambda(t) \mapsto (L_\lambda(\mathbb{R}))^*$.
(iii) $a\lambda \leqslant \mu$ implies $aL_\lambda(\mathbb{R}) \leqslant L_\mu(\mathbb{R})$.
(iv) $(L_\lambda^t, L_\mu^t)_{[\theta]} = L_{\lambda^\theta \mu^{1-\theta}}^t$.

Are these enough? Perhaps (iii) needs to be a little more detailed. For example, one also has:

(v) If $a\lambda(s) \leqslant \mu(s)$ for $s \leqslant t$ and the support of f is contained in the interval $[0,t]$, then $a\|f\|_\lambda \leqslant \|f\|_\mu$.
(vi) If f^* is constant on $[0,u]$ and $a\lambda(s) \leqslant \mu(s)$ for $s \geqslant u$, then $a\|f\|_\lambda \leqslant \|f\|_\mu$.

(C) Recall the remark at the end of Section 4, that only $\log n$ changes between S and T are needed to approximate ℓ_p^n to within a constant. This is clearly best possible. Does it cause the unit ball of the resulting ST-space to have interesting extremal properties of a more geometrical nature amongst polytopes approximating the ball of ℓ_p^n? Unchecked calculations suggest that the number of vertices and faces are both at most exponential (again for $1 < p < \infty$ this is necessary). If they are correct, then we have constructed efficient approximating polytopes in a very explicit way. Perhaps this might have algorithmic uses. It would be interesting to have good estimates for the number of facets of each dimension of the polytope arising from a given ST-sequence.

The restriction of L_λ^n to the subspace generated by the functions $\chi_{[i-1,i)}$ seems to have a polytope as its unit ball when λ is an ST-function (rather than a more general growth function). One can ask similar questions about these polytopes.

(D) It is interesting to define ST-spaces more geometrically, especially if we return to the idea of trying to construct "generic" polytopes. This can be done as follows. Given an affine subspace Y of $\mathbb{R}^n$ not containing zero, define the *canonical extension* of Y to a hyperplane to be the sum of Y and the orthogonal complement of the linear subspace generated by Y. Let the side of this hyperplane containing zero be the *canonical half-space* associated with Y.

Call a set of points $X = \{x_1, \ldots, x_m\}$ in $\mathbb{R}^n$ *compatible* if the canonical half-spaces associated with the (zero-dimensional) affine subspaces $\{x_i\}$ all contain all of X. Suppose we have a compatible set X in general position (for simplicity only; this condition is not really needed) and let Σ be the simplicial complex of all subsets $A \subset X$ such that $\operatorname{conv} A$ is a facet of $\operatorname{conv} X$. If $A \in \Sigma$, denote by $C(A)$ the cone generated by A. Let $C_r = C_r(X)$ be the union of all $C(A)$ such that $|A| = r + 1$ (so that $C(A)$ is r-dimensional).

Given a subset $K \subset C_r$ such that $\operatorname{conv} K$ is a polytope with 0 in its interior and such that $K = C_r \cap \operatorname{conv} K$, we define $SK \subset C_{r+1}$ to be the intersection of C_{r+1} with $\operatorname{conv} K$, and we define TK to be the intersection of C_{r+1} with the intersection of all canonical half-spaces associated with r-facets of $\operatorname{conv} K$ that lie entirely in C_r. This definition agrees with the previous one if X is the set of points $\pm e_i$. It is possible to make precise the sense in which the S and T operations are dual to each other.

Several questions arise immediately. If one takes a suitably well-distributed set of points in the n-sphere and applies an ST-sequence which would have produced an approximation to ℓ_p^n if applied to $\{\pm e_i\}$, then what does one get, or at least what properties can one expect the resulting polytope (or normed space in the symmetric case) to have? (A possible definition of "suitably well-distributed" is that the identity on $\mathbb{R}^n$ can be expressed in the form $y \mapsto \sum c_i \langle x_i, y \rangle x_i$ for positive constants c_i and points x_i in the set.) What happens if one starts with a regular simplex? In this case, does alternating between S and T produce an approximation to a sphere? If so, what can one say about "simplex ℓ_p^n" for $p \neq 2$? (One might, for example, expect that the unit balls of simplex ℓ_p^n and simplex ℓ_q^n were equivalent via a negative multiple of the identity.) Is it possible to define a space given a more general growth function and a compatible set of starting points?

(**E**) One can regard ST-spaces as the result of a kind of interpolation between L_1 and L_∞. We give an indication of how this is done. Given a space $X \in G_t$ and a decreasing function f supported on $[0, \alpha t]$, we calculate its norm in $S_\alpha(X)$ by finding a function g of the form $f \vee \gamma$ restricted to $[0, t]$, where γ is minimal such that $\|g\|_1 = \|f\|_1$. Its norm in $T_\alpha(X)$ can be described in exactly the same way, except that now $\|g\|_\infty = \|f\|_\infty$ (so that $\gamma = 0$). Given a rearrangement-invariant function space V, we could define $V_\alpha(X)$ in the same way. The norm of f in $V_\alpha(X)$ is the norm in X of the restriction of $f \vee \gamma$ to $[0, t]$, where now γ is chosen so that the V-norm of the two functions is the same.

If we have two RI-spaces V and W, and an ST-sequence with corresponding growth function λ, we can replace each S_α by a V_α and each T_α by a W_α. Denote the resulting space by $(V, W)_{[\lambda]}$. Then $(L_1, L_\infty)_{[\lambda]} = L_\lambda$. (We have extended to the whole of $\mathbb{R}_+$—this presents no problems.) We can now take limits as before and define $(V, W)_{[\lambda]}$ for arbitrary growth functions λ.

Some questions that arise out of this definition are the following. What is the relationship between this interpolation method and the complex interpolation method, when $\lambda(t) = t^{1/p}$? It is not hard to show that $(L_\lambda, L_\mu)_{[\nu]} = L_\xi$, where ξ is defined by the differential equation

$$\frac{\xi'(x)}{\xi(x)} = \frac{\nu'(x)}{\nu(x)} \frac{\lambda'(x)}{\lambda(x)} + \left(1 - \frac{\nu'(x)}{\nu(x)}\right) \frac{\mu'(x)}{\mu(x)},$$

so they agree with complex interpolation when V and W are limits of ST-spaces.

Does this method give rise to an interpolation theorem, and is the resulting constant 1? This would be interesting as it is a real method rather than a complex one. Can the method be generalized to interpolation between arbitrary spaces? (Probably it is not hard to generalize at least as far as lattices, but even this is not a triviality.)

(**F**) It seems very likely that it is possible to generalize many of the results of this paper to operator spaces. Do appropriate operator-space versions give operator ℓ_p-spaces, as defined by Pisier? Does this give a means of defining operator Orlicz spaces? Of course the answer to this last question can only be yes if ST-spaces and Orlicz spaces coincide isomorphically.

Acknowledgement

This paper is a revised and expanded version of a chapter in my PhD thesis, which contained all the main results. The thesis was written under the supervision of Dr B. Bollobás at Cambridge University, to whom I am very grateful. I would also like to thank the referee, whose long and detailed report, full of valuable suggestions, reached me less than two weeks after I submitted the paper. As a result, the presentation has been significantly improved and I have been saved from several errors.

References

[BL] J. Bergh and J. Löfström, Interpolation Spaces, an Introduction, Springer, Berlin, 1976.

[G] E. D. Gluskin, *The diameter of the Minkowski compactum is approximately equal to n*, Funct. Anal. Appl. **15** (1981), 72–73.

[LT] J. Lindenstrauss and L. Tzafriri, Classical Banach spaces I: sequence spaces, Springer, Berlin, 1977.

W. Timothy Gowers
Department of Pure Mathematics and Mathematical Statistics
Cambridge University
16 Mill Lane
Cambridge CB2 1SB
United Kingdom
W.T.Gowers@dpmms.cam.ac.uk

A Remark about the Scalar-Plus-Compact Problem

W. TIMOTHY GOWERS

> ABSTRACT. In [GM] a Banach space X was constructed such that every operator from a subspace $Y \subset X$ into the space is of the form $\lambda I_{Y \to X} + S$, where $I_{Y \to X}$ is the inclusion map and S is strictly singular. In this paper we show that there is an operator T from a subspace $Y \subset X$ into X which is not of the form $\lambda I_{Y \to X} + K$ with K compact.

1. Introduction

It is an open problem whether there exists an infinite-dimensional Banach space X such that every bounded linear operator from X to itself is of the form $\lambda I + K$, where λ is a scalar, I is the identity on X and K is a compact operator. The strongest property of a similar nature that has been obtained is that a space may be *hereditarily indecomposable* (see [GM] for this definition and several others throughout the paper), which implies [GM] that every operator on it is of the form $\lambda I + S$, where S is strictly singular, and even [F1] that every operator from a subspace into the space is a strictly singular perturbation of a multiple of the inclusion map. (These results assume complex scalars but several examples are known where the conclusion holds with real scalars.) In this note, we show that the first hereditarily indecomposable space to be discovered [GM], which we shall call X, has a subspace Y such that there is a non-compact strictly singular operator from Y into X. Therefore this operator is not a compact perturbation of a multiple of the inclusion map. Since all we are doing is showing that one particular space does not give an example of a stronger property than that required by the problem, the existence of this note needs some justification, which we shall now provide.

First, if one is trying to solve the problem with an example, then a natural line of attack is to try to construct a hereditarily indecomposable space such that every strictly singular operator is compact. To ensure the second property, a natural sufficient condition is the following: if $u_1 < u_2 < \cdots$ and $v_1 < v_2 < \cdots$

are any pair of normalized block bases such that $(u_n)_1^\infty$ dominates $(v_n)_1^\infty$, then they are actually equivalent. However, if such an example existed, then it would also give an example of the stronger property about maps from subspaces, so the stronger property is worth considering.

Second, the known hereditarily indecomposable spaces (for example [AD, F2, G1, G2, GM, H]) are obvious places to start in any search for a counterexample. Since not much was known about any of them in this respect, this note performs a modest, but necessary function.

Third, the method of proof does not rely very much on the detailed properties of the space X, so it is highly likely that it can be generalized, perhaps even to some very wide class of spaces such as reflexive ones. Indeed, it is the author's belief (but this is just a guess) that every reflexive space has a subspace such that there is a map from the subspace into the space which is not a compact perturbation of the inclusion map.

Nevertheless, since the result of this note is rather specific, we shall assume familiarity with the paper [GM], including its notation (although it is not necessary to have followed everything), and in some places we shall sketch easy arguments rather than proving them in full.

2. Construction of the Subspace and Operator into X

The main properties we shall use of the space X are the following two. Let $f(n) = \log_2(n+1)$. Then, for every $x \in X$,

$$\|x\| \leq \|x\|_\infty \vee \sup\left\{ f(k)^{-1/2} \sum_{i=1}^k \|E_i x\| : k \geq 2, E_1 < \cdots < E_k \right\},$$

which implies that the norm on X is dominated by the norm on Schlumprecht's space defined with the function $\sqrt{f}$.

The second property is that for every $\varepsilon > 0$ and every $m \in \mathbb{N}$ there is a normalized block basis $u_1 < u_2 < \cdots$ of X such that if $a_1, \ldots, a_m$ are scalars and $i_1 < \cdots < i_m$, then, setting $a = \sum_{j=1}^m a_j u_{i_j}$, there exist k and intervals $E_1 < \cdots < E_k$ such that

$$\|a\| \geq f(k)^{-1} \sum_{r=1}^k \|E_r a\| \geq (1+\varepsilon)^{-1} \sum_{j=1}^m |a_j|.$$

(To sketch the proof: let $v_1 < v_2 < \cdots$ be an infinite sequence in X such that every subsequence of length $M >> m$ is a rapidly increasing sequence with constant $1 + \varepsilon/2$. Let each u_i be a block consisting of M/m of the v_js added together.)

Now let $N_1 < N_2 < N_3 < \cdots$ be a sufficiently fast-growing sequence of integers. (It will be clear later that suitable choices of N_i exist.) For each integer s, let $u_1^{(s)} < u_2^{(s)} < u_3^{(s)} < \cdots$ be a block basis satisfying the condition above, with $m = N_s$ and $\varepsilon = 1$. Now let us choose vectors $y_1, y_2, y_3, \ldots$ satisfying the following conditions.

(i) There is some function $\phi : \mathbb{N}^2 \to \mathbb{N}$ such that $y_n = \sum_{s=1}^{\infty} 2^{-s} u_{\phi(s,n)}^{(s)}$ for every n, s.
(ii) Any pair of distinct $u_{\phi(s,n)}^{(s)}$ have disjoint ranges.
(iii) If $m < n$ then $u_{\phi(s,m)}^{(s)} < u_{\phi(s,n)}^{(s)}$.

It is not hard to show that all these properties can be satisfied simultaneously.

Consider a sum of the form $\sum_{i=1}^{N} a_i y_{n_i}$. If $N \leq N_s$, then

$$\left\| \sum_{i=1}^{N} a_i y_{n_i} \right\| \geq 2^{-s} \geq 2^{-(s+1)} \sum_{i=1}^{N} |a_i|,$$

because we can estimate the norm on the left-hand side by isolating the contribution from the block basis $(u_j^{(s)})$ and using the intervals $E_1 < \cdots < E_k$ guaranteed by the condition on this block basis. (Note that we are not simply projecting onto the span of the $u_j^{(s)}$, which would not be allowed as X does not have an unconditional basis.)

Therefore, given any monotone function $\omega : \mathbb{N} \to [4, \infty]$ such that $\omega(n)$ tends to infinity with n, one can choose the sequence $N_1 < N_2 < \cdots$ in such a way that

$$\left\| \sum_{i=1}^{\infty} a_i y_i \right\| \geq \tfrac{1}{4} \sup_{A \subset \mathbb{N}} \omega(|A|)^{-1} \sum_{i \in A} |a_i|$$

whenever the left-hand side makes sense.

Our next aim is to show that if ω is sufficiently slow-growing, then the norm in X of any vector $\sum_{i=1}^{\infty} a_i e_i$ is at most $C \sup_{A \subset \mathbb{N}} \omega(|A|)^{-1} \sum_{i \in A} |a_i|$ for some absolute constant C. This will imply that there is a bounded linear map from Y to X taking y_n to e_n. This map is certainly not compact. Moreover, it follows easily from the above estimate that it is infinitely singular, and hence, since X is hereditarily indecomposable, strictly singular also. Thus, once we have the estimate, the proof is finished.

Let $L(\omega)$ be the space of all scalar sequences $a = (a_1, a_2, \ldots)$ with

$$\|a\| = \sup_{A \subset \mathbb{N}} \omega(|A|)^{-1} \sum_{i \in A} |a_i|.$$

(This space is the dual of a Lorentz sequence space.) Let S be Schlumprecht's space, defined using the function $g(n) = (\log_2(n+1))^{1/2}$. Let S' be the symmetrization of S. That is, $\|a\|_{S'}$ is the supremum of $\|b\|_S$ over all rearrangements b of a. Then we have $\|a\|_X \leq \|a\|_S \leq \|a\|_{S'}$. Therefore, it is enough to show that the formal identity from $L(\omega)$ to S' is continuous.

To prove this, it is enough to consider extreme points in the unit ball of $L(\omega)$. Such a point has a decreasing rearrangement $a = (a_1, a_2, \ldots)$, say, and it is easy to see that whenever $\sum_{i=1}^{n} a_i < \omega(n)$, we must have $a_n = a_{n+1}$. Therefore, a must be of the form

$$\sum_{i=1}^{n_1} \frac{\omega(n_1)}{n_1} e_i + \sum_{i=n_1+1}^{n_2} \frac{\omega(n_2) - \omega(n_1)}{n_2 - n_1} e_i + \sum_{i=n_2+1}^{n_3} \frac{\omega(n_3) - \omega(n_2)}{n_3 - n_2} e_i + \cdots.$$

Furthermore,

$$\sum_{i=n_{r-1}+1}^{n_r} \frac{\omega(n_r) - \omega(n_{r-1})}{n_r - n_{r-1}} e_i = \mathbb{E} \sum_{i=n_{r-1}+1}^{n_r} (\omega(i) - \omega(i-1)) e_{\pi(i)},$$

where the average is over all permutations π of the set $\{n_{r-1}+1, \ldots, n_r\}$, so in fact every extreme point of the ball of $L(\omega)$ has as its decreasing rearrangement the sequence $(\omega(1), \omega(2) - \omega(1), \omega(3) - \omega(2), \ldots)$.

Since the unit vector basis is 1-symmetric in both $L(\omega)$ and S', all that remains is to choose a decreasing sequence $b_m \to 0$ such that $\sum_{m=1}^{\infty} b_m = \infty$ and $\|(b_m)\|_{S'} < \infty$, so that we can set $\omega(n) = \sum_{m=1}^{n} b_m$. The existence of such a sequence is an easy exercise, given that the norm of $\sum_{i=1}^{r} e_i$ in S' is $r/f(r)$.

3. Further Questions

It would be nice of course to solve the whole problem, but if this cannot immediately be done, then to get more of a feel for the technicalities involved, it would be good to obtain results similar to those of this paper for other known hereditarily indecomposable spaces. For some of them the argument carries through with only minor modifications (this is certainly true of [G2] and probably of [F] and [H] as well). However, at least three known spaces present difficulties, each of a different kind. One is the dual of the space X considered here, another is the asymptotic ℓ_1-space constructed by Argyros and Delyanni and a third is the non-reflexive space constructed in [G1]. Some of these difficulties appear to be merely technical, but the problems should still be investigated.

References

[AD] S. Argyros and I. Deliyanni, *Examples of asymptotic ℓ_1 Banach spaces*, preprint (1994).

[F1] V. Ferenczi, *Operators on subspaces of hereditarily indecomposable Banach spaces*, Bull. London Math. Soc. (to appear).

[F2] V. Ferenczi, *A uniformly convex and hereditarily indecomposable Banach space*, Israel J. Math. (to appear).

[G1] W. T. Gowers, *A Banach space not containing c_0, ℓ_1 or a reflexive subspace*, Trans. Amer. Math. Soc. **344** (1994), 407–420.

[G2] W. T. Gowers, *A hereditarily indecomposable space with an asymptotic unconditional basis*, GAFA Israel Seminar 1992-94, Operator Theory Advances and Applications **77**, Birkhäuser, 1995, 111–120.

[GM] W. T. Gowers and B. Maurey, *The unconditional basic sequence problem*, J. Amer. Math. Soc. **6** (1993), 851–874.

[H] P. Habala, *Banach spaces all of whose subspaces fail the Gordon–Lewis property* (submitted).

W. TIMOTHY GOWERS
DEPARTMENT OF PURE MATHEMATICS AND MATHEMATICAL STATISTICS
CAMBRIDGE UNIVERSITY
16 MILL LANE
CAMBRIDGE CB2 1SB
UNITED KINGDOM
   W.T.Gowers@dpmms.cam.ac.uk

Another Low-Technology Estimate in Convex Geometry

GREG KUPERBERG

ABSTRACT. We give a short argument that for some $C > 0$, every n-dimensional Banach ball K admits a 256-round subquotient of dimension at least $Cn/(\log n)$. This is a weak version of Milman's quotient of subspace theorem, which lacks the logarithmic factor.

Let V be a finite-dimensional vector space over $\mathbb{R}$ and let V^* denote the dual vector space. A *symmetric convex body* or *(Banach) ball* is a compact convex set with nonempty interior which is invariant under under $x \mapsto -x$. We define $K^\circ \subset V^*$, the *dual* of a ball $K \subset V$, by

$$K^\circ = \{y \in V^* \big| y(K) \subset [-1,1]\}.$$

A ball K is the unit ball of a unique Banach norm $\|\cdot\|_K$ defined by

$$\|v\|_K = \min\{t \big| v \in tK\}.$$

A ball K is an *ellipsoid* if $\|\cdot\|_K$ is an inner-product norm. Note that all ellipsoids are equivalent under the action of $\mathrm{GL}(V)$.

If V is not given with a volume form, then a volume such as Vol K for $K \subset V$ is undefined. However, some expressions such as (Vol K)(Vol K°) or (Vol K)/(Vol K') for $K, K' \subset V$ are well-defined, because they are independent of the choice of a volume form on V, or equivalently because they are invariant under $\mathrm{GL}(V)$ if a volume form is chosen.

An n-dimensional ball K is r-*semiround* [8] if it contains an ellipsoid E such that

$$(\mathrm{Vol}\ K)/(\mathrm{Vol}\ E) \leq r^n.$$

1991 *Mathematics Subject Classification.* Primary 52A21; Secondary 46B03.

The author was supported by an NSF Postdoctoral Fellowship, grant #DMS-9107908.

It is *r-round* if it contains an ellipsoid E such that $K \subseteq rE$. Santaló's inequality states that if K is an n-dimensional ball and E is an n-dimensional ellipsoid,

$$(\text{Vol } K)(\text{Vol } K^\circ) \leq (\text{Vol } E)(\text{Vol } E^\circ).$$

(Saint-Raymond [7], Ball [1], and Meyer and Pajor [4] have given elementary proofs of Santaló's inequality.) It follows that if K is r-round, then either K or K° is $\sqrt{r}$-semiround.

If K is a ball in a vector space V and W is a subspace, we define $W \cap K$ to be a *slice* of K and the image of K in V/W to be a *projection* of K; they are both balls. Following Milman [5], we define a *subquotient* of K to be a slice of a projection of K. Note that a slice of a projection is also a projection of a slice, so that we could also have called a subquotient a proslice. It follows that a subquotient of a subquotient is a subquotient. Note also that a slice of K is dual to a projection of K°, and therefore a subquotient of K is dual to a proslice (or a subquotient) of K°.

In this paper we prove the following theorem:

THEOREM 1. *Suppose that K is a $(2^{k+1}n)$-dimensional ball which is $(2^{(3/2)^k} \cdot 4)$-semiround, with $k \geq 0$. Then K has a 256-round, n-dimensional subquotient.*

COROLLARY 2. *There exists a constant $C > 0$ such that every n-dimensional ball K admits a 256-round subquotient of dimension at least $Cn/(\log n)$.*

The corollary follows from the theorem of John that every n-dimensional ball is $(\sqrt{n})$-round.

The corollary is a weak version of a celebrated result of Milman [5; 6]:

THEOREM 3 (MILMAN). *For every $C > 1$, there exists $D > 0$, and for every $D < 1$ there exists a C, such that every n-dimensional ball K admits a C-round subquotient of dimension at least Dn.*

However, the argument given here for Theorem 1 is simpler than any known proof of Theorem 3.

Theorem 3 has many consequences in the asymptotic theory of convex bodies, among them a dual of Santalo's inequality:

THEOREM 4 (BOURGAIN, MILMAN). *There exists a $C > 0$ such that for every n and for every n-dimensional ball K,*

$$(\text{Vol } K)(\text{Vol } K^\circ) \geq C^n (\text{Vol } E)(\text{Vol } E^\circ).$$

Theorem 4 is an asymptotic version of Mahler's conjecture, which states that for fixed n, $(\text{Vol } K)(\text{Vol } K^\circ)$ is minimized for a cube. In a previous paper, the author [3] established a weak version of Theorem 4 also, namely that

$$\text{Vol } (K) \text{Vol } (K^\circ) \geq (\log_2 n)^{-n} \text{Vol } (E) \text{Vol } (E^\circ)$$

for $n \geq 4$. That result was the motivation for the present paper.

ANOTHER LOW-TECHNOLOGY ESTIMATE IN CONVEX GEOMETRY   119

The author speculates that there are elementary arguments for both Theorems 3 and 4, which moreoever would establish reasonable values for the arbitrary constants in the statements of these theorems.

The Proof

The proof is a variation of a construction of Kashin [8]. For every k let Ω_k be the volume of the unit ball in $\mathbb{R}^k$; Ω_k is given by the formula

$$\frac{\pi^{k/2}}{\Gamma(\frac{k}{2}+1)}.$$

Let V be an n-dimensional vector space with a distinguished ellipsoid E, to be thought of as a round unit ball in V, so that V is isometric to standard $\mathbb{R}^n$ under $\|\cdot\|_E$. Give V the standard volume structure $d\vec{x}$ on $\mathbb{R}^n$. In particular, Vol $E = \Omega_n$. Endow ∂E, the unit sphere, with the invariant measure μ with total weight 1. If K is some other ball in V, then

$$\text{Vol } K = \Omega_n \int_{\partial E} \|x\|_K^{-n} d\mu$$

and, more generally,

$$\int_K \|x\|_E^k \, d\vec{x} = \frac{n \Omega_n}{n+k} \int_{\partial E} \|x\|_K^{-n-k} d\mu.$$

Let f be a continuous function on ∂E. Let $0 < d < n$ be an integer and consider the space of d-dimensional subspaces of V. This space has a unique probability measure invariant under rotational symmetry. If W is such a subspace chosen at random with respect to this measure, then for any continuous function f,

$$\int_{\partial E} f(x) d\mu = \mathrm{E}\left[\int_{\partial(E \cap W)} f(x) d\mu\right], \qquad (1)$$

where μ denotes the invariant measure of total weight 1 on $E \cap W$ also. In particular, there must be some W for which the integral of f on the right side of equation (1) is less than or equal to that of the left side, which is the average value.

The theorem follows by induction from the case $k = 0$ and from the claim that if K is a $(2n)$-dimensional ball which is r-semiround, then K has an n-dimensional slice K'' such that either K'' or its dual is $(2r)^{2/3}$-semiround. In both cases, we assume that K is r-semiround and has dimension $2n$ and we proceed with a parallel analysis.

There exists an $(n+1)$-dimensional subspace V' of V such that:

$$\int_{\partial E'} \|x\|_K^{-2n} d\mu \leq \frac{\text{Vol } K}{\text{Vol } E} = r^{2n}, \qquad (2)$$

where $E' = E \cap V'$. Let $K' = V' \cap K$. Then

$$\int_{\partial E'} \|x\|_K^{-2n} d\mu = \frac{2n}{(n-1)\Omega_{n+1}} \int_{K'} \|x\|_{E'}^{n-1} d\vec{x}. \tag{3}$$

Let p be a point in K' such that $s = \|p\|_E$ is maximized; in particular K' is s-round Let V'' be the subspace of V' perpendicular to p and define $K'' = V'' \cap K$ and $E'' = V'' \cap E$. The convex hull $S(K'')$ of $K'' \cup \{p, -p\}$ is a double cone with base K'' (or suspension of K''), and $S(K'') \subseteq K'$. We establish an estimate that shows that either s or Vol K'' is small. Let x_0 be a coordinate for V' given by distance from V''. Then

$$\int_{K'} \|x\|_{E'}^{n-1} d\vec{x} \geq \int_{S(K'')} \|x\|_{E'}^{n-1} d\vec{x} > \int_{S(K'')} |x_0|^{n-1} d\vec{x}$$

$$= 2\int_0^s x_0^{n-1} \left(\text{Vol}\left(1 - \frac{x_0}{s}\right)K''\right) dx_0$$

$$= 2(\text{Vol } K'')s^n \int_0^1 t^{n-1}(1-t)^n \, dt$$

$$= (\text{Vol } K'')s^n \frac{2(n-1)!n!}{(2n)!}. \tag{4}$$

We combine equations (2), (3), and (4) with the inequality

$$\frac{\Omega_n 4n(n-1)!n!}{\Omega_{n+1}(n-1)(2n)!} = \frac{2\Gamma(\frac{n+3}{2})(n-2)!n!}{\sqrt{\pi}\Gamma(\frac{n+2}{2})(2n-1)!} > 4^{-n}.$$

(Proof: Let $f(n)$ be the left side. By Stirling's approximation, $f(n)4^n \to 2^{3/2}$ as $n \to \infty$. Since

$$\frac{f(n+2)}{f(n)} = \frac{1}{4}\frac{n^2 + 2n - 3}{4n^2 + 8n + 3} < \frac{1}{16},$$

the limit is approached from above.) The final result is that

$$\frac{\text{Vol } K''}{\text{Vol } E''} \leq (2r)^{2n} s^{-n}.$$

In the case $k = 0$, $r = 8$. Since $E'' \subseteq K''$, Vol $K'' \geq$ Vol E'', which implies that $s \leq 4r^2 = 256$. Since K'' is s-round, it is the desired subquotient of K.

If $k > 1$, then suppose first that $s \leq (2r)^{4/3}$. In this case K'' is $(2r)^{4/3}$-round, which implies by Santaló's inequality that either K'' or K''° is $(2r)^{2/3}$-semiround. On the other hand, if $s \geq (2r)^{4/3}$, then K'' is $(2r)^{2/3}$-semiround. In either case, the induction hypothesis is satisfied.

References

[1] K. Ball. *Isometric problems in ℓ_p and sections of convex sets*. PhD thesis, Trinity College, 1986.

[2] J. Bourgain and V. D. Milman. New volume ratio properties for convex symmetric bodies in $\mathbb{R}^n$. *Invent. Math.*, 88:319–340, 1987.

[3] G. Kuperberg. A low-technology estimate in convex geometry. *Duke Math. J.*, 68:181–183, 1992.

[4] M. Meyer and A. Pajor. *On Santaló's inequality*, volume 1376, pages 261–263. Springer, 1989.

[5] V. D. Milman. Almost Euclidean quotient spaces of subpspaces of finite-dimensional normed spaces. *Proc. Amer. Math. Soc.*, 94:445–449, 1985.

[6] G. Pisier. *The Volume of Convex Bodies and Banach Space Geometry*, volume 94 of *Cambridge Tracts in Mathematics*. Cambridge University Press, 1989.

[7] J. Saint-Raymond. Sur le volume des corps convexes symètriques. Technical report, Universitè P. et M. Curie, Paris, 1981. Sèminaire Initiation à l'Analyse.

[8] S. Szarek. On Kashin's almost Euclidean orthogonal decomposition of ℓ_1^n. *Bull. Acad. Polon. Sci.*, 26:691–694, 1978.

GREG KUPERBERG
DEPARTMENT OF MATHEMATICS
UNIVERSITY OF CALIFORNIA
DAVIS, CA 95616
UNITED STATES OF AMERICA
greg@math.ucdavis.edu

On the Equivalence Between Geometric and Arithmetic Means for Log-Concave Measures

RAFAŁ LATAŁA

ABSTRACT. Let X be a random vector with log-concave distribution in some Banach space. We prove that $\|X\|_p \leq C_p \|X\|_0$ for any $p > 0$, where $\|X\|_p = (E\|X\|^p)^{1/p}$, $\|X\|_0 = \exp E \ln \|X\|$ and C_p are constants depending only on p. We also derive some estimates of log-concave measures of small balls.

Introduction. Let X be a random vector with log-concave distribution (for precise definitions see below). It is known that for any measurable seminorm and $p, q > 0$ the inequality

$$\|X\|_p \leq C_{p,q} \|X\|_q$$

holds with constants $C_{p,q}$ depending only on p and q (see [4], Appendix III). In this paper we show that the above constants can be made independent of q, which is equivalent to the inequality

$$\|X\|_p \leq C_p \|X\|_0, \tag{1}$$

where $\|X\|_0$ is the geometric mean of $\|X\|$. In the particular case in which X is uniformly distributed on some convex compact set in R^n and the seminorm is given by some functional, inequality (1) was established by V. D. Milman and A. Pajor [3]. As a consequence of (1) we prove the result of Ullrich [6] concerning the equivalence of means for sums of independent Steinhaus random variables with vector coefficients, even though these random-variables are not log-concave (Corollary 2).

To prove (1) we derive some estimates of log-concave measures of small balls (Corollary 1), which are of independent interest. In the case of Gaussian random variables they were formulated and established in a weaker version in [5] and completelely proved in [2].

Definitions and Notation. Let E be a complete, separable, metric vector space endowed with its Borel σ-algebra $\mathcal{B}_E$. By μ we denote a log-concave probability measure on $(E, \mathcal{B}_E)$ (for some characterizations, properties and examples, see [1]) i.e. a probability measure with the property that for any Borel subsets A, B and all $0 < \lambda < 1$ we have

$$\mu(\lambda A + (1-\lambda)B) \geq \mu(A)^\lambda \mu(B)^{1-\lambda}.$$

We say that a random vector X with values in E is log-concave if the distribution of X is log-concave. For a random vector X and a measurable seminorm $\|.\|$ on E (i.e. Borel measurable, nonnegative, subadditive and positively homogeneous function on E) we define

$$\|X\|_p = (E\|X\|^p)^{1/p} \text{ for } p > 0$$

and

$$\|X\|_0 = \lim_{p \to 0^+} \|X\|_p = \exp(E \ln \|X\|).$$

Let us begin with the following Lemma from [1].

LEMMA 1. *For any convex, symmetric Borel set B and $k \geq 1$ we have*

$$\mu((kB)^c) \leq \mu(B) \left(\frac{1-\mu(B)}{\mu(B)} \right)^{(k+1)/2}.$$

PROOF. The statement follows immediately from the log-concavity of μ and the inclusion

$$\frac{k-1}{k+1} B + \frac{2}{k+1} (kB)^c \subset B^c. \qquad \square$$

LEMMA 2. *If B is a convex, symmetric Borel set, with $\mu(KB) \geq (1+\delta)\mu(B)$ for some $K > 1$ and $\delta > 0$ then*

$$\mu(tB) \leq Ct\mu(B) \text{ for any } t \in (0,1),$$

where $C = C(K/\delta)$ is a constant depending only on K/δ.

PROOF. Obviously it's enough to prove the result for $t = 1/2n$, $n = 1, 2, \ldots$. So let us fix n and define, for $u \geq 0$,

$$P_u = \{x : \|x\|_B \in (u - 1/2n, u + 1/2n)\},$$

where

$$\|x\|_B = \inf\{t > 0 : x \in tB\}.$$

By simple calculation $\lambda P_u + (1-\lambda)(2n)^{-1} B \subset P_{\lambda u}$, so

$$\mu(P_{\lambda u}) \geq \mu(P_u)^\lambda \mu((2n)^{-1}B)^{1-\lambda} \text{ for } \lambda \in (0,1). \qquad (2)$$

From the assumptions it easily follows that there exists $u \geq 1$ such that $\mu(P_u) \geq \delta\mu(B)/Kn$. Let $\mu((2n)^{-1}B) = \kappa\mu(B)/n$. If $\kappa \leq 2\delta/K$ we are done, so we will assume that $\kappa \geq 2\delta/K$. Then by (2) it follows that $\mu(P_1) \geq \delta\mu(B)/Kn$. The sets $P_{(n-1)/n}, P_{(n-2)/n}, \ldots, P_{1/n}, (2n)^{-1}B$ are disjoint subsets of B, and hence

$$\mu(B) \geq \mu(P_{(n-1)/n}) + \cdots + \mu(P_{1/n}) + \mu((2n)^{-1}B).$$

Using our estimations of $\mu(P_1)$ and $\mu((2n)^{-1}B)$ we obtain by (2)

$$\mu(B) \geq n^{-1}\mu(B)((\delta/K)^{(n-1)/n}\kappa^{1/n} + \cdots + (\delta/K)^{1/n}\kappa^{(n-1)/n} + \kappa) =$$

$$= \frac{\kappa}{n}\mu(B)\frac{1 - \delta/K\kappa}{1 - (\delta/K\kappa)^{1/n}} \geq \frac{\kappa}{2n}\mu(B)\frac{1}{1 - (\delta/K\kappa)^{1/n}}.$$

Therefore

$$\kappa \leq 2n(1 - (\delta/K\kappa)^{1/n}) \leq 2\ln K\kappa/\delta,$$

so that $\kappa \leq C(K/\delta)$ and the lemma follows. □

COROLLARY 1. *For each $b < 1$ there exists a constant C_b such that for every log-concave probability measure μ and every measurable convex, symmetric set B with $\mu(B) \leq b$ we have*

$$\mu(tB) \leq C_b t\mu(B) \text{ for } t \in [0,1].$$

PROOF. If $\mu(B) = 2/3$ then by Lemma 1 $\mu(3B) \geq 5/6 = (1 + 1/4)\mu(B)$, so by Lemma 2 for some constant $\tilde{C}_1$, $\mu(tB) \leq \tilde{C}_1 t\mu(B)$.

If $\mu(B) \in [1/3, 2/3]$ then obviously $\mu(tB) \leq 2\tilde{C}_1 t\mu(B)$.

If $\mu(B) < 1/3$, let K be such that $\mu(KB) = 2/3$. By the above case $\mu(B) \leq \tilde{C}_1 K^{-1}\mu(KB)$, and hence

$$K \leq 2\tilde{C}_1\left(\frac{\mu(KB)}{\mu(B)} - 1\right).$$

So Lemma 2 gives in this case that $\mu(tB) \leq \tilde{C}_2 t\mu(B)$ for some constant $\tilde{C}_2$.

Finally if $\mu(B) > 2/3$, but $\mu(B) \leq b < 1$ then by Lemma 1 for some $K_b < \infty$, $\mu(K_b^{-1}B) \leq 2/3$ and we can use the previous calculations. □

THEOREM 1. *For any $p > 0$ there exists a universal constant C_p, depending only on p such that for any sequence $X_1, \ldots, X_n$ of independent log-concave random vectors and any measurable seminorm $\|.\|$ on E we have*

$$\left\|\sum_{i=1}^n X_i\right\|_p \leq C_p \left\|\sum_{i=1}^n X_i\right\|_0.$$

PROOF. Since a convolution of log-concave measures is also log-concave (see [1]) we may and do assume that $n = 1$. Let

$$M = \inf\{t : P(\|X_1\| \geq t) \leq 2/3\}.$$

Then by Lemma 1 (used for $B = \{x \in E : \|x\| \leq M\}$) it follows easily that $\|X_1\|_p \leq a_p M$ for $p > 0$ and some constants a_p depending only on p. By similar reasoning Corollary 1 yields $\|X_1\|_0 \geq a_0 M$. □

COROLLARY 2. *Let E be a complex Banach space and $X_1, \ldots, X_n$ be a sequence of independent random variables uniformly distributed on the unit circle $\{z \in \mathbb{C} : |z| = 1\}$. Then for any sequence of vectors $v_1, \ldots, v_n \in E$ and any $p > 0$ the following inequality holds:*

$$\left\|\sum v_k X_k\right\|_p \leq K_p \left\|\sum v_k X_k\right\|_0,$$

where K_p is a constant depending only on p.

PROOF. It is enough to prove Corollary for $p \geq 1$. Let $Y_1, \ldots, Y_n$ be a sequence of independent random variables uniformly distributed on the unit disc $\{z : |z| \leq 1\}$. By Theorem 1 we have

$$\left\|\sum v_k Y_k\right\|_p \leq C_p \left\|\sum v_k Y_k\right\|_0. \tag{3}$$

But we may represent Y_k in the form $Y_k = R_k X_k$, where R_k are independent, identically distributed random variables on $[0, 1]$ (with an appropriate distribution), which are independent of X_k. Hence, by taking conditional expectation we obtain

$$\left\|\sum v_k Y_k\right\|_p \geq (ER_1) \left\|\sum v_k X_k\right\|_p. \tag{4}$$

Finally let us observe that for any $u, v \in E$ the function $f(z) = \ln \|u + zv\|$ is subharmonic on $\mathbb{C}$, so $g(r) = E \ln \|u + rv X_1\|$ is nondecreasing on $[0, \infty)$ and therefore

$$\left\|\sum v_k X_k\right\|_0 \geq \left\|\sum v_k Y_k\right\|_0. \tag{5}$$

The corollary follows from (3), (4) and (5). □

Acknowledgements

This paper was prepared during the author's stay at MSRI during the spring of 1996. The idea of the proof of Corollary 2 was suggested by Prof. S. Kwapień.

References

[1] C. Borell, "Convex measures on locally convex spaces", *Ark. Math.* **12** (1974), 239–252.

[2] P. Hitczenko, S. Kwapień, W. V. Li, G. Schechtman, T. Schlumprecht, and J. Zinn, "Hypercontractivity and comparison of moments of iterated maxima and minima of independent random variables", *Electron. J. Probab.* **3** (1998), 26pp. (electronic).'

[3] V. D. Milman and A. Pajor, "Isotropic positions and inertia ellipsoids and zonoids of the unit balls of a normed n-dimensional space", pp. 64–104 in *Geometric aspects of functional analysis: Israel Seminar* (GAFA), 1987-88, edited by J. Lindenstrauss and V. D. Milman, Lecture Notes in Math. **1376**, Springer, Berlin 1989.

[4] V. D. Milman and G. Schechtman, *Asymptotic theory of finite dimensional normed spaces*, Lecture Notes in Math. **1200**, Springer, Berlin, 1986.

[5] S. Szarek, "Conditional numbers of random matrices", *J. Complexity* **7** (1991), 131–149.

[6] D. Ullrich, "An extension of Kahane–Khinchine inequality in a Banach space", *Israel J. Math.* **62** (1988), 56–62.

RAFAŁ LATAŁA
INSTITUTE OF MATHEMATICS
WARSAW UNIVERSITY
BANACHA 2
02-097 WARSAW
POLAND
    rlatala@mimuw.edu.pl

On the Constant in the Reverse Brunn–Minkowski Inequality for p-Convex Balls

ALEXANDER E. LITVAK

ABSTRACT. This note is devoted to the study of the dependence on p of the constant in the reverse Brunn–Minkowski inequality for p-convex balls (that is, p-convex symmetric bodies). We will show that this constant is estimated as $c^{1/p} \le C(p) \le C^{ln(2/p)/p}$, for absolute constants $c > 1$ and $C > 1$.

Let $K \subset \mathbb{R}^n$ and $0 < p \le 1$. K is called a p-convex set if for any $\lambda, \mu \in (0,1)$ such that $\lambda^p + \mu^p = 1$ and for any points $x, y \in K$ the point $\lambda x + \mu y$ belongs to K. We will call a p-convex compact centrally symmetric body a p-ball.

Recall that a p-norm on real vector space X is a map $\|\cdot\| : X \to \mathbb{R}^+$ satisfying these conditions:

(1) $\|x\| > 0$ for all $x \ne 0$.
(2) $\|tx\| = |t|\|x\|$ for all $t \in \mathbb{R}$ and $x \in X$.
(3) $\|x + y\|^p \le \|x\|^p + \|y\|^p$ for all $x, y \in X$.

Note that the unit ball of p-normed space is a p-ball and, vice versa, the gauge of p-ball is a p-norm.

Recently, J. Bastero, J. Bernués, and A. Peña [BBP] extended the reverse Brunn–Minkowski inequality, which was discovered by V. Milman [M], to the class of p-convex balls. They proved the following result:

THEOREM 0. *Let $0 < p \le 1$. There exists a constant $C = C(p) \ge 1$ such that for all $n \ge 1$ and all p-balls $A_1, A_2 \subset \mathbb{R}^n$, there exists a linear operator $u : \mathbb{R}^n \to \mathbb{R}^n$ with $|\det(u)| = 1$ and*

$$|uA_1 + A_2|^{1/n} \le C\bigl(|A_1|^{1/n} + |A_2|^{1/n}\bigr), \tag{1}$$

where $|A|$ denotes the volume of body A.

This research was supported by Grant No. 92–00285 from United States–Israel Binational Science Foundation (BSF).

Their proof yields an estimate $C(p) \leq C^{ln(2/p)/p^2}$.

We will obtain a much better estimate for $C(p)$:

THEOREM 1. *There exist absolute constants $c > 1$ and $C > 1$ such that the constant $C(p)$ in (1) satisfies*

$$c^{1/p} \leq C(p) \leq C^{ln(2/p)/p}.$$

The proof of Theorem 0 [BBP] was based on an estimate of the entropy numbers (see also [Pi]). We use the same idea, but obtain the better dependence of the constant on p.

Let us recall the definitions of the Kolmogorov and entropy numbers. Let $U : X \to Y$ be an operator between two Banach spaces. Let $k > 0$ be an integer. The Kolmogorov numbers are defined by the following formula

$$d_k(U) = \inf \{ \|Q_S U\| \mid S \subset Y, \ \dim S = k \},$$

where $Q_S : Y \to Y/S$ is a quotient map. For any subsets K_1, K_2 of Y denote by $N(K_1, K_2)$ the smallest number N such that there are N points $y_1, \ldots, y_N$ in Y such that

$$K_1 \subset \bigcup_{i=1}^{N} (y_i + K_2).$$

Denote the unit ball of the space X (Y) by B_X (B_Y) and define the entropy numbers by

$$e_k(U) = \inf \{ \varepsilon > 0 \mid N(UB_X, \varepsilon B_Y) \leq 2^{k-1} \}.$$

For p-convex balls $B_1, B_2 \subset \mathbb{R}^n$, with $0 < p \leq 1$, we will denote the identity operator from $(\mathbb{R}^n, \|\cdot\|_1)$ to $(\mathbb{R}^n, \|\cdot\|_2)$ by $B_1 \to B_2$, where $\|\cdot\|_i$ ($i = 1, 2$) is the p-norm whose unit ball is B_i.

THEOREM 2. *Given $\alpha > 1/p - 1/2$, there exists a constant $C = C(\alpha, p)$ such that, for any n and any p-convex ball $B \subset \mathbb{R}^n$, there exists an ellipsoid $D \subset \mathbb{R}^n$ such that, for every $1 \leq k \leq n$,*

$$\max\{d_k(D \to B), e_k(B \to D)\} \leq C(n/k)^\alpha .$$

Moreover, there is an absolute constant c such that

$$C(\alpha, p) \leq \left(\frac{2}{p}\right)^{c/p} \left(\frac{1}{1-\delta}\right)^{8/\delta} \quad \text{for } \alpha > \frac{3(1-p)}{2p}, \ \delta = \frac{3(1-p)}{2p\alpha}, \ p \leq \frac{1}{2} \quad (2)$$

and

$$C(\alpha, p) \leq \left(\frac{2}{p}\right)^{c/p^2} \left(\frac{1}{1-\varepsilon}\right)^{\frac{2}{\varepsilon p^2}} \quad \text{for } \alpha > \frac{1}{p} - \frac{1}{2}, \ \varepsilon = \frac{1/p - 1/2}{\alpha}. \quad (3)$$

REMARK 1. In fact, in [BBP] Theorem 2 was proved with estimate (3). Using this result we prove estimate (2).

In the following $C(\alpha, p)$ will denote the best possible constant from Theorem 2. The main point of the proof is the following lemma.

LEMMA 1. *Let $p, q, \theta \in (0,1)$ such that $1/q - 1 = (1/p - 1)(1-\theta)$ and $\gamma = \alpha(1-\theta)$. Then*
$$C(\alpha, p) \leq 2^{1/p} 2^{1/(1-\theta)} (e/(1-\theta))^\alpha C_{p\theta}^{1/(1-\theta)} C(\gamma, q)^{1/(1-\theta)},$$
where
$$C_{p\theta} = \frac{\Gamma(1 + (1-p)/p)}{\Gamma(1 + \theta(1-p)/p)\Gamma(1 + (1-\theta)(1-p)/p)}, \quad \Gamma \text{ is the gamma function.}$$

For the reader's convenience we postpone the proof of this lemma.

PROOF OF THEOREM 2. Take $q = 1/2$, $1-\theta = p/(1-p)$. Then $C_{p\theta} = (1-p)/p$ and, consequently, by Lemma 1,
$$C(\alpha, p) \leq c \left(\frac{e}{p}\right)^\alpha 2^{2/p} \left(\frac{1}{p}\right)^{1/p} C\left(\frac{\alpha p}{1-p}, \frac{1}{2}\right).$$

Inequality (3) implies
$$C\left(\frac{\alpha p}{1-p}, \frac{1}{2}\right) \leq c \left(\frac{1}{1-\delta}\right)^{8/\delta}, \quad \text{where } \delta = \frac{3(1-p)}{2p\alpha}.$$

Thus for $\alpha > 3(1-p)/(2p)$ and $p \leq 1/2$ we obtain
$$C(\alpha, p) \leq \left(\frac{2}{p}\right)^{c/p} \left(\frac{1}{1-\delta}\right)^{8/\delta}. \quad \square$$

PROOF OF THEOREM 1. By B. Carl's theorem ([C], or see Theorem 5.2 of [Pi]) for any operator u between Banach spaces the following inequality holds
$$\sup_{k \leq n} k^\alpha e_k(u) \leq \rho_\alpha \sup_{k \leq n} k^\alpha d_k(u).$$

One can check that Carl's proof works in the p-convex case also and gives
$$\rho_\alpha \leq C^{1/p} (C\alpha)^{C\alpha}$$
for some absolute constant C. Let us fix $\alpha = 2/p$. Then, by Theorem 2, we have that for any p-convex body K there exists an ellipsoid D such that
$$\max\{e_n(D \to B), e_n(B \to D)\} \leq C^{ln(2/p)/p}.$$

The standard argument [Pi] gives the upper estimate for C_p.

To show the lower bound we use the following example. Let B_p^n be a unit ball in the space l_p^n and B_2^n be a unit ball in the space l_2^n. Denote
$$A = \frac{|B_2^n|^{1/n}}{|B_p^n|^{1/n}} = \frac{\Gamma(3/2)\Gamma^{1/n}(1+n/p)}{\Gamma^{1/n}(1+n/2)\Gamma(1+1/p)} \geq C_0 \frac{n^{1/p - 1/2}}{\sqrt{1/p}},$$
where C_0 is an absolute constant.

Consider a body
$$K = AB_p^n.$$
We are going to estimate from below
$$\frac{|UB_2^n + K|^{1/n}}{|UB_2^n|^{1/n} + |K|^{1/n}} = \frac{|UB_2^n + K|^{1/n}}{2|B_2^n|^{1/n}}$$
for an arbitrary operator $U : \mathbb{R}^n \to \mathbb{R}^n$ with $|\det U| = 1$.

To simplify the sum of bodies in the example let us use the Steiner symmetrization with respect to vectors from the canonical basis of $\mathbb{R}^n$ (see, e.g., [BLM], for precise definitions). Usually the Steiner symmetrization is defined for convex bodies, but if we take the unit ball of l_p^n and any coordinate vector then we have the similar situation. The following properties of the Steiner symmetrization are well-known (and can be directly checked):

(i) It preserves volume.
(ii) The symmetrization of sum of two bodies contains sum of symmetrizations of these bodies.
(iii) Given an ellipsoid UB_2^n, a consecutive application of the Steiner symmetrizations with respect to all vectors from the canonical basis results in the ellipsoid VB_2^n, where V is a diagonal operator (depending on U).

That means that in our example it is enough to consider a diagonal operator U with $|\det U| = 1$.

Let $b \in (0,1)$ and P_1 be the orthogonal projection on a coordinate subspace of dimension $n-1$. Then direct computations give for every $r > 0$
$$|UB_2^n + rB_p^n| \geq 2 \int_0^{rb_p} |P_1 UB_2^n + br P_1 B_p^n| \, dx \geq 2rb_p |P_1 UB_2^n + br P_1 B_p^n|,$$
where $b_p = (p(1-b))^{1/p}$. Since $P_1 K = AB_p^{n-1}$, by induction arguments one has
$$|UB_2^n + K| \geq \left(2Ab^{(k-1)/2} b_p\right)^k |P_k UB_2^n + b^k P_k K|,$$
where P_k is the orthogonal projection on an arbitrary $(n-k)$-dimensional coordinate subspace of $\mathbb{R}^n$. Choosing $b = \exp(-2/(kp))$, P_k such that $|P_k UB_2^n| \geq |B_2^{n-k}|$ and $k = [n/2]$ we get
$$C(p) \geq \frac{|UB_2^n + K|^{1/n}}{2|B_2^n|^{1/n}} \geq \tfrac{1}{2} \left(2Ae^{-1/p} (2/k)^{1/p}\right)^{k/n} \left(\frac{|B_2^{n-k}|}{|B_2^n|}\right)^{1/n}$$
$$\geq c_1 \sqrt{p^{1/2} (4/e)^{1/p}}$$
for sufficiently large n and an absolute constant c_1. That gives the result for p small enough, namely, $p \leq c_2$, where c_2 is an absolute constant. For $p \in (c_2, 1]$ the result follows from the convex case. □

To prove Lemma 1 we will use the Lions–Peetre interpolation [BL, K] with parameters $(\theta, 1)$.

Let us recall some definitions.

Let X be a quasi-normed space with an equivalent quasi-norms $\|\cdot\|_0$ and $\|\cdot\|_1$. Let $X_i = (X, \|\cdot\|_i)$.

Define $K(t, x) = \inf\{\|x_0\|_0 + t\|x_1\|_1 \mid x = x_0 + x_1\}$ and

$$\|x\|_{\theta,1} = \theta(1-\theta) \int_0^{+\infty} \frac{K(t,x)}{t^{1+\theta}} dt,$$

for $\theta \in (0, 1)$.

The interpolation space $(X_0, X_1)_{\theta,1}$ is the space $(X, \|\cdot\|_{\theta,1})$.

CLAIM 1. *Let $\|\cdot\|_0 = \|\cdot\|_1 = \|\cdot\|$ be p-norms on space X. Then*

$$\frac{1}{C_{p\theta}}\|x\| \leq \|x\|_{\theta,1} \leq \|x\|$$

for every $x \in X$, with $C_{p\theta}$ as in Lemma 1.

PROOF. $\|x\|_{\theta,1} \leq \|x\|$ since

$$\inf\{\|x_0\|_0 + t\|x_1\|_1 \mid x = x_0 + x_1\} \leq \min(1,t)\|x\|$$

and

$$\|x\|_{\theta,1} = \theta(1-\theta) \int_0^{+\infty} \frac{K(t,x)}{t^{1+\theta}} dt \leq \theta(1-\theta) \int_0^{+\infty} \frac{\min(1,t)}{t^{1+\theta}}\|x\|dt = \|x\|.$$

By p-convexity of the norm $\|\cdot\|$ for $a = \|y\|/\|x\| \leq 1$ we have

$$\frac{\|y\| + t\|x - y\|}{\|x\|} \geq a + t(1-a^p)^{1/p} \geq \frac{t}{(1+t^s)^{1/s}}, \quad \text{where } s = \frac{p}{1-p}.$$

Hence

$$K(t,x) = \inf\{\|x_0\|_0 + t\|x_1\|_1 \mid x = x_0 + x_1\} \geq \|x\|\frac{t}{(1+t^s)^{1/s}}$$

and

$$\frac{\|x\|_{\theta,1}}{\|x\|} \geq \theta(1-\theta) \int_0^{+\infty} \frac{dt}{(1+t^s)^{1/s}t^\theta} = B\left(\frac{1-\theta}{s}, \frac{\theta}{s}\right) \frac{\theta(1-s)}{s}$$

$$= \frac{(\theta/s)\Gamma(\theta/s)((1-\theta)/s)\Gamma((1-\theta)/s)}{(1/s)\Gamma(1/s)} = \frac{1}{C_{p\theta}},$$

where $B(x, y)$ is the beta function. This proves the claim. □

CLAIM 2. *Let $\|\cdot\|_0 = \|\cdot\|_1 = \|\cdot\|$ be norms on X. Then $\|x\|_{\theta,1} = \|x\|$ for every $x \in X$.*

PROOF. In case of norm $K(t,x) = \min(1,t)\|x\|$. So, $\|x\|_{\theta,1} = \|x\|$. □

The next statement is standard (see [BL] or [K]).

CLAIM 3. *Let X_i, Y_i ($i = 0, 1$) be quasi-normed spaces. Let $T : X_i \to Y_i$ ($i = 0, 1$) be a linear operator. Then*

$$\|T : (X_0, X_1)_{\theta,1} \to (Y_0, Y_1)_{\theta,1}\| \le \|T : X_0 \to Y_0\|^{1-\theta} \|T : X_1 \to Y_1\|^{\theta}.$$

CLAIM 4. *Let X_i ($i = 0, 1$) be quasi-normed spaces. Then for every $N \ge 1$,*

$$\left(l_1^N(X_0), l_1^N(X_1)\right)_{\theta,1} = l_1^N\left((X_0, X_1)_{\theta,1}\right)$$

with equal norms.

PROOF. The conclusion of this claim follows from the equality

$$K(t, x = (x_1, x_2, \ldots, x_N), l_1^N(X_0), l_1^N(X_1)) = \sum_{i=1}^N K(t, x_i, X_0, X_1). \quad \square$$

CLAIM 5. *Let X_i ($i = 0, 1$) be quasi-normed spaces, Y be a p-normed space. Let $T : X_i$ ($i = 0, 1$) $\to Y$ be a linear operator. Then for every $k_0, k_1 \ge 1$*

$$d_{k_0+k_1-1}\left(T : (X_0, X_1)_{\theta,1} \to Y\right) \le C_{p\theta} d_{k_0}^{1-\theta}(T : X_0 \to Y) d_{k_1}^{\theta}(T : X_1 \to Y).$$

PROOF. As in the convex case [P], fix $\varepsilon > 0$. Consider a subspace $S_i \subset Y$ ($i = 0, 1$) such that $\dim S_i < k_i$ and

$$\|Q_{S_i} T : X_i \to Y/S_i\| \le (1 + \varepsilon) d_{k_i}(T : X_i \to Y).$$

Let $S = \operatorname{span}(S_0, S_1) \subset Y$. Then $\dim S < k_0 + k_1 - 1$ and

$$\|Q_S T : X_i \to Y/S\| \le \|Q_{S_i} T : X_i \to Y/S_i\|.$$

Note that quotient space of a p-normed space is again a p-normed one. Because of this, and by Claims 1 and 3,

$$\|Q_S T : (X_0, X_1)_{\theta,1} \to Y/S\| \le C_{p\theta} \|Q_S T : (X_0, X_1)_{\theta,1} \to (Y/S, Y/S)_{\theta,1}\|$$
$$\le C_{p\theta} \|Q_S T : X_0 \to Y/S\|^{1-\theta} \|Q_S T : X_1 \to Y/S\|^{\theta}$$
$$\le C_{p\theta} \|Q_{S_0} T : X_0 \to Y/S_0\|^{1-\theta} \|Q_{S_1} T : X_1 \to Y/S_1\|^{\theta}$$
$$\le C_{p\theta} (1+\varepsilon)^2 d_{k_0}(T : X_0 \to Y)^{1-\theta} d_{k_1}(T : X_1 \to Y)^{\theta}.$$

This completes the proof. $\square$

PROOF OF LEMMA 1.

Step 1. Let D be an optimal ellipsoid such that

$$d_k(D \to B) \le C(\alpha, p)(n/k)^{\alpha} \quad \text{and} \quad e_k(B \to D) \le C(\alpha, p)(n/k)^{\alpha}$$

for every $1 \le k \le n$.
 Let $\lambda = C(\alpha, p)(n/k)^{\alpha}$.

Step 2. Now denote the body $(B, D)_{\theta,1}$ by B_θ. By Claim 5 (applied for $k_0 = 1$), for every $1 \leq k \leq n$ we have

$$d_k(B_\theta \to B) \leq C_{p\theta}\|B \to B\|^{1-\theta}(d_k(D \to B))^\theta \leq C_{p\theta}\lambda^\theta.$$

It follows from the definition of entropy numbers that B is covered by 2^{k-1} translates of λD with centers in $\mathbb{R}^n$. Replacing λD with $2\lambda D$ we can choose these centers in B. Therefore there are 2^{k-1} points $x_i \in B$ $(1 \leq i \leq 2^{k-1})$ such that

$$B \subset \bigcup_{i=1}^{2^{k-1}} (x_i + 2\lambda D).$$

This means that for any $z \in B$ there is some $x_i \in B$ such that $\|z - x_i\|_D \leq 2\lambda$. Also, by p-convexity, $\|z - x_i\|_B \leq 2^{1/p}$. By taking the operator $u_x : \mathbb{R} \to X$, $u_x t = tx$ for some fixed x, and applying Claim 3 (or see [BL], [BS]) it is clear that

$$\|x\|_{B_\theta} \leq \|x\|_B^{1-\theta}\|x\|_D^\theta.$$

Hence, for any $z \in B$ there exists $x_i \in B$ such that

$$\|z - x_i\|_{B_\theta} \leq (2^{1/p})^{1-\theta}(2\lambda)^\theta,$$

that is,

$$e_k(B \to B_\theta) \leq 2^{(1-\theta)/p}(2\lambda)^\theta.$$

Thus, we obtain

$$d_k(B_\theta \to B) \leq C_{p\theta}\lambda^\theta \quad \text{and} \quad e_k(B \to B_\theta) \leq 2^\theta 2^{(1-\theta)/p}\lambda^\theta$$

for every $1 \leq k \leq n$.

LEMMA 2. *Let $B \subset \mathbb{R}^n$ be a p-convex ball and $D \subset \mathbb{R}^n$ be a convex body. Let $0 < \theta < 1$ and $B_\theta = (B, D)_{\theta,1}$. Then there exists a q-convex body B^q such that $B_\theta \subset B^q \subset 2^{1/q}B_\theta$, where $1/q - 1 = (1/p - 1)(1 - \theta)$.*

PROOF. Take the operator $U : l_1^2(\mathbb{R}^n) \to \mathbb{R}^n$ defined by $U((x, y)) = x + y$. Since

$$\|x + y\|_B \leq 2^{1/p-1}(\|x\|_B + \|y\|_B) \quad \text{and} \quad \|x + y\|_D \leq (\|x\|_D + \|y\|_D)$$

and by Claims 3, 4 we have

$$\|x + y\|_{B_\theta} \leq 2^{(1-\theta)(1/p-1)}(\|x\|_{B_\theta} + \|y\|_{B_\theta}).$$

But by the Aoki–Rolewicz theorem for every quasi-norm $\|\cdot\|$ with the constant C in the quasi-triangle inequality there exists a q-norm

$$\|\cdot\|_q = \inf\left\{\left(\sum_{i=1}^n \|x_i\|^q\right)^{1/q} \,\bigg|\, n > 0,\, x = \sum_{i=1}^n x_i\right\}$$

such that $\|x\|_q \leq \|x\| \leq 2C\|x\|_q$ with q satisfying $2^{1/q-1} = C$ ([KPR, R]; see also [K], p.47).

Thus, $B_\theta \subset B^q \subset 2^{1/q}B_\theta$, where B^q is a unit ball of q-norm $\|\cdot\|_q$. □

REMARK 2. Essentially, Lemma 2 goes back to Theorem 5.6.2 of [BL]. However, the particular case that we need is simpler and we are able to estimate the constant of equivalence.

Note that Lemma 2 can be easily extended to the more general case:

LEMMA 2'. *Let $B_i \subset \mathbb{R}^n$ be a p_i-convex bodies for $i = 0, 1$ and $B_\theta = (B_0, B_1)_{\theta,1}$. Then there exists a q-convex body B^q such that $B_\theta \subset B^q \subset 2^{1/q} B_\theta$, where*

$$\frac{1}{q} = \frac{1-\theta}{p_0} + \frac{\theta}{p_1}.$$

REMARK 3. N. Kalton pointed out to us that the interpolation body $(B, D)_{\theta,1}$ between a p-convex B and an ellipsoid D is equivalent to some q-convex body for any $q \in (0, 1]$ satisfying

$$1/q - 1/2 > (1/p - 1/2)(1 - \theta).$$

To prove this result one have to use methods of [Kal] and [KT]. Certainly, with growing q the constant of equivalence becomes worse.

Step 3. By definition of $C(\alpha, p)$ for B^q from Lemma 2 and $\gamma = \alpha(1 - \theta)$ there exists an ellipsoid D_1 such that for every $1 \leq k \leq n$

$$d_k(D_1 \to B^q) \leq C(\gamma, q)(n/k)^\gamma \text{ and } e_k(B^q \to D_1) \leq C(\gamma, q)(n/k)^\gamma.$$

By the ideal property of the numbers d_k, e_k and because of the inclusion $B_\theta \subset B^q \subset 2^{1/q} B_\theta$, for every $1 \leq k \leq n$

$$d_k(D_1 \to B_\theta) \leq 2^{1/q} C(\gamma, q)(n/k)^\gamma \text{ and } e_k(B_\theta \to D_1) \leq C(\gamma, q)(n/k)^\gamma.$$

Step 4. Let $a = 1 + [k(1 - \theta)]$. Using multiplicative properties of the numbers d_k, e_k we get

$$d_k(D_1 \to B) \leq d_{k+1-a}(D_1 \to B_\theta) d_a(B_\theta \to B)$$

$$\leq C_{p\theta} \lambda^\theta 2^{1/q} C(\gamma, q)(n/k)^\gamma \left(\frac{1}{(1-\theta)^{1-\theta} \theta^\theta}\right)^\alpha$$

$$\leq C(\alpha, p)^\theta \left(\frac{e}{1-\theta}\right)^{\alpha(1-\theta)} C_{p\theta} 2^{1/q} C(\gamma, q)(n/k)^\alpha$$

and

$$e_k(B \to D_1) \leq e_{k+1-a}(B \to B_\theta) e_a(B_\theta \to D_1)$$

$$\leq 2^\theta 2^{(1-\theta)/p} \lambda^\theta C(\gamma, q)(n/k)^\gamma \left(\frac{1}{(1-\theta)^{1-\theta} \theta^\theta}\right)^\alpha$$

$$\leq C(\alpha, p)^\theta \left(\frac{e}{1-\theta}\right)^{\alpha(1-\theta)} 2^\theta 2^{(1-\theta)/p} C(\gamma, q)(n/k)^\alpha.$$

By the minimality of $C(\alpha, p)$ and since $1/q \leq 1 + (1-\theta)/p$ we have

$$C(\alpha, p) \leq C(\alpha, p)^\theta \left(\frac{e}{1-\theta}\right)^{\alpha(1-\theta)} C_{p\theta} 2^{1-\theta/p} 2 C(\gamma, q)(n/k)^\alpha.$$

That proves Lemma 1. □

Acknowledgement

I want to thank Prof. E. Gluskin and Prof. V. Milman for their great help, and to thank Prof. N. Kalton for his useful remarks.

References

[BBP] J. Bastero, J. Bernués, and A. Peña, *An extension of Milman's reverse Brunn–Minkowski inequality*, GAFA **5**:3 (1995), 572–581.

[BS] C. Bennett and R. Sharpley, *Interpolation of Operators*, Academic Press, Orlando, 1988.

[BL] J. Bergh and J. Löfström, *Interpolation spaces, an introduction*, Springer, Berlin, 1976.

[BLM] J. Bourgain, J. Lindenstrauss, and V. Milman, *Estimates related to Steiner symmetrizations*, Geometrical aspects of functional analysis, Israel Seminar, 1987–88, Lect. Notes in Math. 1376, Springer, 264–273, 1989.

[C] B. Carl, *Entropy numbers, s-numbers, and eigenvalue problems*, J. Funct. Anal. **41**:3 (1981), 290–306.

[Kal] N. J. Kalton, *Convexity, type and the three space problem*, Studia Math. **69** (1981), 247–287.

[KPR] N. J. Kalton, N. T. Peck, and J. W. Roberts, *An F-space sampler*, London Mathematical Society Lecture Note Series, 89, Cambridge University Press, Cambridge, 1984.

[KT] N. Kalton and Sik-Chung Tam, *Factorization theorems for quasi-normed spaces*, Houston J. Math. **19** (1993), 301-317.

[K] H. König, *Eigenvalue distribution of compact operators*, Birkhäuser, 1986.

[M] V. D. Milman, *Inégalité de Brunn–Minkowsky inverse et applications á la théorie locale des espaces normés*, C. R. Acad. Sci. Paris Sér. 1 **302** (1986), 25–28.

[P] A. Pietch, *Operator ideals*, North-Holland, Berlin, 1979.

[Pi] G. Pisier, *The volume of convex bodies and Banach space geometry*, Cambridge University Press, Cambridge, 1989.

[R] S. Rolewicz, *Metric linear spaces*, Monografie Matematyczne 56, PWN, Warsaw, 1972.

ALEXANDER E. LITVAK
DEPARTMENT OF MATHEMATICS
TEL AVIV UNIVERSITY
RAYMOND AND BEVERLY SACKLER FACULTY OF EXACT SCIENCES
RAMAT AVIV, TEL AVIV 69978
ISRAEL
alexandr@math.tau.ac.il
Current address: Department of Mathematical Sciences, 632 CAB, University of Alberta, Edmonton, AB, Canada T6G 2G1 (alexandr@math.ualberta.ca)

The Extension of the Finite-Dimensional Version of Krivine's Theorem to Quasi-Normed Spaces

ALEXANDER E. LITVAK

ABSTRACT. In 1980 D. Amir and V. D. Milman gave a quantitative finite-dimensional version of Krivine's theorem. We extend their version of the Krivine's theorem to the quasi-convex setting and provide a quantitative version for p-convex norms.

Recently, a number of results of the Local Theory have been extended to the quasi-normed spaces. There are several works [Kal1, Kal2, D, GL, KT, GK, BBP1, BBP2, M2] where such results as Dvoretzky–Rogers lemma [DvR], Dvoretzky theorem [Dv1, Dv2], Milman's subspace-quotient theorem [M1], Krivine's theorem [Kr], Pisier's abstract version of Grotendick's theorem [P1, P2], Gluskin's theorem on Minkowski compactum [G], Milman's reverse Brunn–Minkowski inequality [M3], and Milman's isomorphic regularization theorem [M4] are extended to quasi-normed spaces after they were established for normed spaces. It is somewhat surprising since the first proofs of these facts substantially used convexity and duality.

In [AM2] D. Amir and V. D. Milman proved the local version of Krivine's theorem (see also [Gow], [MS]). They studied quantitative estimates appearing in this theorem. We extend their result to the q- and quasi-normed spaces.

Recall that a quasi-norm on a real vector space X is a map $\|\cdot\| : X \to \mathbb{R}^+$ satisfying these conditions:

(1) $\|x\| > 0$ for all $x \neq 0$.
(2) $\|tx\| = |t|\|x\|$ for all $t \in \mathbb{R}$ and $x \in X$.
(3) There exists $C \geq 1$ such that $\|x + y\| \leq C(\|x\| + \|y\|)$ for all $x, y \in X$.

If (3) is substituted by

(3a) $\|x + y\|^q \leq \|x\|^q + \|y\|^q$ for all $x, y \in X$, for some fixed $q \in (0, 1]$,

---

This research was supported by Grant No. 92–00285 from United States–Israel Binational Science Foundation (BSF).

then $\|\cdot\|$ is called a q-norm on X. Note that 1-norm is the usual norm. It is obvious that every q-norm is a quasi-norm with $C = 2^{1/q-1}$. However, not every quasi-norm is q-norm for some q. Moreover, it is even not necessary continuous. It can be shown by the following simple example. Let f be a positive function on the Euclidean sphere S^{n-1} defined by

$$f(x) = \begin{cases} |x| & \text{for } x \in A, \\ 2|x| & \text{otherwise.} \end{cases}$$

Here A is a subset of S^{n-1} such that both A and $S^{n-1} \setminus A$ are dense in S^{n-1}. Denote $\|x\| = |x| f(x/|x|)$. Because f is not continuous it is clear that $\|\cdot\|$ is not q-norm for any q though it is the quasi-norm.

The next lemma is the Aoki–Rolewicz Theorem ([KPR, R]; see also [K, p. 47]).

LEMMA 1. *Let $\|\cdot\|$ be a quasi-norm with the constant C in the quasi-triangle inequality. Then there exists a q-norm $\|\cdot\|$ for which*

$$\|x\|_q \leq \|x\| \leq 2C\|x\|_q$$

with q satisfying $2^{1/q-1} = C$. This q-norm can be defined as follows

$$\|x\|_q = \inf\left\{ \left(\sum_{i=1}^n \|x_i\|^q \right)^{1/q} : n > 0, \ x = \sum_{i=1}^n x_i \right\}.$$

We refer to [KPR] for further properties of the quasi- and q-norms.

THEOREM 1. *Let $\{e_i\}_1^n$ be a unit vector basis in $\mathbb{R}^n$, $\|\cdot\|_p$ be a l_p-norm on $\mathbb{R}^n$, i.e., $\|\sum_{i=1}^n a_i e_i\|_p = \left(\sum_i |a_i|^p \right)^{1/p}$, for $0 < p < \infty$. Let $\|\cdot\|$ be a q-norm on $\mathbb{R}^n$ such that*

$$C_1^{-1}\|x\|_p \leq \|x\| \leq C_2\|x\|_p \qquad (1)$$

for every $x \in \mathbb{R}^n$. Then for every $\varepsilon > 0$ and $C = C_1 C_2$ there exists a block sequence $u_1, u_2, \ldots, u_m$ of $e_1, e_2, \ldots, e_n$ which satisfies

$$(1-\varepsilon)\left(\sum_{i=1}^m |a_i|^p \right)^{1/p} \leq \left\| \sum_{i=1}^m a_i u_i \right\| \leq (1+\varepsilon)\left(\sum_{i=1}^m |a_i|^p \right)^{1/p} \qquad (2)$$

for all $a_1, a_2, \ldots, a_m$ and $m \geq C(\varepsilon, p, q) (n/\log n)^\nu$, where

$$\nu = \frac{\alpha \varepsilon_0}{\varepsilon_0 + p + \alpha \varepsilon_0} \quad \text{for } p < 1 \quad \text{and} \quad \nu = \frac{\varepsilon_0}{2\varepsilon_0 + 1} \quad \text{for } p \geq 1;$$

$$\alpha = \min\{p, q\}, \quad \varepsilon_0 = \left(\frac{q\varepsilon/2}{1 + C^q 12^{q/p}} \right)^{p/q}.$$

REMARK 1. If $p \geq 1$ in this theorem, then we have the well-known finite-dimensional version of Krivine's theorem with some modifications concerning change of the usual norm to the q-norm. In this case for small enough q we get $\varepsilon_0 \approx (q\varepsilon/4)^{p/q}$ and $\nu \approx \varepsilon_0$.

The case $p < 1$ is more interesting. We get an extension of the finite-dimensional version of Krivine's theorem. To provide an intuition for the behavior of the constant in the theorem we point out that for small enough p and q with $p = q$ we can take $\varepsilon_0 \approx q\varepsilon/30$ and $\nu \approx \varepsilon_0$.

REMARK 2. By Lemma 1 in the case of quasi-norm with the constant C_0 the inequality (2) is substituted with

$$(1-\varepsilon)\left(\sum_{i=1}^m |a_i|^p\right)^{1/p} \leq \left\|\sum_{i=1}^m a_i u_i\right\| \leq 2(1+\varepsilon)C_0 \left(\sum_{i=1}^m |a_i|^p\right)^{1/p}.$$

Due to the example above, we can not remove the constant C_0 in this inequality.

The proof of the theorem consists of two lemmas.

LEMMA 2. *For every $\eta > 0$ there exists a constant $C(\eta) > 0$ such that if $\|\cdot\|$ is a q-norm on $\mathbb{R}^n$ satisfying (1) then there exists a block sequence $y_1, y_2, \ldots, y_k$ of $e_1, e_2, \ldots, e_n$ which is $(1+\eta)$-symmetric and $k \geq C(\eta, q, p)n/\log n$.*

LEMMA 3. *If $y_1, y_2, \ldots, y_k$ is a 1-symmetric sequence in a normed space satisfying*

$$C_1^{-1}\|a\|_p \leq \left\|\sum_{i=1}^k a_i y_i\right\| \leq C_2 \|a\|_p$$

for all $a = (a_1, a_2, \ldots, a_k) \in \mathbb{R}^k$ then for every $\varepsilon > 0$ there exists a block sequence $u_1, u_2, \ldots, u_m$ of $y_1, y_2, \ldots, y_k$ such that

$$(1-\varepsilon)\|a\|_p \leq \left\|\sum_{i=1}^m a_i u_i\right\| \leq (1+\varepsilon)\|a\|_p$$

for all $a = (a_1, a_2, \ldots, a_m) \in \mathbb{R}^m$, where $m \geq C(p,q)\varepsilon^{p/q}k^\nu$,

$$\nu = \frac{\alpha \varepsilon_0}{\varepsilon_0 + p + \alpha \varepsilon_0} \text{ for } p < 1 \text{ and } \nu = \frac{\varepsilon_0}{2\varepsilon_0 + 1} \text{ for } p \geq 1,$$

$$\alpha = \min\{p, q\}, \quad \varepsilon_0 = \left(\frac{q\varepsilon}{1 + C^q 12^{q/p}}\right)^{p/q}.$$

At first, D. Amir and V. D. Milman ([AM2]; see also [MS]) proved Lemma 2 for $q = 1$, $p \geq 1$ with the estimate $k \geq C(\eta, q, p)n^{1/3}$. Their proof can be modified to obtain result for $0 < p < \infty$, $q \leq 1$. Afterwards, W. T. Gowers [Gow] showed that the estimate of k can be improved to $k \geq C(\eta, q, p)n/\ln n$. In fact, he gave two different, though similar, proofs for cases $p = 1$ and $p > 1$. The proof given for case $p = 1$ strongly used the convexity of the norm and the fact that p is equal to 1. However, the method used for $p > 1$ actually works for every $0 < p < \infty$ and even for q-norms. Let us recall the idea of W. T. Gowers. First we will introduce some definition.

Let Ω be the group $\{-1,1\}^n \times S_n$, where S_n is the permutation group. Let Ψ be the group $\{-1,1\}^k \times S_k$. For

$$b = \sum_{i=1}^n b_i e_i \in \mathbb{R}^n, \quad a = \sum_{i=1}^k a_i e_i \in \mathbb{R}^k, \quad (\varepsilon, \pi) \in \Omega, \quad (\eta, \sigma) \in \Psi$$

set

$$b_{\varepsilon\pi} = \sum_{i=1}^n \varepsilon_i b_i e_{\pi(i)}, \quad a_{\eta\sigma} = \sum_{i=1}^k \eta_i a_i e_{\sigma(i)}.$$

Let $h \cdot k = n$. For $i \leq k$, $j \leq h$ put

$$e_{ij} = e_{(i-1)h+j}, \quad \varepsilon_{ij} = \varepsilon_{(i-1)h+j}, \quad \pi_{ij} = \pi((i-1)h+j).$$

Define an action of Ψ on Ω by

$$\Psi_{\eta\sigma}((\varepsilon,\pi)) = (\varepsilon^1, \pi^1), \quad \text{where} \quad \varepsilon_{ij}^1 = \eta_i \varepsilon_{\sigma(i)j}, \quad \pi_{ij}^1 = \pi_{\sigma(i)j}.$$

For any $(\varepsilon, \pi) \in \Omega$ define the operator

$$\Phi_{\varepsilon\pi} : \mathbb{R}^k \to \mathbb{R}^n \quad \text{by} \quad \Phi_{\varepsilon\pi}\left(\sum_{i=1}^k a_i e_i\right) = \sum_{i=1}^k \sum_{j=1}^h \varepsilon_{ij} a_i e_{\pi_{ij}}.$$

For every $a \in \mathbb{R}^k$ by M_a denote the median of $\Phi_{\varepsilon\pi}(a)$ taken over Ω. Finally, let $A = \{a \in l_p^k : \|a\|_p \leq 1, a_1 \geq a_2 \geq \cdots \geq a_k \geq 0\}$.

The following claim, which W. T. Gowers proved for case $p > 1$ and $q = 1$, is the main step in the proof of Lemma 2.

CLAIM 1. *Let $\|\cdot\|$ be a q-norm on $\mathbb{R}^n$ satisfying $\|x\|_p \leq \|x\| \leq B\|x\|_p$. There is a constant $C_0 = C(p,q,\delta,B)$ such that given $\lambda > 0$ for every $a \in A$*

$$\mathbf{Prob}_\Omega \left\{ \exists (\eta, \sigma) : \left|\|\Phi_{\varepsilon\pi}(a_{\eta\sigma})\|^q - M_a^q\right|^{1/q} > \frac{1}{2^{1/q}} \delta \|a\|_p h^{1/p} \right\} < 1/N$$

with $k = C_0 \frac{n}{\lambda \log n}$ and $N = k^\lambda$.

The proof of this claim can be equally well applied for all $0 < p < \infty$ and $0 < q \leq 1$. The only change that we have to do is to replace the triangle inequality

$$\left|\|x\| - \|y\|\right| \leq \|x - y\| \quad \text{by} \quad \left|\|x\|^q - \|y\|^q\right|^{1/q} \leq \|x - y\|.$$

The following two claims are technical and can be proved using ideas of [Gow] with small changes, connected with replacing $p \geq 1$ by $p < 1$ and the norm by q-norm.

CLAIM 2. *Let $0 < p < \infty$ and $\delta > 0$. There exist a constant λ, depending on p and δ only, such that for every integer k the set A contains a δ-net K of cardinality k^λ.*

CLAIM 3. *Let $\|\cdot\|$ be a q-norm on $\mathbb{R}^n$ satisfying $\|x\|_p \leq \|x\| \leq B\|x\|_p$. If there is $(\varepsilon, \pi) \in \Omega$ such that for every a in some δ-net K of A*

$$\left|\|\Phi_{\varepsilon\pi}(a_{\eta\sigma})\|^q - \|\Phi_{\varepsilon\pi}(a_{\eta_1\sigma_1})\|^q\right|^{1/q} \leq \delta \|a\|_p h^{1/p}$$

for every $(\eta, \sigma), (\eta_1, \sigma_1) \in \Psi$ then the block basis

$$\{\Phi_{\varepsilon\pi}(e_i)\}_{i=1}^k$$

of $(\mathbb{R}^k, \|\cdot\|)$ is $(1 + 6(B\delta)^q)^{1/q}$-symmetric.

These three claims imply Lemma 2 in the standard way (see [Gow] for the details).

PROOF OF LEMMA 3. Our method of proof is close to the method used in [AM1], but our notation follows that of [MS, chapter 10].

First, we will give the Krivine's construction of block basis. Let a and N be some integers which will be specified later. Let us introduce some set of numbers $\{\lambda_j\}_J$. We will say that set

$$\{B_{j,i}\}_{j \in J, i \in I}$$

(if card $I = 1$ then we have only one index j) is $\{\lambda_j\}_J$-set if

(1) $B_{j,i} \subset \{1, \ldots, n\}$ for every $j \in J, i \in I$,
(2) $B_{j,i}$ are mutually disjoint,
(3) card $B_{j,i} = \lambda_j$ for every $j \in J$, $i \in I$.

Let us fix some $\{[\rho^j]\}$-set

$$\{A_{j,s}\}_{0 \leq j \leq N-1, 1 \leq s \leq m}$$

for $\rho = 1 + 1/a$.

For $0 \leq j \leq N - 1$ and $1 \leq s \leq m$, define

$$Y_{j,s} = \sum_{i \in A_{j,s}} y_i$$

and

$$z_s = \sum_{j=0}^{N-1} \rho^{(N-j)/p} Y_{j,s}.$$

Clearly, $\|z_1\| = \|z_2\| = \cdots = \|z_m\|$. The integer m will be defined from

$$k \approx m \sum_{j=0}^{N-1} [\rho^{(N-j)/p}] \approx m\rho^N (\rho - 1)^{-1} = ma\left(\frac{a+1}{a}\right)^N.$$

Finally, we define the block sequence $\{u_s\}_{s=1}^m$ by

$$u_s = z_s / \|z_s\|.$$

Now, as in [MS], we will establish the necessary estimates.

Fix $N, M \in \{T+1, T+2, \ldots, m\}$ and $t_s \in \{0, \ldots, T\}$ for $s \in \{1, \ldots, m\}$ such that
$$\sum_{s=1}^{M} \rho^{-t_s} = 1 + \eta, \quad \text{with } |\eta| = 1.$$
Then
$$\sum_{s=1}^{M} \rho^{-t_s/p} z_s = \sum_{s=1}^{M} \sum_{j=0}^{N-1} \rho^{(N-j-t_s)/p} Y_{j,s}$$
$$= \sum_{i=0}^{N-1+T} \rho^{(N-i)/p} \sum_{\substack{s \leq M,\ j \leq N-1 \\ j+t_s=i}} \sum_{l \in A_{j,s}} y_l = \sum_{i=0}^{N-1+T} \rho^{(N-i)/p} \sum_{l \in B_i} y_l$$
for some $\{a_i\}$-set $\{B_i\}_{i=0}^{N-1+T}$, where
$$a_i = \sum_{\substack{s \leq M,\ j \leq N-1 \\ j+t_s=i}} [\rho^{i-t_s}], \quad \text{for } 0 \leq i \leq N-1+T.$$

Therefore, we can choose a vector z which has the same structure as z_s (i.e., $z = \sum_{j=0}^{N-1} \rho^{(N-j)/p} \sum_{i \in A_j} y_i$ for some $\{[\rho^j]\}$-set $\{A_j\}_{0 \leq j \leq N-1}$) such that the difference Δ is
$$\Delta = \sum_{s=1}^{M} \rho^{-t_s/p} z_s - z = \sum_{s=1}^{N-1} \rho^{(N-i)/p} \sum_{l \in C_i} y_l + \sum_{s=N}^{N-1+T} \rho^{(N-i)/p} \sum_{l \in C_i} y_l$$
for some $\{b_j\}$-set $\{C_j\}_{i=0}^{N-1+T}$, where
$$b_j = \begin{cases} |[\rho^j - a_j]| & \text{for } 0 \leq j \leq N-1, \\ a_j & \text{for } N \leq j \leq N-1+T. \end{cases}$$

Using techniques from [MS, pp. 66-67] we obtain
$$\|\Delta\| \leq C_2 \rho^{N/p} (4T + N|\eta| + NM\rho^{-T})^{1/p} \quad \text{and} \quad \|z\| \geq (1/C_1) \rho^{N/p} (N/2)^{1/p}.$$
Hence
$$\left| \left\| \sum_{s=1}^{M} \rho^{-t_s/p} u_s \right\|^q - 1 \right| \leq \left\| \sum_{s=1}^{M} \rho^{-t_s/p} u_s - \frac{z}{\|z\|} \right\|^q$$
$$= \left(\frac{\|\Delta\|}{\|z\|} \right)^q \leq (C_1 C_2)^q \left(\frac{8T}{N} + 2|\eta| + 2M\rho^{-T} \right)^{q/p}.$$
Thus
$$\left| \left\| \sum_{s=1}^{M} \rho^{-t_s/p} u_s \right\|^q - 1 \right| \leq C^q (12\varepsilon_0)^{q/p},$$
provided $T \leq N\varepsilon_0$, $|\eta| \leq \varepsilon_0$, and $M\rho^{-T} \leq m\rho^{-T} \leq \varepsilon_0$, for some ε_0. Assume $T = [N\varepsilon_0]$.

Case 1: $p < 1$.

Let $\sum_{s=1}^{m} |\alpha_s|^p = 1$ and $a_s = |\alpha_s|$. Let $\alpha = \min\{p, q\}$ and $\delta = \varepsilon_0^{1/p}/m^{1/\alpha}$. Take $\beta_s = \rho^{-t_s/p}$ or $\beta_s = 0$, $t_s \in \{0, 1, \ldots, T\}$ such that $|a_s - \beta_s| \leq \delta$ for every s. It is possible if $\rho^{-T/p} \leq \delta$ and $1 - \rho^{-1/p} \leq \delta$. Since $p \leq 1$ it is enough to take a such that it satisfies following the inequalities

$$\left(\frac{a}{a+1}\right)^{[N\varepsilon_0]} \leq \delta^p = \frac{\varepsilon_0}{m^{p/\alpha}} \quad \text{and} \quad \delta \geq \frac{1}{p(a+1)}.$$

Take $a = [1/(\delta p)] = \left[m^{1/\alpha}/(p\varepsilon_0^{1/p})\right]$. Thus $\delta \geq \frac{1}{p(a+1)}$,

$$\left|\sum \rho^{-t_s} - 1\right| = \left|\sum \beta_s^p - 1\right| \leq \left|\sum (a_s + \delta)^p - 1\right|$$
$$\leq \left|\sum (a_s^p + \delta^p) - 1\right| = \delta^p m \leq \varepsilon_0$$

and

$$\left\|\sum_{s=1}^{m} \beta_s u_s\right\|^q - \left\|\sum_{s=1}^{m} \alpha_s u_s\right\|^q \leq \left\|\sum_{s=1}^{m} |\beta_s - a_s| u_s\right\|^q$$
$$\leq \delta^q \left\|\sum_{s=1}^{m} u_s\right\|^q \leq \delta^q m \leq \varepsilon_0^{q/p}.$$

Hence

$$\left|\left\|\sum_{s=1}^{m} \alpha_s u_s\right\|^q - 1\right| \leq \varepsilon_0^{q/p}(1 + C^q 12^{q/p}),$$

if $m^{p/\alpha} \leq \varepsilon_0(\frac{1+a}{a})^{[N\varepsilon_0]}$ and $ma(\frac{1+a}{a})^N \leq k$, when $a = \left[\frac{m^{1/\alpha}}{p\varepsilon_0^{1/p}}\right]$. Choose N such that $\left(\frac{a}{1+a}\right)^{N\varepsilon_0}$ is of the order $\varepsilon_0/m^{p/\alpha}$. Then

$$m \frac{m^{1/\alpha}}{p\varepsilon_0^{1/p}} \left(\frac{m^{p/\alpha}}{\varepsilon_0}\right)^{1/\varepsilon_0} = \frac{m^{1+1/\alpha+p/(\alpha\varepsilon_0)}}{\varepsilon_0^{1/p} p\varepsilon_0^{1/\varepsilon_0}} \sim k.$$

Thus, since $1/\alpha \geq \max\{1/p, 1/q\}$,

$$m \sim \varepsilon_0 (pk)^{\frac{\alpha\varepsilon_0}{\varepsilon_0+p+\alpha\varepsilon_0}} \sim \varepsilon_0 k^{\frac{\alpha\varepsilon_0}{\varepsilon_0+p+\alpha\varepsilon_0}}$$

and for $\varepsilon_1 = \varepsilon_0^{q/p}\left(1 + c^q 12^{q/p}\right)$

$$(1 - \varepsilon_1)^{1/q} \|(\alpha_s)\|_p \leq \left\|\sum \alpha_s u_s\right\| \leq (1 + \varepsilon_1)^{1/q} \|(\alpha_s)\|_p$$

holds. For ε_1 small enough ($\varepsilon_1 < 2^q - 1$) we obtain $1 - \varepsilon_1/q \leq (1 - \varepsilon_1)^{1/q}$ and $1 + 2\varepsilon_1/q \geq (1 + \varepsilon_1)^{1/q}$. Take $\varepsilon = 2\varepsilon_1/q$, then

$$\varepsilon_0 = \left(\frac{q\varepsilon/2}{1 + C^q 12^{q/p}}\right)^{p/q}$$

and

$$m \geq C(p, q)\varepsilon^{p/q} k^{\frac{\alpha\varepsilon_0}{\varepsilon_0+p+\alpha\varepsilon_0}}.$$

Case 2: $p \geq 1$. We use the same idea. Let $\sum_{s=1}^{m} |\alpha_s|^p = 1$ and $a_s = |\alpha_s|$. Let $\delta = \varepsilon_0/(C^p m)$. Take $\beta_s = \rho^{-t_s/p}$ or $\beta_s = 0$, $t_s \in \{0, 1, \ldots, T\}$ such that $|a_s^p - \beta_s^p| \leq \delta$ for every s. It is possible if $\rho^{-T} \leq \delta$ and $1 - \rho^{-1} \leq \delta$. These two conditions are met if

$$\left(\frac{a}{a+1}\right)^{[N\varepsilon_0]} \leq \delta = \frac{\varepsilon_0}{C^p m} \quad \text{and} \quad \delta \geq \frac{1}{a+1}.$$

Take $a = [1/\delta] = [C^p m/\varepsilon_0]$. Thus

$$\left|\sum \rho^{-t_s} - 1\right| = \left|\sum \beta_s^p - 1\right| \leq \left|\sum (a_s^p + \delta) - 1\right| = \delta m \leq \varepsilon_0.$$

Since

$$\left\|\sum_{s=1}^{m} u_s\right\| \leq C_1 C_2 \frac{\|\sum_{s=1}^{m} u_s\|_p}{\|z\|_p} \leq C_1 C_2 \left(\frac{m \sum \rho^{N-j}[\rho^j]}{\|z\|_p^p}\right)^{1/p} = C m^{1/p}$$

and

$$|\beta_s - a_s| \leq |\beta_s^p - a_s^p|^{1/p} \leq \delta^{1/p},$$

we obtain

$$\left|\left\|\sum_{s=1}^{m} \beta_s u_s\right\|^q - \left\|\sum_{s=1}^{m} \alpha_s u_s\right\|^q\right| \leq \left\|\sum_{s=1}^{m} |\beta_s - a_s| u_s\right\|^q$$

$$\leq \delta^{q/p} \left\|\sum_{s=1}^{m} u_s\right\|^q \leq \delta^{q/p} C^q m^{q/p} \leq \varepsilon_0^{q/p}.$$

Hence

$$\left|\left\|\sum_{s=1}^{m} \alpha_s u_s\right\|^q - 1\right| \leq \varepsilon_0^{q/p}(1 + C^q 12^{q/p}),$$

if $m \leq \frac{\varepsilon_0}{C^p}(\frac{1+a}{a})^{[N\varepsilon_0]}$ and $m a(\frac{1+a}{a})^N \leq k$, when $a = [C^p m/\varepsilon_0]$. Choose N such that $(\frac{a}{1+a})^{N\varepsilon_0}$ is of the order $\varepsilon_0/(C^p m)$. Then

$$m \frac{C^p m}{\varepsilon_0} \left(\frac{C^p m}{\varepsilon_0}\right)^{1/\varepsilon_0} = \left(\frac{C^p}{\varepsilon_0}\right)^{1+1/\varepsilon_0} m^{2+1/\varepsilon_0} \sim k.$$

Thus

$$m \geq \frac{\varepsilon_0}{C^p} k^{\frac{\varepsilon_0}{2\varepsilon_0 + 1}}$$

and, for $\varepsilon_1 = \varepsilon_0^{q/p}\left(1 + C^q 12^{q/p}\right)$,

$$(1 - \varepsilon_1)^{1/q} \|(\alpha_s)\|_p \leq \left\|\sum \alpha_s u_s\right\| \leq (1 + \varepsilon_1)^{1/q} \|(\alpha_s)\|_p$$

holds. For ε_1 small enough ($\varepsilon_1 < 2^q - 1$) we obtain $1 - \varepsilon_1/q \leq (1 - \varepsilon_1)^{1/q}$ and $1 + 2\varepsilon_1/q \geq (1 + \varepsilon_1)^{1/q}$. Take $\varepsilon = 2\varepsilon_1/q$, then

$$\varepsilon_0 = \left(\frac{q\varepsilon/2}{1 + C^q 12^{q/p}}\right)^{p/q}.$$

and
$$m \geq C(p,q)\varepsilon^{p/q} k^{\frac{\varepsilon_0}{2\varepsilon_0+1}}.$$
□

Acknowledgement

I thank Prof. V. Milman for his guidance and encouragement, and the referee for his helpful remarks.

References

[AM1] D. Amir and V. D. Milman, *Unconditional and symmetric sets in n-dimensional normed spaces*, Isr. J. Math. **37** (1980), 3–20.

[AM2] D. Amir and V. D. Milman, *A quantitative finite-dimensional Krivine theorem*, Isr. J. Math. **50** (1985), 1–12.

[BBP1] J. Bastero, J. Bernués, and A. Peña, *An extension of Milman's reverse Brunn–Minkowski inequality*, GAFA **5**:3 (1995), 572–581.

[BBP2] J. Bastero, J. Bernués, and A. Peña, *The theorem of Caratheodory and Gluskin for $0 < p < 1$*, Proc. Amer. Math. Soc. **123**:1 (1995), 141–144.

[D] S. J. Dilworth, *The dimension of Euclidean subspaces of quasi-normed spaces*, Math. Proc. Camb. Phil. Soc. **97** (1985), 311–320.

[Dv1] A. Dvoretzky, *A theorem on convex bodies and applications to Banach spaces*, Proc. Nat. Acad. Sci. USA **45** (1959), 223–226.

[Dv2] A. Dvoretzky, *Some results on convex bodies and Banach spaces*, Proc. Symp. Linear spaces, Jerusalem, 123–160, 1961.

[DvR] A. Dvoretzky and C. Rogers, *Absolute and unconditional convergence in normed linear spaces*, Proc. Nat. Acad. Sci. USA **36** (1950), 192–197.

[G] E. Gluskin, *The diameter of the Minkowski compactum is approximately equal to n*, Functional Anal. Appl. **15** (1981), 72–73.

[GK] Y. Gordon and N. J. Kalton, *Local structure theory for quasi-normed spaces*, Bull. Sci. Math. **118** (1994), 441–453.

[GL] Y. Gordon, D. R. Lewis, *Dvoretzky's theorem for quasi-normed spaces*, Illinois J. Math. **35** (1991), 250-259.

[Gow] W. T. Gowers, *Symmetric block bases in finite-dimensional normed spaces*, Isr. J. Math. **68** (1989), 193–219.

[Kal1] N. J. Kalton, *The convexity type of a quasi-Banach space*, unpublished note, 1977.

[Kal2] N. J. Kalton, *Convexity, type and the three space problem*, Studia Math. **69** (1981), 247–287.

[KPR] N. J. Kalton, N. T. Peck, and J. W. Roberts, *An F-space sampler*, London Mathematical Society Lecture Note Series, 89, Cambridge University Press, Cambridge, 1984.

[KT] N. Kalton and Sik-Chung Tam, *Factorization theorems for quasi-normed spaces*, Houston J. Math. **19** (1993), 301-317.

[K] H. König, *Eigenvalue distribution of compact operators*, Birkhäuser, 1986.

[Kr] J. L. Krivine, *Sous-espaces de dimension finie des espaces de Banach réticulés*, Ann. Math. **104** (1976), 1–29.

[M1] V. D. Milman, *Almost Euclidean quotient spaces of subspaces of a finite dimensional normed space*, Proc. Amer. Math. Soc. **94**:3 (1985), 445–449.

[M2] V. Milman, *Isomorphic Euclidean regularization of quasi-norms in $\mathbb{R}^n$*, C. R. Acad. Sci. Paris **321** (1996), 879-884.

[M3] V. D. Milman, *Inégalité de Brunn–Minkowsky inverse et applications á la théorie locale des espaces normés*, C. R. Acad. Sci. Paris Sér. 1 **302** (1986), 25–28.

[M4] V. Milman, *Some applications of duality relations*, GAFA Seminar 1989–90, Lecture Notes in Math. 1469, Springer, 1991, 13–40.

[MS] V. Milman, G. Schechtman, *Asymptotic theory of finite dimensional normed spaces*, Lecture Notes in Math. 1200, Springer, 1986.

[P1] G. Pisier, *Un théorème sur les opérateurs entre espaces de Banach qui se factorisent par un espace de Hilbert*, Ann. École Norm. Sup. **13** (1980), 23–43.

[P2] G. Pisier, *Factorization of linear operators and geometry of Banach spaces*, NSF-CBMS Regional Conference Series 60, AMS, Providence, 1986.

[R] S. Rolewicz, *Metric linear spaces*, Monografie Matematyczne 56, PWN, Warsaw, 1972.

ALEXANDER E. LITVAK
DEPARTMENT OF MATHEMATICS
TEL AVIV UNIVERSITY
RAYMOND AND BEVERLY SACKLER FACULTY OF EXACT SCIENCES
RAMAT AVIV, TEL AVIV 69978
ISRAEL
  alexandr@math.tau.ac.il

  Current address: Department of Mathematical Sciences, 632 CAB, University of Alberta, Edmonton, AB, Canada T6G 2G1 (alexandr@math.ualberta.ca)

A Note on Gowers' Dichotomy Theorem

BERNARD MAUREY

ABSTRACT. We present a direct proof, slightly different from the original, for an important special case of Gowers' general dichotomy result: If X is an arbitrary infinite dimensional Banach space, either X has a subspace with unconditional basis, or X contains a hereditarily indecomposable subspace.

The first example of dichotomy related to the topic discussed in this note is the classical combinatorial result of Ramsey: for every set A of pairs of integers, there exists an infinite subset M of $\mathbb{N}$ such that, either every pair $\{m_1, m_2\}$ from M is in A, or no pair from M is in the set A. There exist various generalizations to "infinite Ramsey theorems" for sets of finite or infinite sequences of integers, beginning with the result of Nash-Williams [NW]: for any set A of finite increasing sequences of integers, there exists an infinite subset M of $\mathbb{N}$ such that either no finite sequence from M is in A, or every infinite increasing sequence from M has some initial segment in A (although it does not look so at the first glance, notice that the result is symmetric in A and A^c, the complementary set of A; for further developments, see also [GP], [E]). The first naive attempt to generalize this result to a vector space setting would be to ask the following question: given a normed space X with a basis, and a set A of finite sequences of blocks in X (i.e., finite sequences of vectors $(x_1, \ldots, x_k)$ where $x_1, \ldots, x_k \in X$ are successive linear combinations from the given basis), does there exist a vector subspace Y of X spanned by a block basis, such that either every infinite sequence of blocks from Y has some initial segment in A, or no finite sequence of blocks from Y belongs to A, up to some obviously necessary perturbation involving the norm of X. It turns out that the answer to this question is negative, as a consequence of the existence of *distortable spaces*, like Tsirelson's space [T]. A correct vector generalization requires a more delicate statement, which in particular is not symmetric in A and A^c. Gowers' dichotomy theorem is such a result; in its first form [G1], this theorem is about sets of finite sequences of blocks in a normed space, and it was later extended in [G2] to analytic sets of infinite sequences of blocks. We will not state these general results here, in particular we will not describe the very interesting "vector game" that seems necessary for expressing Gowers' theorem. The first striking application of this result (probably the one for which the combinatorial result was proved) is an application to the *unconditional basic sequence problem*. This problem asks whether it is possible to find in a given Banach space X an infinite unconditional basic sequence (x_n), or, in equivalent terms, an infinite dimensional subspace Y of X with an unconditional basis. The answer is negative for some spaces X, as was shown in [GM1]. Furthermore, the example in [GM1] has a property which seems rather extreme in the direction opposite to an unconditional behaviour: this space X is *Hereditarily Indecomposable*, or H.I. for short. This means that no vector subspace of X is the topological direct sum of two infinite dimensional subspaces (and of course X is infinite dimensional).

It was natural to investigate more closely the connection between the failure of the unconditional basic sequence property and the H.I. property. Was it just accidental if the first example of a space not containing any infinite unconditional sequence was actually a H.I. space? Gowers' result completely clarifies the situation.

THEOREM 1 (SPECIAL CASE OF GOWERS' DICHOTOMY THEOREM). *Let X be an arbitrary infinite dimensional Banach space. Either X has a subspace with unconditional basis, or X contains a H.I. subspace.*

Let us mention that there exist non trivial examples of non H.I. spaces not containing any infinite unconditional basic sequence (see [GM2]; on the other hand, trivial examples of this situation are simply obtained by considering spaces of the form $X \oplus X$, with an H.I. space X). Recall that Theorem 1 above, together with the results by Komorowski and Tomczak [KT] gave a positive solution to the *homogeneous Banach space problem*, which appeared in Banach's book [B] sixty years before: if a Banach space X is isomorphic to all its infinite dimensional closed subspaces, then X is isomorphic to the Hilbert space ℓ_2.

The purpose of this note is to present a variant for a direct proof of this important special case of Gowers' general dichotomy result. It is of course not essentially different from the original argument in [G1], and the attentive reader will easily detect several steps here that are very similar to some parts of [G1], for example our Lemma 2 below and its Corollary. Our main intention is to give a more geometric exposition. We shall try to gather all the easy geometric information that we need before embarking for the central part of the argument, which is the combinatorial part.

We begin with some notation and definitions. For any normed space X we denote by $S(X)$ the unit sphere of X. The notation Y, Z, or U, V will be used for infinite dimensional vector subspaces of X, and E, F, G for finite dimensional subspaces of X. Given a real number $C \geq 1$, a finite or infinite sequence (e_n) of non zero vectors in a normed space X is called C-*unconditional* if

$$\left\|\sum \varepsilon_i a_i e_i\right\| \leq C \left\|\sum a_i e_i\right\|$$

for any sequence of signs $\varepsilon_i = \pm 1$ and any finitely supported sequence (a_i) of scalars. An infinite sequence (e_n) of non zero vectors is called *unconditional* if it is C-unconditional for some C. It is usual to normalize the sequence (e_n) by the condition $\|e_n\| = 1$ for each integer n, but this is unimportant here.

An infinite dimensional normed space X is called *Hereditarily Indecomposable* (in short H.I.) if for any infinite dimensional vector subspaces Y and Z of X,

(1) $$\inf\{\|y - z\| : y \in S(Y), z \in S(Z)\} = 0.$$

It is easy to check that this property is equivalent to the fact that no subspace of X is the topological direct sum of two infinite dimensional subspaces Y and Z. Property (1) says that the *angle* between any two infinite dimensional subspaces of X is equal to 0. This notion of angle will be discussed with more details below.

In order to compare easily the H.I. property and the unconditionality property, we rephrase unconditionality in terms of angle of subspaces. Saying that (e_n) is C-unconditional is of course equivalent to saying that

$$\left\|\sum a_i e_i\right\| \leq C \left\|\sum \varepsilon_i a_i e_i\right\|$$

for all signs (ε_i) and all scalars (a_i) (we just moved the signs to the other side). For any finite subset K of the set of indices, let E_K denote the linear span of $(e_k)_{k \in K}$. Consider a linear combination $\sum \varepsilon_i a_i e_i$, let $I = \{i : \varepsilon_i = 1\}$ and $J = \{i : \varepsilon_i = -1\}$. Letting $x = \sum_{i \in I} a_i e_i \in E_I$ and $y = \sum_{i \in J} a_i e_i \in E_J$ we may restate the above inequality as

$$\|x + y\| \leq C \|x - y\|$$

for all $x \in E_I$ and $y \in E_J$, whenever I and J are disjoint. This is again an angle property. There are however several ways for measuring the angle between two subspaces, and we want to introduce two of them. For any L, M finite or infinite dimensional subspaces of X, we denote by $a(L, M)$ the measure of the angle between L and M given by

$$a(L, M) = \inf\{\|x - y\| : x \in S(L), y \in S(M)\}.$$

This expression is symmetric, decreasing in L and M, and (Lipschitz-) continuous for the metric $\delta(L, M)$ given by the Hausdorff distance between the unit spheres $S(L)$ and $S(M)$,

$$\delta(L, M) = \max\{\sup\{d(x, S(M)) : x \in S(L)\}, \sup\{d(y, S(L)) : y \in S(M)\}\}.$$

An equivalent expression for the angle is

$$b(L, M) = \inf\{\inf\{d(x, M) : x \in S(L)\}, \inf\{d(y, L) : y \in S(M)\}\}.$$

It is clear that $b(L,M) \leq a(L,M)$; in the other direction we have $a(L,M) \leq 2b(L,M)$. To see this, let $b > b(L,M)$, and assume for example that $b(L,M) = \inf\{d(x,M) : x \in S(L)\}$. Let $x \in S(L)$ and $u \in M$ be such that $\|x-u\| < b$, hence $1-b < \|u\| < 1+b$. Letting $u' = u/\|u\|$, we have $\|u-u'\| = |\|u\|-1| < b$, and $d(x, S(M)) \leq \|x - u'\| < 2b$.

According to the above discussion, we see that a sequence $(e_n)_{n \in \mathbb{N}}$ of non zero vectors in X is unconditional iff there exists $\beta > 0$ such that

(2) $$b(\mathrm{span}\{e_n : n \in I\}, \mathrm{span}\{e_n : n \in J\}) \geq \beta$$

whenever I and J are finite disjoint subsets of $\mathbb{N}$. The relations between β and the unconditional constant of the sequence (e_n) are as follows: given $\beta > 0$ with the above property, the sequence (e_n) is C-unconditional with $C \leq 2/\beta$. Conversely, if the sequence (e_n) is C-unconditional, then (2) is true with $\beta \geq 2/(C+1)$.

Let us check these two facts. Suppose first that (2) is true for some $\beta > 0$. If I and J are disjoint, and $x \in E_I$, $y \in E_J$, we see that

$$\|x+y\| \leq \frac{2}{\beta} \|x-y\|,$$

proving that (e_n) is $2/\beta$-unconditional. Indeed, suppose that $\|x\| = 1 \geq \|y\|$; we know that $\|x-y\| \geq b(E_I, E_J) \geq \beta$ and $\|x+y\| \leq 2$, and the inequality above follows by homogeneity. Conversely, if (e_n) is C-unconditional, the projection $P_I : \sum a_i e_i \to \sum_{i \in I} a_i e_i$ on E_I has norm $\leq (C+1)/2$ for any subset I of the set of indices, and this implies that $\|x\| \leq \frac{C+1}{2}\|x-y\|$, hence we may choose $\beta = 2/(C+1)$.

The following easy technical Lemma will be used in the proof of Theorem 2 below.

LEMMA 1. *Assume that E, E' are finite dimensional subspaces of X, M any subspace of X and Z an infinite dimensional subspace of X. We have*

$$\sup_{U \subset Z} a(E' + U, M) \leq \sup_{U \subset Z} a(E + U, M) + 2\delta(E', E),$$

where the supremum above runs over all infinite dimensional subspaces U of Z.

PROOF. Let

$$s > \sup_{V \subset Z} a(E + V, M), \quad \delta = \delta(E', E),$$

$t > 1$ and let U be any infinite dimensional subspace of Z. By a standard argument, we may find an infinite dimensional subspace $U' \subset U$ such that $t\|e + u'\| \geq \|e\|$ for every $e \in E$ and $u' \in U'$ (we intersect U with the kernels of a finite set of functionals forming a t^{-1}-norming set for E). By assumption we have $a(E + U', M) < s$, hence we can find $e + u' \in S(E + U')$ and $y \in S(M)$ such that $\|(e + u') - y\| < s$. We know then that $\|e\| \leq t$, thus there exists

$e' \in E'$ such that $\|e' - e\| \le t\delta$. Now $1 - t\delta \le \|e' + u'\| \le 1 + t\delta$ and we can find $x \in S(E' + U')$ such that $\|x - (e' + u')\| \le t\delta$. Finally,
$$a(E' + U, M) \le a(E' + U', M) \le \|x - y\| < s + 2t\delta,$$
ending the proof. $\square$

So far we did not say if our normed spaces are real or complex, and everything above applies to both cases. In the complex case however, it is customary to define the *complex unconditional constant* by replacing in the definition above the signs $\varepsilon_i = \pm 1$ by arbitrary complex numbers of modulus one. This makes no essential difference, because a sequence of vectors in a complex normed space is complex-unconditional iff it is real-unconditional, except that the complex unconditional constant may differ from the real constant by some factor (less than 3 say). We shall therefore work with the real definition of the unconditional constant.

We introduce the intermediate notion of a $HI(\varepsilon)$ space. Given $\varepsilon > 0$, an infinite dimensional normed space X will be called a $HI(\varepsilon)$ space if for every infinite dimensional subspaces Y and Z of X we have
$$a(Y, Z) \le \varepsilon.$$
Obviously, a normed space X is H.I. iff it is $HI(\varepsilon)$ for every $\varepsilon > 0$.

THEOREM 2. *Let X be an infinite dimensional normed space. For each $\varepsilon > 0$, either X contains an infinite sequence with unconditional constant $\le 4/\varepsilon$, or X contains a $HI(\varepsilon)$ subspace Z.*

Of course, when X does not contain any infinite sequence with unconditional constant $\le 4/\varepsilon$, this implies that *every* infinite dimensional subspace Y of X contains a $HI(\varepsilon)$ subspace. Theorem 2 implies Theorem 1 by a simple diagonalization procedure that already appears in [G1]: assume that X does not contain any infinite unconditional sequence; by Theorem 2, every subspace Y of X contains for each $\varepsilon > 0$ a subspace Z which is $HI(\varepsilon)$. Taking successively $\varepsilon = 2^{-n}$, we construct a decreasing sequence (Z_n), where Z_n is a $HI(2^{-n})$ subspace of X. Let Z be a subspace obtained from the sequence (Z_n) by the diagonal procedure. For every n, this space Z is contained in Z_n up to finitely many dimensions, therefore Z is $HI(\varepsilon)$ for every $\varepsilon > 0$, so Z is H.I.

PROOF OF THEOREM 2. We may clearly restrict our attention to separable spaces X. Let (E, F) be a couple of finite dimensional subspaces of X and let Z be an infinite dimensional subspace of X. We set
$$A(E, F, Z) = \sup_{U, V \subset Z} a(E + U, F + V),$$
where the supremum is taken over all infinite dimensional subspaces U and V of Z. It follows from Lemma 1 that $A(E', F', Z) \le A(E, F, Z) + 2\delta(E', E) +$

$2\delta(F', F)$ for all finite dimensional subspaces E' and F'. We will keep $\varepsilon > 0$ fixed throughout the proof.

We introduce a convenient terminology, inspired by [GP]. We say that the couple (E, F) *accepts* the subspace Z if

$$A(E, F, Z) < \varepsilon.$$

This perhaps unnatural strict inequality is necessary for approximation reasons. Indeed, we get from Lemma 1 that when (E, F) accepts a subspace Z of X, then (E', F') also accepts Z provided $\delta(E', E)$ and $\delta(F', F)$ are small enough. When (E, F) accepts Z, we know that $a(E + U, F + V) < \varepsilon$ for all infinite dimensional subspaces U and V of Z, and except for the small technicality just mentioned, this is exactly the idea that the reader should keep in mind. Before going any further, let us notice that when the couple $(\{0\}, \{0\})$ accepts a subspace Z, then Z is $HI(\varepsilon)$ (actually, Z is then $HI(\varepsilon')$ for some $\varepsilon' < \varepsilon$). Acceptance is clearly symmetric: (F, E) accepts Z iff (E, F) accepts Z. If (E, F) accepts Z, it also accepts every $Z' \subset Z$ (obvious) and every $Z + G$, when $\dim G < +\infty$; this last fact is easy: given two infinite dimensional subspaces U, V of $Z + G$, we may consider the two infinite dimensional subspaces $U' = U \cap Z$ and $V' = V \cap Z$ of Z; since (E, F) accepts Z, we have

$$a(E + U, F + V) \leq a(E + U', F + V') \leq A(E, F, Z) < \varepsilon.$$

Notice that what we just did was proving the equality $A(E, F, Z) = A(E, F, Z + G)$, which is one of the main ingredients for the proof: we are dealing here with a function of Z that does not depend upon changing finitely many dimensions.

We say that a couple $\tau = (E, F)$ *rejects* Z if no subspace $Z' \subset Z$ is accepted by τ. Rejection is also symmetric, and saying that (E, F) rejects some subspace Z (or simply: does not accept Z) implies that

$$a(E, F) \geq \varepsilon$$

because $a(E, F) \geq a(E + U, F + V)$ for all U, V, hence $a(E, F) \geq A(E, F, Z)$ for every Z. This yields $b(E, F) \geq \frac{1}{2}a(E, F) \geq \varepsilon/2$ and will be used in connection with the property (2) for $\beta = \varepsilon/2$, in order to produce an upper bound $4/\varepsilon$ for the unconditional constant. This notion of rejection will therefore be the tool for constructing inductively subspaces with an angle bounded away from 0; the strength of the rejection hypothesis will allow the induction to run. Observe that when a couple τ rejects a subspace Z, it is clearly true by definition that τ rejects every subspace Z' of Z, and τ also rejects "supspaces" of Z of the form $Z + G$, when G is finite dimensional (otherwise, τ would accept some $Z' \subset Z + G$, hence also accept $Z'' = Z' \cap Z$, contradicting the fact that τ rejects Z); combining the above observations, we see that when τ accepts or rejects Z, the same is true for every Z' such that $Z' \subset Z + G$, when G is any finite dimensional subspace of X. This simple remark is the basis for our first step. Since X was assumed separable, we may select a countable family $\mathcal{E}$, dense in the set of finite dimensional subspaces of X (for the Hausdorff metric of spheres).

CLAIM 1. *There exists an infinite dimensional subspace Z_0 of X such that for every couple (E, F) with $E, F \in \mathcal{E}$ and every rational α in $(0, \varepsilon)$, either $A(E, F, Z_0) < \alpha$ or, for every infinite dimensional subspace Z' of Z_0, we have $A(E, F, Z') \geq \alpha$.*

PROOF. We use a very usual diagonal argument. Let $(\sigma_n)_{n \geq 1}$, with $\sigma_n = (E_n, F_n, \alpha_n)$ be a listing of all triples (E, F, α) such that $E, F \in \mathcal{E}$ and α is a rational number in $(0, \varepsilon)$. We construct a decreasing sequence $(X_n)_{n \geq 0}$ of subspaces of X in the following way: $X_0 = X$, and if $A(E_{n+1}, F_{n+1}, Z') \geq \alpha_{n+1}$ for every subspace Z' of X_n, we simply let $X_{n+1} = X_n$. Otherwise, there exists a subspace of X_n, which we call X_{n+1}, such that $A(E_{n+1}, F_{n+1}, X_{n+1}) < \alpha_{n+1}$. We consider then a diagonal infinite dimensional subspace Z_0 which is the linear span of a sequence $(z_n)_{n \geq 1}$ built by picking inductively z_{n+1} in X_{n+1} and not in the linear span of $z_1, z_2, \ldots, z_n$. For each integer $n \geq 1$, we see that $Z_0 \subset X_n + G_n$ for some finite dimensional subspace G_n, and either

$$A(E_n, F_n, Z_0) \leq A(E_n, F_n, X_n + G_n) = A(E_n, F_n, X_n) < \alpha_n,$$

or for every subspace Z' of Z_0, $A(E_n, F_n, Z') = A(E_n, F_n, Z' \cap X_n) \geq \alpha_n$. □

By an easy approximation argument, we can state a version of Claim 1 above that will apply to any couple τ, and not only to those from the dense subset $\mathcal{E}$. Let (E, F) be an arbitrary couple. If (E, F) does not reject Z_0, it accepts some $Z' \subset Z_0$ and we may choose a rational α in $(0, \varepsilon)$ such that $A(E, F, Z') < \alpha$; let β be rational and $0 < \beta < (\varepsilon - \alpha)/8$; let $E', F' \in \mathcal{E}$ be such that $\delta(E', E) < \beta$ and $\delta(F', F) < \beta$. This implies by Lemma 1 that $A(E', F', Z') < \alpha + 4\beta < \varepsilon$. But then by Claim 1 it follows that $A(E', F', Z_0) < \alpha + 4\beta$; by approximation again (E, F) accepts Z_0. Finally:

CLAIM 2. *For each couple (E, F) of finite dimensional subspaces of X, either (E, F) rejects Z_0 or (E, F) accepts Z_0.*

From now on the whole construction will be performed inside our "stabilizing" subspace Z_0. Here is where the dichotomy really starts. There are two possibilities: either the couple $(\{0\}, \{0\})$ accepts Z_0, or it rejects. As was mentioned before, saying that $(\{0\}, \{0\})$ accepts Z_0 implies that Z_0 is $HI(\varepsilon)$. Suppose now that $(\{0\}, \{0\})$ rejects Z_0; we will find in Z_0 a sequence $(e_k)_{k \geq 1}$ with unconditional constant $C \leq 4/\varepsilon$. This will be done in the following manner: we will choose the sequence (e_k) of non zero vectors in such a way that for each $n \geq 1$ and for all disjoint sets $I, J \subset \{1, \ldots, n\}$, the couple (E_I, E_J) rejects Z_0 (as before, we denote by E_K the linear span of $\{e_k : k \in K\}$). The next Lemma and its Corollary give the tool for constructing the next vector e_{n+1} of our unconditional sequence, when $e_1, \ldots, e_n$ are already selected.

LEMMA 2. *If (E, F) rejects Z_0, then for every infinite dimensional subspace Z' of Z_0 there exists a further infinite dimensional subspace $U' \subset Z'$ such that for every finite dimensional subspace E' of U', the couple $(E + E', F)$ rejects Z_0.*

PROOF. Otherwise, there exists $Z' \subset Z_0$ such that, for every subspace $U' \subset Z'$, there exists $E' \subset U'$ such that $(E + E', F)$ does not reject Z_0. We know that $(E + E', F)$ accepts Z_0 by Claim 2; for every subspace $V' \subset Z'$ we have, since $E + U' = E + E' + U'$,

$$a(E + U', F + V') = a(E + E' + U', F + V') \leq A(E + E', F, Z_0) < \varepsilon,$$

which implies that (E, F) accepts Z', therefore (E, F) accepts Z_0 by Claim 2, contrary to the initial hypothesis. □

COROLLARY 1. *Suppose that $(E_\alpha, F_\alpha)_{\alpha \in A}$ is a finite family of couples, and that (E_α, F_α) rejects Z_0 for each $\alpha \in A$. For every infinite dimensional subspace Z'' of Z_0 there exists a further infinite dimensional subspace $U'' \subset Z''$ such that for every finite dimensional subspace E' of U'', the couple $(E_\alpha + E', F_\alpha)$ rejects Z_0 for each $\alpha \in A$.*

PROOF. We set $A = \{\alpha_1, \ldots, \alpha_p\}$. Let $Z'' = Z'_0$ be a subspace of Z_0. By Lemma 2, there exists $U' = Z'_1 \subset Z'_0$ such that for every $E' \subset Z'_1$, $(E_{\alpha_1} + E', F_{\alpha_1})$ rejects Z_0. We apply again Lemma 2, this time to the couple $(E_{\alpha_2}, F_{\alpha_2})$, with $Z' = Z'_1$, and so on until $U'' = Z'_p \subset Z'_{p-1} \subset \ldots \subset Z'$ is reached. □

The Corollary will be applied in the following weakened form; the notation $[z]$ stands for the line $\mathbb{R}z$ or $\mathbb{C}z$ generated by a non zero vector z:

Suppose that $(E_\alpha, F_\alpha)_{\alpha \in A}$ is a finite family of couples, and that (E_α, F_α) rejects Z_0 for each $\alpha \in A$. For every infinite dimensional subspace Z' of Z_0 there exists a non zero vector $z \in Z'$ such that the couple $(E_\alpha + [z], F_\alpha)$ rejects Z_0 for each $\alpha \in A$.

Let us finish the proof of the Theorem. Recall that E_K denotes the linear span of $(e_k)_{k \in K}$. Assuming that $(\{0\}, \{0\})$ rejects Z_0, we build by induction a sequence $(e_k)_{k=1}^\infty$ of non zero vectors in Z_0, such that for every integer $n \geq 1$ and every partition (I, J) of $\{1, \ldots, n\}$, the couple (E_I, E_J) rejects Z_0. Assuming that $e_1, \ldots, e_n$ are already selected, let us call *partition of length n* any couple of the form (E_I, E_J), for some partition (I, J) of $\{1, \ldots, n\}$. Consider the finite list A_n of all partitions (E_α, F_α) of length n, where $\alpha \in A_n$. Our induction hypothesis is that for every $\alpha \in A_n$, the couple (E_α, F_α) rejects Z_0; let Z' be an infinite dimensional subspace of Z_0 such that $Z' \cap \text{span}\{e_1, \ldots, e_n\} = \{0\}$. By the Corollary, we can find a non zero vector $z \in Z'$ such that for every $\alpha \in A_n$, the couple $(E_\alpha + [z], F_\alpha)$ rejects Z_0; observe that (F_α, E_α) also belongs to the list, hence $(F_\alpha + [z], E_\alpha)$ also rejects Z_0. We choose now $e_{n+1} = z$. It is clear that with this choice, (E_I, E_J) rejects Z_0 for every partition of length $n+1$. This implies that the infinite sequence $(e_k)_{k \geq 1}$ satisfies property (2) with the constant $\beta = \varepsilon/2$, thus this sequence is $4/\varepsilon$-unconditional. □

REMARK. It is possible to obtain directly a H.I. space, without passing through the intermediate stage of $HI(\varepsilon)$ spaces, by replacing the study of couples by that of triples (E, F, ε), for ε varying. The first version of this paper was indeed written in this way, but the referee said (and was probably right about it) that the earlier version in [M] gave a clearer view of the combinatorics, by dealing first with a countable situation (a countable vector space over $\mathbb{Q}$) and treating the boring approximation afterwards. This version is a sort of midpoint between the two, which perhaps only adds the disadvantages of both...

References

[B] S. Banach, Théorie des opérations linéaires, Warszawa, 1932.

[E] E. Ellentuck, *A new proof that analytic sets are Ramsey*, J. Symbolic Logic **39** (1974), 163–165.

[GP] F. Galvin and K. Prikry, *Borel sets and Ramsey's theorem*, J. Symbolic Logic **38** (1973), 193–198.

[G1] W. T. Gowers, *A new dichotomy for Banach spaces*, Geom. Funct. Anal. **6** (1996), 1083–1093.

[G2] W. T. Gowers, *Analytic sets and games in Banach spaces*, preprint IHES M/94/42.

[G3] W. T. Gowers, *Recent results in the theory of infinite dimensional Banach spaces*, Proceedings ICM 1994, Birkhäuser 1995.

[GM1] W. T. Gowers and B. Maurey, *The unconditional basic sequence problem*, Jour. Amer. Math. Soc. **6** (1993), 851–874.

[GM2] W. T. Gowers and B. Maurey, *Banach spaces with small spaces of operators*, Math. Ann. **307** (1997), 543–568.

[KT] R. Komorowski and N. Tomczak-Jaegermann, *Banach spaces without local unconditional structure*, Israel J. Math. **89** (1995), 205–226.

[M] B. Maurey, *Quelques progrès dans la compréhension de la dimension infinie*, Journée Annuelle 1994, Société Mathématique de France.

[NW] C. St J. A. Nash-Williams, *On well quasi-ordering transfinite sequences*, Proc. Cam. Phil. Soc. **61** (1965), 33–39.

[T] B. S. Tsirelson, *Not every Banach space contains ℓ_p or c_0*, Funct. Anal. Appl. **8** (1974), 139–141.

BERNARD MAUREY
EQUIPE D'ANALYSE ET MATHÉMATIQUES APPLIQUÉES
UNIVERSITÉ DE MARNE LA VALLÉE
5 BOULEVARD DESCARTES, CHAMPS SUR MARNE
77454 MARNE LA VALLÉE CEDEX 2
FRANCE
  maurey@math.univ-mlv.fr

An "Isomorphic" Version of Dvoretzky's Theorem, II

VITALI MILMAN AND GIDEON SCHECHTMAN

ABSTRACT. A different proof is given to the result announced in [MS2]: For each $1 \leq k < n$ we give an upper bound on the minimal distance of a k-dimensional subspace of an arbitrary n-dimensional normed space to the Hilbert space of dimension k. The result is best possible up to a multiplicative universal constant.

Our main result is the following extension of Dvoretzky's theorem (from the range $1 < k < c \log n$ to $c \log n \leq k < n$), first announced in [MS2, Theorem 2]. As is remarked in that paper, except for the absolute constant involved the result is best possible.

THEOREM. *There exists a $K > 0$ such that, for every n and every $\log n \leq k < n$, any n-dimensional normed space, X, contains a k-dimensional subspace, Y, satisfying*

$$d(Y, \ell_2^k) \leq K \sqrt{\frac{k}{\log(1+n/k)}}.$$

In particular, if $\log n \leq k \leq n^{1-K\varepsilon^2}$, there exists a k-dimensional subspace Y (of an arbitrary n-dimensional normed space X) with

$$d(Y, \ell_2^k) \leq \frac{K}{\varepsilon} \sqrt{\frac{k}{\log n}}.$$

Jesus Bastero pointed out to us that the proof of the theorem in [MS2] works only in the range $k \leq cn/\log n$. Here we give a different proof which corrects this oversight. The main addition is a computation due to E. Gluskin (see the proof of the Theorem in [Gl1] and the remark following the proof of Theorem 2 in [Gl2]). In the next lemma we single out what we need from Gluskin's argument and sketch Gluskin's proof.

Partially supported by BSF and NSF. Part of this research was carried on at MSRI..

GLUSKIN'S LEMMA. *Let $1 \leq k \leq n/2$ and let $\nu_{n,k}$ denote the normalized Haar measure on the Grassmannian of k-dimensional subspaces of $\mathbb{R}^n$. Then, for some absolute positive constant c,*

$$\nu_{n,k}\left(\left\{E : \exists x \in E \text{ with } \|x\|_\infty < c\sqrt{\frac{\log(1+n/k)}{n}}\|x\|_2\right\}\right) < 1/2.$$

PROOF. Let $g_{i,j}$, for $i = 1,\ldots,k$ and $j = 1,\ldots,n$, be independent standard Gaussian variables. The invariance of the Gaussian measure under the orthogonal group implies that the conclusion of the lemma is equivalent to

$$\text{Prob}\left(\inf_{x \in S^{k-1}} \max_{1 \leq j \leq n} \frac{|\sum_{i=1}^k x_i g_{i,j}|}{\left(\sum_{l=1}^n \left(\sum_{i=1}^k x_i g_{i,l}\right)^2\right)^{1/2}} < c\sqrt{\frac{\log(1+n/k)}{n}}\right) < 1/2.$$

As is well known, the variable $\left(\sum_{j=1}^n \left(\sum_{i=1}^k x_i g_{i,j}\right)^2\right)^{1/2}$ is well concentrated near the constant $\|x\|_2 = \left(\sum_{i=1}^k x_i^2\right)^{1/2}\sqrt{n}$. In particular,

$$\text{Prob}\left(\exists x \in S^{k-1} \text{ with } \left(\sum_{j=1}^n \left(\sum_{i=1}^k x_i g_{i,j}\right)^2\right)^{1/2} > 2\sqrt{n}\right) < 1/4$$

if n is large enough. It is thus enough to prove that

$$\text{Prob}\left(\inf_{x \in S^{k-1}} \max_{1 \leq j \leq n} \left|\sum_{i=1}^k x_i g_{i,j}\right| < 2c\sqrt{\log(1+n/k)}\right) < 1/4.$$

The left-hand side here is clearly dominated by

$$\text{Prob}\left(\inf_{x \in S^{k-1}} \left(\sum_{j=1}^{2k} \left|\sum_{i=1}^k x_i g_{i,j}\right|^{*2}\right)^{1/2} < 2c\sqrt{2k\log(1+n/k)}\right)$$

where $\{a_i^*\}$ denotes the decreasing rearrangement of the sequence $\{|a_i|\}$. To estimate the last probability we use the usual deviation inequalities for Lipschitz functions of Gaussian vectors (see, for example, [MS1] or [Pi]). The only two facts one should notice are that the norm $\|a\| = (\sum_{j=1}^{2k} a_j^{*2})^{1/2}$ on $\mathbb{R}^n$ is dominated by the Euclidean norm, i.e. that the Lipschitz constant of the function $\|a\| = (\sum_{j=1}^{2k} a_j^{*2})^{1/2}$ on $\mathbb{R}^n$ is at most one and that the expectation of $\|(g_1,\ldots,g_n)\|$ is larger than $c_1(2k)^{1/2}\sqrt{\log(1+n/k)}$ for some absolute constant c_1. Then, for an appropriate c,

$$\text{Prob}\left(\inf_{x \in S^{k-1}} \left(\sum_{j=1}^{2k} \left|\sum_{i=1}^k x_i g_{i,j}\right|^{*2}\right)^{1/2} < 2c\sqrt{2k\log(1+n/k)}\right)$$
$$< \exp(-k\log(1+n/k)) < 1/4$$

if n is large enough. □

PROOF OF THE THEOREM. The proof follows in parts the proof in [MS2]. For completeness we shall repeat these parts. In what follows $0 < \eta, K < \infty$ denote absolute constants, not necessarily the same in each instance. By a result of Bourgain and Szarek ([BS, Theorem 2]; but refer to [MS2; Remark 4] for an explanation why we need only a much simpler form of their result), we may assume without loss of generality that there exists a subspace, $Z \subseteq X$, with $m = \dim Z > n/2$ and $\alpha \|x\|_{\ell_\infty^m} \leq \|x\|_Z \leq \|x\|_{\ell_2^m}$ for all $x \in Z$ for some absolute constant $\alpha > 0$.

Let M denote the median of $\|x\|_Z$ over $S^{m-1} = S^{n-1} \cap Z$. Fix k as in the statement of the theorem. If $M > K\sqrt{\frac{k}{n}}$ then, by [Mi] (see also [FLM] or [MS1, Theorem 4.2]), Z and thus X contains a k-dimensional subspace, Y, satisfying $d(Y, \ell_2^k) \leq 2$. Also, for each k, with probability $> 1 - e^{-\eta k}$

$$(*) \qquad \|x\| \leq 2\left(M + K\sqrt{\frac{k}{n}}\right)\|x\|_2.$$

This again follows by the usual deviation inequalities. Let us refresh the reader's memory: Let $\mathcal{M}$ be a $\frac{1}{2}$-net in the sphere of a fixed k-dimensional subspace, Y_0, of Z with $|\mathcal{M}| \leq 6^k$ (see [MS1, Lemma 2.6]). Denoting by ν the Haar measure on the orthogonal group $O(m)$ we get,

$$\nu\left(\left\{U \; ; \; \|Ux\|_Z \geq M + K\sqrt{\frac{k}{n}} \text{ for some } x \in \mathcal{M}\right\}\right) \leq \exp(k \log 6 - \eta K^2 k).$$

Thus, a successive approximation argument gives that, with probability larger than $1 - e^{-\eta k}$, a k-dimensional subspace E of Z satisfies

$$\|x\| \leq 2\left(M + K\sqrt{\frac{k}{n}}\right)\|x\|_2, \quad \text{for all } x \in E.$$

By Gluskin's Lemma, for $k < n/4$,

$$\nu_{m,k}\left(\left\{E \; ; \; \exists x \in E \text{ with } \|x\| < c\alpha\sqrt{\frac{\log(1+n/k)}{n}}\|x\|_2\right\}\right)$$
$$\leq \nu_{m,k}\left(\left\{E \; ; \; \exists x \in E \text{ with } \|x\|_\infty < c\sqrt{\frac{\log(1+m/k)}{m}}\|x\|_2\right\}\right) < 1/2.$$

That is, with probability larger than $1/2$ a k-dimensional subspace E satisfies

$$\|x\| \geq c\alpha\sqrt{\frac{\log(1+n/k)}{n}}\|x\|_2, \quad \text{for all } x \in E.$$

Combining this with $(*)$, we conclude that, if $M \leq K\sqrt{k/n}$, there exists a k-dimensional subspace whose distance to Euclidean space is smaller than

$$4K\sqrt{\frac{k}{n}} \bigg/ c\alpha\sqrt{\frac{\log(1+n/k)}{n}} = K'\sqrt{k/\log(1+n/k)}. \qquad \square$$

REMARK. Can one show that the conclusion of the Theorem holds for a random subspace Y? The only obstacle in the proof here and also in [MS2] is the use of the result from [BS]. Michael Schmuckenschläger showed us how to overcome this obstacle in the proof of [MS2]: Instead of using [BS] one can use [ScSc, Proposition 4.11], which says that any multiple of the ℓ_∞^n unit ball has larger or equal Gaussian measure than the same multiple of the unit ball of any other norm on $\mathbb{R}^n$ whose ellipsoid of maximal volume is S^{n-1}. This can replace the first inequality on the last line of [MS2, p. 542] (with $m = n$ and $\alpha = 1$. The change from the Gaussian measure to the spherical measure is standard.) It follows that at least for $k \leq cn/\log n$ the answer to the question above is positive.

What is the "isomorphic" version of Dvoretzky's Theorem for spaces with non-trivial cotype? It is known that in this case one has a version of Dvoretzky's theorem with a much better dependence of the dimension of the Euclidean section on the dimension of the space ([FLM] or see [MS1, 9.6]). We do not know if one can extend this theorem in a similar "isomorphic" way as the theorem above. The proposition below gives such an extension under the additional assumption that the space also has non-trivial type. Recall that it is a major open problem whether an n-dimensional normed space with non-trivial cotype has a subspace of dimension $[n/2]$ which is of type 2 (with the type 2 constant depending on the cotype and the cotype constant only) or at least of some non-trivial type. If this open problem has a positive solution, the next proposition would imply the desired "isomorphic" cotype case of the theorem. The proof we sketch here (as well as the statement of the proposition) uses quite a lot of background material (which can be found in [MS1]) and is intended for experts.

PROPOSITION. *For every n and every $n^{2/q} \leq k < n/2$, any n-dimensional normed space, X, contains a k-dimensional subspace, Y, satisfying $d(Y, \ell_2^k) \leq Kk^{1/2}/n^{1/q}$. Here K depends on $q < \infty$, the cotype q constant of X and the norm of the Rademacher projection in $L_2(X)$ only. Up to the exact value of the constant involved the result is best possible and is attained for $X = \ell_q^n$.*

SKETCH OF PROOF. We use the notations of [MS1]. We first find an operator $T : \ell_2^n \to X$ for which

$$\ell(T)\ell(T^{-1*}) \leq Kn,$$

where K depends on the norm of the Rademacher projection in $L_2(X)$ only (see [MS1, 15.4.1]).

Next we use the "lower bound theorem" of Milman [MS1, 4.8] to find an $3n/4$ dimensional subspace $E \subseteq \ell_2^n$ for which

$$\|(T_{|E})^{-1}\| \leq C\ell(T^{-1*})/\sqrt{n},$$

for an absolute constant C.

By a theorem of Figiel and Tomczak (see [FT] or [MS1, 15.6]), there exists a further subspace $F \subseteq E$ of dimension larger than $n/2$ for which

$$\|T_{|F}\| \leq Kn^{-1/2}n^{1/2-1/q}\ell(T_{|E}) \leq Kn^{-1/q}\ell(T),$$

for K depending on q and the cotype q constant of X only.

This reduces the problem to the following: Given a norm $\|\cdot\|$ on $\mathbb{R}^{n/2}$ for which

$$C^{-1}\|x\|_2 \leq \|x\| \leq Kn^{1/2-1/q}\|x\|_2$$

for all x, for constants C, K depending only on p, q and the type p and cotype q constants of X, and for which $M = \int_{S^{n/2-1}} \|x\| = 1$, find a subspace Y of dimension k as required in the statement of the proposition. Since we have to take care of the upper bound only, this can be accomplished by the usual "concentration" method as described in the first few chapters of [MS1].

The fact that, for some absolute constant η and for all k, ℓ_q^n does not have k-dimensional subspaces of distance smaller than $\eta q^{-1/2}k^{1/2}/n^{1/q}$ to ℓ_2^k follows from the method developed in [BDGJN] (or see [MS1, 5.6]). One just need to replace the constant 2 in [MS1, 5.6] by a general constant d and follow the proof to get a lower bound on d. □

References

[BDGJN] G. Bennett, L. E. Dor, V. Goodman, W. B. Johnson, and C. M. Newman, *On uncomplemented subspaces of L_p, $1 < p < 2$*, Israel J. Math. **26** (1977), 178–187.

[BS] J. Bourgain and S. J. Szarek, *The Banach–Mazur distance to the cube and the Dvoretzky–Rogers factorization*, Israel J. Math. **62** (1988), 169–180.

[FLM] T. Figiel, J. Lindenstrauss, and V. D. Milman, *The dimension of almost spherical sections of convex bodies*, Acta. Math. **139** (1977), 53–94.

[FT] T. Figiel and N. Tomczak-Jaegermann, *Projections onto Hilbertian subspaces of Banach spaces*, Israel J. Math. **33** (1979), 155–171.

[Gl1] E. D. Gluskin, *The octahedron is badly approximated by random subspaces*, Funct. Anal. Appl. **20** (1986), 11–16.

[Gl2] E. D. Gluskin, *Extremal properties of orthogonal parallelepipeds and their applications to the geometry of Banach spaces*, Math. USSR Sbornik **64** (1989), 85–96.

[Mi] V. D. Milman, *A new proof of the theorem of A. Dvoretzky on sections of convex bodies*, Funct. Anal. Appl. **5** (1971), 28–37 (translated from Russian).

[MS1] V. D. Milman and G. Schechtman, *Asymptotic theory of finite dimensional normed spaces*, Lect. Notes in Math. **1200** (1986), Springer-Verlag, Berlin.

[MS2] V .D. Milman and G. Schechtman, *An "isomorphic" version of Dvoretzky's theorem*, C. R. Acad. Sci. Paris Sér. I **321** (1995), 541–544.

[Pi] G. Pisier, *The volume of convex bodies and Banach space geometry* (1989), Cambridge Univ. Press.

[ScSc] G. Schechtman and M. Schmuckenschläger, *A concentration inequality for harmonic measures on the sphere*, Operator Theory: Advances and Applications **77** (1995), 255–273.

VITALI MILMAN
DEPARTMENT OF MATHEMATICS
TEL AVIV UNIVERSITY
RAYMOND AND BEVERLY SACKLER FACULTY OF EXACT SCIENCES
RAMAT AVIV, TEL AVIV 69978
ISRAEL
   vitali@math.tau.ac.il

GIDEON SCHECHTMAN
DEPARTMENT OF THEORETICAL MATHEMATICS
THE WEIZMANN INSTITUTE OF SCIENCE
REHOVOT
ISRAEL
   gideon@wisdom.weizmann.ac.il

Asymptotic Versions of Operators and Operator Ideals

VITALI MILMAN AND ROY WAGNER

ABSTRACT. The goal of this note is to introduce new classes of operator ideals, and, moreover, a new way of constructing such classes through an application to operators of the asymptotic structure recently introduced by Maurey, Milman, and Tomczak-Jaegermann in *Op. Th. Adv. Appl.* **77** (1995), 149–175.

1. Preliminaries

1.1. Notation. We follow standard Banach-space theory notation, as outlined in [LTz].

Throughout this note X will be an infinite dimensional Banach-space with a shrinking basis $\{e_i\}_{i=1}^\infty$. The notation $[X]_n$ will stand for the head subspace (span$\{e_i\}_{i=1}^n$) and $[X]_{>n}$ for the tail subspace (the closure of span$\{e_i\}_{i=n+1}^\infty$). P_n and $P_{>n}$ are the coordinate-orthogonal projections on these subspaces respectively.

A basis $\{e_i\}$ is *equivalent* to a basis $\{f_i\}$ if

$$C_2 \left\| \sum a_i f_i \right\| \leq \left\| \sum a_i e_i \right\| \leq C_1 \left\| \sum a_i f_i \right\|$$

for any scalars $\{a_i\}$. The equivalence is quantified by the ratio C_1/C_2; the closer it is to 1, the better the equivalence.

A vector is called *a block* if it has *finite support*, that is, if it is a finite linear combination of elements of the basis. The blocks v and w are said to be *consecutive* ($v < w$) if the *support* of v (the set of elements of the basis that form v as a linear combination) ends before the support of w begins. $S(X)^n_<$ is the collection of all n-tuples of consecutive normalised blocks; thus $S(X)^1_<$ means normalised finite support vectors.

1.2. An intuitive introduction to asymptotic structure.
The language of asymptotic structure has been introduced to study the essentially infinite dimensional structure of Banach-spaces, and to help bridge between finite dimensional and infinite dimensional theories. This approach does generalise spreading models, but takes an essentially different view. Formally introduced in [MMiT], it has already been studied, extended and applied in [KOS], [OTW], [T], [W1] and [W2]; it is closely related to the new surge of results in infinite dimensional Banach-space theory, and especially to [G] and [MiT]. For formal arguments, the most convenient terminology is the game terminology coming from [G], presented below. However, we choose to preface this by a less rigorous intuitive introduction.

The main idea behind this theory is a stabilisation at infinity of finite dimensional objects (subspaces, restrictions of operators), which repeatedly appear arbitrarily far and arbitrarily spread out along the basis. Piping these stabilised objects together gives rise to infinite dimensional notions: *asymptotic versions* of a Banach-space X or of an operator acting on X.

To define this structure we first have to choose a frame of reference in the form of a family of subspaces, $\mathcal{B}(X)$. It is most convenient to choose $\mathcal{B}(X)$ such that the intersection of any two subspaces from $\mathcal{B}(X)$ is in $\mathcal{B}(X)$. The family of tail subspaces is such a family; so is the family of finite codimensional subspaces, but here we will work with the former. The construction proceeds as follows.

Fix n and ε. Consider the tail subspace $[X]_{>N_1}$ for some "very large" N_1, and take a normalised vector in this tail subspace. Consider now a further tail subspace $[X]_{>N_2}$, with N_2 "very large", depending on the choice of x_1; choose again any normalised vector, x_2 in $[X]_{>N_2}$. After n steps we have a sequence of n vectors, belonging to a chain of tail subspaces, each subspace chosen 'far enough' with respect to the previous vectors.

The span of a sequence in X, $E = \text{span}\{x_1, \ldots, x_n\}$, is called ε-*permissible* if we can produce by the above process vectors $\{y_i\}_{i=1}^n$, which are $(1+\varepsilon)$-equivalent to the basis of E, regardless of the choice of tail subspaces.

Now we can explain how far is "far enough". The choice of the tail subspaces $[X]_{>N_i}$ is such that no matter what normalised blocks are chosen inside them, they will always form ε-permissible spaces. The existence of such "far enough" choices of tail subspaces is proved by a compactness argument.

We can now consider basic sequences which are $(1+\varepsilon)$-equivalent to ε-permissible sequences for every ε. These will be called n dimensional *asymptotic spaces*. Our ε-permissible sequences are $(1+\varepsilon)$-realizations in X of asymptotic spaces.

Finally, a Banach-space whose every head-subspace is an asymptotic space of X is called an *asymptotic version* of X.

The same construction can be made for an operator T as well. In this case, we would like to stabilise not only the domain (which is an asymptotic space), but also the image and action of the operator. More precisely, our permissible sequences will now be sequences $\{x_i\}_{i=1}^n$, such that we can find arbitrarily far and arbitrarily spread out sequences in our space, which are closely equivalent to $\{x_i\}_{i=1}^n$, and on top of that, are mapped by T to sequences closely equivalent to $\{T(x_i)\}_{i=1}^n$. The assumption that X has a shrinking basis promises that $\{T(x_i)\}_{i=1}^n$ are close to successive blocks. In turn this means that the asymptotic version of T, viewed as an operator from $[x_i]_{i=1}^n$ to $\left[T(x_i)/\|T(x_i)\|\right]_{i=1}^n$, is always a formally diagonal operator.

1.3. The Game. The following game (up to slight formalities) is used in [MMiT] to define asymptotic spaces. The game terminology comes from [G].

DEFINITION 1. This is a game for two players. One is the subspace player, $\mathcal{S}$, and the other is the vector player, $\mathcal{V}$. The "board" of the game consists of a Banach-space with a basis, a natural number n, and two subsets of $S(X)_<^n$: Φ and Σ. Player $\mathcal{S}$ begins, and they play n turns. In the first turn player $\mathcal{S}$ chooses a tail subspace, $[X]_{>m_1}$. Player $\mathcal{V}$ then chooses a normalised block in this subspace, $x_1 \in [X]_{>m_1}$. In the k-th turn, player $\mathcal{S}$ chooses a tail subspace $[X]_{>m_k}$. Player $\mathcal{V}$ then chooses a normalised block, x_k, such that $x_k \in [X]_{>m_k}$ and $x_k > x_{k-1}$.

$\mathcal{V}$ wins if it produces a sequence of vectors in Φ.

$\mathcal{S}$ wins if it forces the choices of $\mathcal{V}$ to be in Σ.

Note that in this game it is not always true that one player wins and the other loses. Furthermore, in some cases, we are only interested in the winning prospects of one player, and therefore may ignore either Φ or Σ.

If $\mathcal{V}$ has a *winning strategy* in this game for Φ, that is a recipe for producing sequences in Φ considering any possible moves of $\mathcal{S}$, we call Φ *an asymptotic set of length* n. Formally this means:

$$\forall m_1 \; \exists x_1 \in X_{>m_1} \; \forall m_2 \; \exists x_2 \in X_{>m_2} \ldots \forall m_n \; \exists x_n \in X_{>m_n}$$

$$\text{such that } (x_1, \ldots, x_n) \in \Phi.$$

Note that this generalises an earlier notion of an asymptotic set for $n = 1$ (in this context of tail subspaces, rather than block subspaces; compare [GM]).

If $\mathcal{S}$ has a winning strategy in this game for the collection Σ, we call Σ *an admission set of length* n. Formally this means:

$$\exists m_1 \; \forall x_1 \in X_{>m_1} \; \exists m_2 \; \forall x_2 \in X_{>m_2} \ldots \exists m_n \; \forall x_n \in X_{>m_n}$$

$$\text{such that } (x_1, \ldots, x_n) \in \Sigma$$

This terminology comes from admissibility criteria in the study of Tsirelson's space and its variants. An admission set contains all vector sequences beginning "far enough" and spread out "far enough" — for some interpretation of the term "enough".

When the context is clear, we will omit the length, and simply write "an asymptotic (admission) set".

REMARKS 2. (i) Note that a collection containing an admission set is an admission set, and that a collection containing an asymptotic set is an asymptotic set.

(ii) It is also useful to note that if $\mathcal{V}$ has a winning strategy for Φ, and $\mathcal{S}$ has a winning strategy for Σ, then playing these strategies against each other will necessarily produce sequences in $\Phi \cap \Sigma$.

In fact, in such a case $\mathcal{V}$ has a winning strategy for $\Phi \cap \Sigma$. Player $\mathcal{V}$'s strategy is as follows; player $\mathcal{V}$ plays his winning strategy for Φ, while pretending that the subspace dictated to him in each turn is the intersection of the subspace actually chosen by $\mathcal{S}$ with the subspace arising from the winning strategy for Σ.

The following is simply formal negation.

LEMMA 3. *Let $\Sigma \subseteq S(X)_<^n$. Then either Σ is an admission set, or Σ^c is an asymptotic set. These options are mutually exclusive.*

Tail subspaces of a Banach-space form a filter. This allows us to demonstrate filter (cofilter) behaviour for admission (asymptotic) sets.

LEMMA 4. (i) *Let $\Sigma_1, \ldots, \Sigma_k \subseteq S(X)_<^n$ be admission sets. Then $\bigcap_{j=1}^k \Sigma_j$ is also an admission set.*
(ii) *Let $\Phi_1, \ldots, \Phi_k \subseteq S(X)_<^n$, such that $\bigcup_{j=1}^k \Phi_j$ is asymptotic. Then, for some $1 \leq j \leq k$, Φ_j is asymptotic.*

PROOF. (i) Suppose at any turn of the game player $\mathcal{S}$ has to choose $[X]_{>m_1}$ in order to win for Σ_1, $[X]_{>m_2}$ in order to win for $\Sigma_2, \ldots$, and $[X]_{>m_k}$ in order to win for Σ_k. If player $\mathcal{S}$ chooses $[X]_{>m}$, where $m = \max\{m_1, m_2, \ldots, m_k\}$, the vector sequence chosen by player $\mathcal{V}$ will have to be in $\bigcap_{j=1}^k \Sigma_j$.

(ii) Suppose for all $1 \leq j \leq k$, Φ_j is not asymptotic. Then Lemma 3 implies that Φ_j^c are all admission sets. By part 1 of the proof, $\bigcap_{j=1}^k \Phi_j^c$ is an admission set. Therefore, using Lemma 3 again, $\bigcup_{j=1}^k \Phi_j$ is not asymptotic, in contradiction. □

1.4. Asymptotic versions. The following notions come from [MMiT].

DEFINITION 5. *An n-dimensional Banach space F with a basis $\{f_i\}_{i=1}^n$ is called an asymptotic space of a Banach-space X, if for every $\varepsilon > 0$ the set of all sequences in $S(X)_<^n$, which are $(1+\varepsilon)$-equivalent to $\{f_i\}_i$, is an asymptotic set.*

A space $\tilde{X}$ is an asymptotic version of X, if all the spaces $[\tilde{X}]_n$ are asymptotic spaces of X.

We use $\{X\}_\infty$ to denote the collection of asymptotic versions of a space X, and $\{X\}_n$ to denote the collection of its n-dimensional asymptotic spaces.

REMARK 6. The existence of asymptotic versions and spaces for every Banach-space X is elementary, as observed in [MMiT]. Spreading models make obvious examples of asymptotic versions. Existence of some special asymptotic versions is dealt with in [MMiT]. In our existence theorem below we will prove that we can extract close realizations of an asymptotic version out of any sequence of increasingly long asymptotic sets.

1.5. König's Unendlichkeitslemma. The following combinatorial lemma comes from [Kö]. Its proof is an elementary exercise. A *rooted tree* is simply a connected tree with some vertex labelled as *root*. The root allows us to consider the *level* of vertices in the tree.

LEMMA 7. *A rooted tree has an infinite branch emanating from the root if*

(i) *there are vertices arbitrarily far from the root, and*
(ii) *the set of vertices with any fixed distance to the root is finite.*

This Lemma will be used to allow us to find asymptotic versions not only inside the whole space, but also inside asymptotic subsets. This in turn will be used to show, that if the collections of sequences with a certain property is asymptotic, an asymptotic version with the said property can be extracted.

2. Asymptotic Versions of Operators

DEFINITION 8. Let $T \in \mathcal{L}(X)$. Define $\tilde{T}$, *an asymptotic version of* T, to be a formally diagonal operator between asymptotic versions $\tilde{Y}, \tilde{Z} \in \{\tilde{X}\}_\infty$, such that for every $n \in \mathbb{N}$ and every $\varepsilon > 0$, the following set is asymptotic:

All sequences in $S(X)^n_<$, which are $(1+\varepsilon)$-equivalent to the basis of $[\tilde{Y}]_n$, and whose images under T are $(1+\varepsilon)$-equivalent to the images of the basis of $[\tilde{Y}]_n$ under $\tilde{T}$.

A collection of such asymptotic sets, for arbitrarily small ε's and arbitrarily large n's will be referred to as *asymptotic sets realizing the asymptotic version* $\tilde{T}$.

The set of all asymptotic versions of an operator T will be denoted $\{T\}_\infty$.

Note that the asymptotic versions of the identity operator correspond simply to asymptotic versions of the space.

We are about to prove the basic existence theorem for asymptotic versions of operators. In order to formulate it, we need the following technical terminology.

DEFINITION 9. A *truncation (of length k)* of a given collection, $\Sigma \subseteq S(X)^n_<$, is the collection of sequences of the leading k blocks from all sequences in Σ.

A truncation of an asymptotic set is obviously asymptotic.

THEOREM 10 (THE EXISTENCE THEOREM). *Let X be a space with a shrinking basis. For every operator $T \in \mathcal{L}(X)$ and every sequence $\{\Phi_n\}_{n=1}^\infty$ of increasingly long asymptotic sets, there exists $\tilde{T} \in \{T\}_\infty$ realized by* **truncated** *subsets of the Φ_n's.*

Before we prove the theorem, let us isolate the part of the proof which requires a shrinking basis.

LEMMA 11. *Let X be a space with a shrinking basis, and let $T \in \mathcal{L}(X)$. Let Σ be an admission set. The collection of all sequences, which T maps ε-close to sequences from Σ, is also an admission set.*

PROOF. As a first step we have to verify that, if a block is supported far enough, its image under T will be — up to an $\varepsilon/2$-perturbation — the first block of some sequence from Σ. This sounds reasonable, recalling that Σ is an admission set, and so any block which is supported sufficiently far can play the first vector of a sequence from Σ.

To verify, let $[X]_{>n_1}$ be the first step in the winning strategy of player $\mathcal{S}$ for Σ. We know that all blocks x_1 supported far enough have $\|P_{n_1}(T(x_1))\| < \varepsilon/2$. Indeed, if this weren't the case, we would have $\|P_{n_1}(T(x_1^k))\| \geq \varepsilon/2$ for a sequence $\{x_1^k\}_{k=1}^{\infty}$ of normalised vectors supported increasingly far (and hence weakly null). This would imply that one of the bounded functionals $P_{\{e_i\}}(T(x))$, $1 \leq i \leq n_1$, does not go to zero when applied to a sequence of weakly null vectors — a contradiction. Therefore, as long as x_1 is supported far enough, $T(x_1)$ fits (up to $\varepsilon/2$) as the first vector from a sequence in Σ. The first step is accomplished.

Let now $[X]_{>n_2}$ be the second step in the winning strategy of player $\mathcal{S}$ for Σ, given that player $\mathcal{V}$ has just chosen the appropriate perturbation of $T(x_1)$. The above reasoning shows that if x_2 is supported far enough, it's image under T will be supported (up to an $\varepsilon/4$ perturbation) after n_2, that is, far enough to fit as the second vector of a sequence from Σ, which begins with our slight perturbation of $T(x_1)$.

Repeating this argument we see, that if a sequence of vectors is sufficiently spread out (each vector is supported sufficiently far with respect to its predecessors), the image of that sequence will be a sequence of essentially consecutive vectors, ε-close to a sequence from Σ. We have thus proved that our collection is indeed an admission collection, and we're through. □

REMARK 12. In the proof we will apply the last Lemma to the collection of close realizations of asymptotic spaces in X. The paper [MMiT] explains that the collection $\Sigma_{n,\varepsilon}(X)$ of all block sequences of length n, which are $(1+\varepsilon)$-equivalent to asymptotic spaces, is indeed an admission set (we will point out a proof for this claim in Remark 13 below).

PROOF OF THE EXISTENCE THEOREM. The proof will split into three parts. First we have to make sure that we can restrict T to operate between asymptotic spaces. To do this we will extract from the Φ_n's asymptotic subsets of block sequences, whose normalised images under T are closely equivalent to asymptotic spaces of X (this is where we use the last Lemma and the shrinking property). Then we will use a compactness argument and Lemma 4 to extract asymptotic subsets of block sequences, where the norm of linear combinations, the normalised image under T and the action of T are almost fixed. Finally we will pipe such asymptotic sets of different lengths together by means of Lemma 7, to realize an asymptotic version of T.

First step: For every $\delta > 0$ and for every asymptotic set Ψ of length n, there is an asymptotic subset, Ψ', of sequences whose normalised images under T are $(1+\delta)$-equivalent to elements of $\{X\}_n$.

Proof of first step: We want to show that the following subset of Ψ is asymptotic:

The intersection of Ψ with the collection of block sequences which T maps (up to $\delta/2$) into the admission collection $\Sigma_{n,\delta/2}(X)$ from Remark 12.

Ψ is already an asymptotic set. By the Lemma 11, the sequences mapped close to $\Sigma_{n,\delta/2}(X)$ form an admission set. Remark 2 says that their intersection must indeed be an asymptotic set.

Second step: Let M be the compactum of n-dimensional spaces with a normalised basis and a basic constant not worse than that of X (the metric on this space is given by equivalence of bases). Consider a finite covering $\{V_i\}_i$ of M. Consider a finite covering $\{I_k\}_k$ of the cube $[0, \|T\|]^n$. For every asymptotic set Φ of length n there is an asymptotic subset, Φ', of block-sequences with the following additional properties:

(i) The sequences in Φ' are contained in some fixed V_{i_0}.
(ii) The normalised images under T of sequences from Φ' are contained in some fixed W_{j_0}.
(iii) For all $\{x_i\}_{i=1}^n \in \Phi'$, the sequences $\{\|T(x_i)\|\}_{i=1}^n$ are contained in some fixed I_{k_0}.

Proof of second step: This is an easy application of Lemma 4 and the fact that the covering is finite. Our covering induces a covering of $M \times M \times [0, \|T\|]^n$. Each cell of this covering contains a subset of Φ of the form

$$\Phi_{i,j,k} = \left\{ \{x_i\}_{i=1}^n : \{x_i\}_{i=1}^n \in V_i, \; \left\{ \frac{T(x_i)}{\|T(x_i)\|} \right\}_{i=1}^n \in W_j, \; \{\|T(x_i)\|\}_{i=1}^n \in I_k \right\}.$$

By Lemma 4 one of those collections must be an asymptotic set.

Third step: For every operator $T \in \mathcal{L}(X)$ and every collection $\{\Phi_n\}_{n=1}^\infty$ of asymptotic sets of increasing lengths, there exists a formally diagonal operator $\tilde{T} \in \{T\}_\infty$, realized by truncated subsets of the Φ_n's.

Proof of third step: Fix a positive sequence converging to zero, $\{\varepsilon_n\}_n$. Next choose inductively open finite coverings as in step 2 above with cells of diameter less than ε_n. Our aim is that the cells of the coverings correspond to the vertices of a tree; we want a cell from the $(n-1)$-th covering to "split" into cells of the n-th covcering. To achieve that, we consider the projection from the n-order space to the $(n-1)$-order space that takes $(x_1, \ldots, x_n; y_1, \ldots, y_n; \lambda_1, \ldots, \lambda_n)$ to $(x_1, \ldots, x_{n-1}; y_1, \ldots, y_{n-1}; \lambda_1, \ldots, \lambda_{n-1})$. We require that this projection maps the covering of the n-order space to a refinement of the covering of the $(n-1)$-order space.

If we choose ε_1 large enough (say $\varepsilon_1 = \|T\|$), The cells of such coverings do form an infinite rooted tree in an obvious way.

Take the collection $\{\Phi_n\}_n$ of asymptotic sets. Fix $k \leq n$. Applying step 2 to the k-truncation of Φ_n, we get that there's a vertex (i.e. a covering cell) which contains an asymptotic subset of the k-truncation of Φ_n. Note that if a certain vertex contains such an asymptotic subset, so must all the predecessors of that vertex (a truncation of an asymptotic set is an asymptotic set). This fact means that the vertices containing such asymptotic sets form a connected rooted subtree. Since k and n were arbitrary, our rooted subtree intersects every level of the original tree. We therefore have that our subtree satisfies both properties of Lemma 7, and must have an infinite branch.

Let $\{\Phi'_m\}_m$ be the truncated asymptotic subsets contained in the vertices of such an infinite branch. By step 1, we may assume that T sends sequences from Φ'_m to increasingly good realizations of asymptotic spaces of X. Thus, for every m, the truncated asymptotic subsets $\{\Phi'_m\}_m$ of $\{\Phi_n\}_n$ have the following properties:

(i) The sequences of blocks from $\{\Phi'_m\}_m$ are $(1+\varepsilon_m)$-equivalent to the basis of $[Y]_m$ for some fixed $Y \in \{X\}_\infty$.
(ii) The normalised images under T of all sequences in $\{\Phi'_m\}_m$ are $(1+\varepsilon_m)$-equivalent to the basis of $[Z]_m$ for some fixed $Z \in \{X\}_\infty$.
(iii) For any $(x_i)_i \in \{\Phi'_m\}_m$, the sequence $\|T(x_i)\|_i$ is fixed up to ε_m.

Hence, the result of this process is a sequence of asymptotic sets realizing a formally diagonal asymptotic version of T. □

REMARK 13. Note that since we may start with any collection of asymptotic sets with increasing lengths, we may choose to extract our asymptotic version of T from asymptotic sets realizing a given asymptotic version of X. We thus have for any operator $T \in \mathcal{L}(X)$ and for any $\tilde{X} \in \{X\}_\infty$ an asymptotic version of T whose domain is $\tilde{X}$.

The proof of the existence theorem also implies that for every positive ε and natural n the collection $\Sigma_{n,\varepsilon}(X)$ (introduced in Remark 12 and used in the proof of step 1) is indeed an admission set.

If this were not the case, by Lemma 3 the complement of $\Sigma_{n,\varepsilon}(X)$ would be an asymptotic set, and by its own definition it could not contain sequences $(1+\varepsilon)$-equivalent to any n-dimensional asymptotic space. Then, using the proof of the existence theorem for the *identity* operator (this does not require the first step of the proof or the assumption that $\Sigma_{n,\varepsilon}(X)$ is an admission set), we extract asymptotic subsets realizing some fixed space. This space must (by definition) be asymptotic — a contradiction.

The same fact was proved in [MMiT, 1.5], by a different compactness argument.

3. Asymptotic Versions of Operator Ideals

3.1. Compact operators.

PROPOSITION 14. *If T is compact then $\{T\}_\infty = \{0\}$. If T is not compact then $\{T\}_\infty$ contains a non-compact operator.*

PROOF. If T is compact, take asymptotic sets, $\{\Phi_n\}_n$, realizing some asymptotic version of the operator. Let $\mathcal{V}$ play his winning strategy for Φ_n, and let $\mathcal{S}$ play tail subspaces $[X]_{>n}$ with n such that $\|T\|_{[X]_{>n}} < \varepsilon$.

The block-sequence resulting from this game will still be an approximate realization of the above asymptotic version. This shows that any asymptotic version of T can be approximated arbitrarily well by operators with norm smaller than any positive ε. Therefore the only asymptotic version of T is zero.

If T is not compact, there is an $\varepsilon > 0$ such that the set

$$\Phi_1 = \{x \in S(X)^1_< : \|T(x)\| \geq \varepsilon\}$$

is asymptotic (of length 1). For any n the collection Φ_n of sequences of n successive elements from Φ_1 is obviously asymptotic.

Using the existence theorem, extract from Φ_n asymptotic subsets realizing some asymptotic version of T. The norm of this asymptotic version will not be smaller than ε on any element of the basis of its domain, and will therefore be non-compact. □

Asymptotic versions induce a seminorm on operators, through the formula:

$$\|\|T\|\| = \sup \|\tilde{T}\|$$

where the supremum is taken over all asymptotic versions of T and the double-bar norm is the usual operator norm.

It is interesting to note that this gives a way of looking at the Calkin algebras $\mathcal{L}(X)/\mathcal{K}(X)$.

PROPOSITION 15. *Suppose X is a Banach space with a shrinking basis, where the norm of all tail projections is exactly 1 (the last property can be achieved by renorming, see* [LTz]*). Then the norm of the image of an operator T in the Calkin algebra is equal to $\|\|T\|\|$.*

PROOF. One direction is clear. If K is a compact operator on X, the norm of $T + K$ is at least the supremum of norms of asymptotic versions of $T + K$. The latter, by the proof of Proposition 14, are the same as asymptotic versions of T.

For the other direction we will show that for every T there exist compact operators K such that the norm of $T + K$ is almost achieved by asymptotic versions of $T + K$, which, again, are the same as asymptotic versions of T.

We will perturb T by a compact operator K, such that the set Φ of normalised blocks mapped by $T+K$ to vectors of norm greater than $\|T+K\|-\varepsilon$ will become an asymptotic set of length 1. We can then extract an asymptotic version of

$T + K$ from asymptotic sets containing only sequences of elements in Φ. Such asymptotic versions will almost achieve the norm of $T + K$, as required.

Take λ to be (up to ε) the largest such that

$$\{x \in S(X)^1_< : \|T(x)\| \geq \lambda\}$$

is asymptotic. By this we mean that the set

$$\{x \in S(X)^1_< : \|T(x)\| \geq \lambda + \varepsilon\}$$

does not have elements in some tail subspace $[X]_{>m}$. Consider $T' = T - T \circ P_m$, which is a compact perturbation of T. By our assumption on the basis we have $\|T'\| \leq \lambda + \varepsilon$, and the set

$$\{x \in S(X)^1_< : \|T'(x)\| \geq \lambda\}$$

is still asymptotic. The proof is now complete. $\square$

3.2. Uniformly singular and asymptotically uniformly singular operators.

DEFINITION 16. An operator T on a sequence space X is *asymptotically uniformly singular* if for every ε there exist $n(\varepsilon, T)$ and an admission set Σ_n of length n, such that any sequence in Σ_n contains a normalised vector, which T send ε-close to zero. Informally we will say, that T has an *almost kernel* in any sequence from Σ_n. T is called *uniformly singular* if it satisfies the above definition with $\Sigma_n = S(X)^n_<$.

In other words, an operator T is asymptotically uniformly singular, if , when restricted to the span of a sequence from Σ_n, T^{-1} is either not defined or has norm larger than $1/\varepsilon$.

An operator ideal close to the ideal of uniformly singular operators was defined in [Mi], and called σ_0. The difference is that the original definition referred to all n dimensional subspaces, rather than just block subspaces, as we read here.

PROPOSITION 17. *Operators on a Banach-space X, which are asymptotically uniformly singular with respect to a given basis, form a Banach space with the usual operator norm, and a two sided ideal of $\mathcal{L}(X)$.*

PROOF. Let T be asymptotically uniformly singular, and let S be bounded. ST is obviously asymptotically uniformly singular. Indeed, $n(\varepsilon, ST) \leq n(\varepsilon/\|S\|, T)$, and the admission sets for ST are the same as those for T.

To see that TS is asymptotically uniformly singular we use Lemma 11, and find admission sets Σ', whose normalised image under S are essentially contained in the admission sets Σ used in the definition of asymptotic uniform singularity for the operator T. TS will take some normalised block from the span of any sequence from Σ' to a vector with arbitrarily small norm. Indeed, S takes (essentially) Σ' to Σ, where T has an "almost kernel".

To show that the sum of two asymptotically uniformly singular operators is also asymptotically uniformly singular, we need to use the following claim:

If T is asymptotically uniformly singular, then for every ε and every k there exists an $N(k, \varepsilon, T)$, and an admission set Σ of length N, such that every sequence from Σ has a k-dimensional block subspace on which T has norm less than ε.

This claim is standardly proved by taking concatenations of the asymptotic sets from the definition of asymptotic uniform singularity for T. It is easier to think here of the uniformly singular case. Suppose T sends some normalised vector from any n_ε-dimensional block subspace into an ε-neighbourhood of zero. Then in any $\sum_{i=1}^{k} n_{\varepsilon_i}$-dimensional block subspace there is a k-dimensional block subspace, on which the norm of T is at most $C \sum_{i=1}^{k} \varepsilon_i$ (where C is the basic constant of X).

To complete the proof of the proposition, fix $\varepsilon > 0$ and consider asymptotically uniformly singular operators, T and S. From the definition of asymptotic uniform singularity produce an admission set Σ, of length $n(\varepsilon/2, S)$, such that S takes some normalised block in the span of any Σ-admissible sequence to a vector with norm less than $\varepsilon/2$. Take the admission set Ψ of length $N(n, \varepsilon/2, T)$ from the above claim. It is possible to extract an admission subset $\Psi' \subseteq \Psi$, such that any n consecutive blocks of a sequence in Ψ' are also in Σ (similarly to Remark 2).

We therefore have that in any block sequence in Ψ' there is an n-dimensional block subspace where T has norm less than $\varepsilon/2$. This subspace must be essentially the span of a sequence from Σ, so it contains a normalised vector, whose image under S has norm less than $\varepsilon/2$. This means that $T+S$ sends this vector ε-close to zero, and hence $T+S$ is asymptotically uniformly singular.

The fact that asymptotically uniformly singular operators form a closed subspace of $\mathcal{L}(X)$ is straightforward. □

REMARKS 18. (i) Every asymptotically uniformly singular operator T is actually uniformly singular on some block subspace.

Indeed, From Gowers' combinatorial lemma (in [G], see also [W1]) it follows that there is a block subspace Y, where $\Sigma_{n(\varepsilon,T)} = S(Y)_<^{n(\varepsilon,T)}$.

In a diagonal subspace we can find an n for every ε, such that any sequence of the form $e_n < x_1 < \ldots < x_n$ has a normalised block whose image under T has norm less than ε.

It is easy to see that on this block subspace T is uniformly singular; indeed, every sequence of n blocks contains a sequence of $[n/2]$ blocks supported after the $[n/2]$-th basic element.

(ii) It is not true, however, that the ideals of asymptotically uniformly singular and uniformly singular operators coincide.

Consider the following example: Let X be the ℓ_2 sum of increasingly long ℓ_2^n's, and let Y be their ℓ_3 sum. The formal identity from X to Y is not uniformly singular, but is asymptotically uniformly singular.

(iii) It is easy to extend Proposition 17 and show that asymptotically uniformly singular operators form a two sided operator ideal with the operator norm, when restricting our attention to the category of Banach spaces with shrinking bases. Uniformly singular operators will only form a left-sided ideal in the (shrinking) basis context. This is because a bounded operator multiplied to the right of a uniformly singular operator does not have to preserve blocks.

(iv) It is worth noting that in the Gowers-Maurey space (from [GM]), all operators are in fact uniformly singular perturbations of a scalar operator (this follows from Lemma 22 and Lemma 3 in [GM]). This point is even more interesting in light of Corollary 21 below, and the Remark which follows it.

The following proposition claims a strong dichotomy in the asymptotic structure of operators: either it contains an isomorphism, or it is composed only of uniformly singular operators.

PROPOSITION 19. *If T is an asymptotically uniformly singular operator then all operators in $\{T\}_\infty$ are uniformly singular. If T is not asymptotically uniformly singular, $\{T\}_\infty$ contains an isomorphism.*

PROOF. Let T be asymptotically uniformly singular. Take asymptotic sets Φ_n realizing an asymptotic version of T. Let player $\mathcal{V}$ play the winning strategy for Φ_n, while $\mathcal{S}$ plays the winning strategy for the admission sets Σ_n from the definition of asymptotic uniform singularity for T. The resulting vector sequences must approximately realize the above asymptotic version, but must also contain an "almost kernel" for T.

Keeping in mind that a restriction of an asymptotic version to a block subspace is still an asymptotic version, we conclude that T has an "almost kernel" on every block subspace of the appropriate dimension. Therefore any asymptotic version of T is uniformly singular.

Suppose T is not asymptotically uniformly singular. Then for some $\varepsilon > 0$ the sets Φ_n of all sequences in $S(X)_<^n$, on which T is an isomorphism with $\|T^{-1}\| \leq 1/\varepsilon$, are asymptotic. Indeed, if they weren't, by Lemma 3, for some ε and for every n, Φ_n^c would be admission sets, and then T would be asymptotically uniformly singular, in contradiction. Using the existence theorem, extract from Φ_n asymptotic subsets Φ_n', which realize an asymptotic version of T. This asymptotic version is an isomorphism. □

REMARK 20. Note that the proof shows that if T is asymptotically uniformly singular, all its asymptotic versions will be uniformly singular operators with the same $n(\varepsilon)$ as T.

We offer here an application of the theory developed above. Recall that an *asymptotic* ℓ_p space is a space, all whose asymptotic versions are isomorphic to ℓ_p.

COROLLARY 21. *Every asymptotically uniformly singular operator from an asymptotic ℓ_p space X to itself is compact.*

PROOF. Consider the set $\mathcal{AUS}(X) \subset \mathcal{L}(X)$ of asymptotically uniformly singular operators. The asymptotic versions of these operators, by Proposition 19 above, are all uniformly singular operators in $\mathcal{L}(\ell_p)$. In particular they are all asymptotically uniformly singular, and therefore belong to some proper closed two-sided ideal. It is well known [GoMaF] that the only proper closed two sided operator ideal in $\mathcal{L}(\ell_p)$ is the ideal of compact operators, but let us sketch a very simple proof for this particular context:

All asymptotic versions of T are diagonal uniformly singular operators in $\mathcal{L}(\ell_p)$. It is clear then, that given any $\varepsilon > 0$, only finitely many entries on the diagonal are larger than ε; otherwise, restricting to the span of the basic elements corresponding to the entries larger than ε we get an isomorphism. Therefore the entries on the diagonal go to zero, and the operator is compact.

We can now complete the proof of the corollary, using Proposition 14 once more.
$$\mathcal{AUS}(X) = \{T \in \mathcal{L}(X) : \{T\}_\infty \subseteq \mathcal{US}(\ell_p)\}$$
$$= \{T \in \mathcal{L}(X) : \{T\}_\infty \subseteq \mathcal{K}(\ell_p)\} = \mathcal{K}(X).$$
where $\mathcal{K}(Z)$ is the set of all compact operators on Z, and $\mathcal{US}(Z)$ is the set of all uniformly singular operators on Z. □

REMARK 22. Note that, by this corollary, if an asymptotic-ℓ_p space with a shrinking basis has the property of the Gowers-Maurey space from the last point of Remark 18, then all bounded linear operators on this space will be compact perturbations of scalar operators. Whether such a space exists is an important open question.

3.3. A general theorem. We conclude with a theorem which explains that the above instances form part of a more general phenomenon. When referring to an operator ideal we invoke the categorical algebraic definition from [P]. An *injective* operator ideal J has the property that if $T : X \to Y$ is in J, then the same operator with a revised range, $T : X \to \overline{\text{Im}}(T)$, is also in J. The following theorem states that the "asymptotic preimage" in $\mathcal{L}(X)$ of an injective ideal is itself an ideal in the algebra $\mathcal{L}(X)$. The preimage may be trivial, but previous examples show this needn't be the case.

THEOREM 23. *Let J be an injective operator ideal, and let X be a space with a shrinking basis. The set of operators $J' = \{T \in \mathcal{L}(X) : \{T\}_\infty \subseteq J\}$ is an operator ideal in $\mathcal{L}(X)$.*

PROOF. If we multiply an operator $S \in J'$ with an operator $T \in \mathcal{L}(X)$, an asymptotic version of the product will always be a product of asymptotic versions.

Indeed, take the asymptotic sets realizing an asymptotic version, $\tilde{R}$, of $R = ST$ (or $R = TS$). Extract by the existence theorem asymptotic subsets, which realize an asymptotic version $\tilde{T}$ of T. Recall that the set of sequences which approximately realize asymptotic versions of S is an admission set (consult Remark 13). Thus Lemma 11 and the proof of the existence theorem allow to extract asymptotic subsets of sequences whose normalised images under T realize an asymptotic version $\tilde{S}$ of S. The product of these asymptotic versions, $\tilde{S}\tilde{T} \in J$, is realized by the same asymptotic subsets as well. But these asymptotic subsets must still realize $\tilde{R}$. Therefore $\tilde{R} \in J$, and R must be in J'.

Let T and S be in J'. Let $R = S + T$, and find asymptotic sets realizing $\tilde{R}$, an asymptotic version of R. Extract asymptotic subsets realizing asymptotic versions of S and T, $\tilde{S}$ and $\tilde{T}$ respectively. Note that we cannot say that $\tilde{R} = \tilde{S} + \tilde{T}$; in fact $\tilde{S}$ and $\tilde{T}$ may even have different ranges. However, we trivially have

$$\|\tilde{R}(x)\| \leq \|\tilde{S}(x)\| + \|\tilde{T}(x)\|. \tag{3.1}$$

This is enough in order to prove $\tilde{R} \in J$. Indeed, we can write

$$\tilde{R} = P \circ (i_1 \circ \tilde{S} + i_2 \circ \tilde{T}),$$

where

$$\tilde{R}: \tilde{X} \to \tilde{W}, \quad \tilde{S}: \tilde{X} \to \tilde{Y}, \quad \tilde{T}: \tilde{X} \to \tilde{Z},$$
$$i_1: \tilde{Y} \to \tilde{Y} \oplus \tilde{Z}, \quad i_2: \tilde{Z} \to \tilde{Y} \oplus \tilde{Z},$$
$$i_1(y) = (y, 0), \quad i_2(z) = (0, z),$$

and $P: \overline{\mathrm{Im}}(i_1 \circ \tilde{S} + i_2 \circ \tilde{T}) \to \tilde{W}$ is defined by the equation $P\big((i_1 \circ \tilde{S} + i_2 \circ \tilde{T})(x)\big) = \tilde{R}(x)$, and extended by continuity. Inequality (3.1) assures that P is well defined and continuous.

Now $\tilde{T}$ and $\tilde{S}$ are in J, so $i_1 \circ \tilde{S} + i_2 \circ \tilde{T}$ is also in J. By injectivity, we are allowed to modify the range as we compose with P, and still maintain that the result $\tilde{R}$ is in J. Thus $R \in J'$, and we conclude that J' is an ideal in $\mathcal{L}(X)$. $\square$

References

[G] Gowers, W. T., *A new dichotomy for Banach spaces (research announcement)*, GAFA **6** (1996), 1083–1093.

[GM] Gowers, W. T. and Maurey, B., *The unconditional basic sequence problem*, J. Amer. Math. Soc. **6** (1993), 851–874.

[GoMaF] Gohberg, I. C., Markus, A. S. and Feldmann, I. A., *Normally solvable operators and ideals associated with them* (Russian), Bul. Akad. Štiince RSS Moldove. **10** (76) (1960), 51–70.

[LTz] Lindenstaruss J. and Tzafriri, L., *Classical Banach spaces*, Vol. I, Ergeb. Math. Grenzgeb. **92**, Springer, Berlin.

[Mi] Milman V., *Spectrum of continuous bounded functions on the unit sphere of a Banach space*, Funct. Anal. Appl. **3** (1969), 67–79 (translated from Russian).

[MMiT] Maurey B., Milman V. and Tomczak-Jaegermann N., *Asymptotic infinite-dimensional theory of Banach spaces*, Operator Theory: Advances and Applications **77** (1995), 149–175.

[MiT] Milman V. and Tomczak-Jaegermann N., *Asymptotic ℓ_p spaces and bounded distortion*, Contemporary Math. **144** (1993), Bor-Luh Lin and W. B. Johnson eds., Amer. Math. Soc., 173–195.

[Kö] König, D., *Theorie der endlichen und unendlichen Graphen*, Akademische Verlagsgesellschaft, Leipzig, 1932; reprinted by Chelsea, New-York, 1950, page 81.

[KOS] Knaust, D., Odell, E. and Schlumprecht, Th., *On asymptotic structure, the Szlenk index and UKK properties in Banach spaces*, preprint.

[OTW] Odell, E., Tomczak-Jaegermann N. and Wagner, R., *Proximity to ℓ_1 and distortion in asymptotic ℓ_1 spaces*, J. Functional Analysis **150** (1997), 101–145.

[P] Pietsch, A., *Operator Ideals*, North-Holland Mathematical Library **20**, North-Holland publishing company, 1980.

[T] Tomczak-Jaegermann, N., *Banach spaces of type p contain arbitrarily distortable subspaces*, GAFA **6** (1996), 1074–1082.

[W1] Wagner, R., *Gowers' dichotomy for asymptotic structure*, Proc. Amer. Math. Soc. **124** (1996), 3089–3095.

[W2] Wagner, R., *Finite high order games and an inductive approach towards Gowers' dichotomy*, preprint.

VITALI MILMAN
DEPARTMENT OF MATHEMATICS
TEL AVIV UNIVERSITY
RAYMOND AND BEVERLY SACKLER FACULTY OF EXACT SCIENCES
RAMAT AVIV, TEL AVIV 69978
ISRAEL
vitali@math.tau.ac.il

ROY WAGNER
D.P.M.M.S.
UNIVERSITY OF CAMBRIDGE
16 MILL LANE
CAMBRIDGE CB2 1SB
UNITED KINGDOM
r.wagner@dpmms.cam.ac.uk

Metric Entropy of the Grassmann Manifold

ALAIN PAJOR

ABSTRACT. The knowledge of the metric entropy of precompact subsets of operators on finite dimensional Euclidean space is important in particular in the probabilistic methods developed by E. D. Gluskin and S. Szarek for constructing certain random Banach spaces. We give a new argument for estimating the metric entropy of some subsets such as the Grassmann manifold equipped with natural metrics. Here, the Grassmann manifold is thought of as the set of orthogonal projection of given rank.

1. Introduction and Notation

Let A be a precompact subset of a metric space (X, τ). An ε-net of A is a subset Λ of X such that any point x of A can be approximated by a point y of Λ such that $\tau(x,y) < \varepsilon$. The smallest cardinality of an ε-net of A is called the *covering number* of A and is denoted by $N(A, \tau, \varepsilon)$. The metric entropy (shortly the entropy) is the function $\log N(A, \tau, \cdot)$.

When X is a d-dimensional normed space equipped with the metric associated to its norm $\|\cdot\|$, we will denote by $N(A, \|\cdot\|, \varepsilon)$ the covering number of a subset A of X and by $B(X)$ the unit ball of X. The metric entropy of a ball $A = rB(X)$ of radius r is computed by volumic method (see [MS] or [P]): for $\varepsilon \in]0, r]$,

$$\left(\frac{r}{\varepsilon}\right)^d \leq N(rB(X), \|\cdot\|, \varepsilon) \leq \left(3\frac{r}{\varepsilon}\right)^d. \tag{1}$$

The space $\mathbb{R}^n$ is equipped with its canonical Euclidean structure and denoted by ℓ_2^n. Its unit ball is denoted by B_2^n, the Euclidean norm by $|\,.\,|$ and the scalar product by $(\,.\,,\,.\,)$. For any linear operator T between two Euclidean spaces and any p, $1 \leq p \leq \infty$, let

$$\sigma_p(T) = \left(\sum_{i \geq 1} |s_i(T)|^p\right)^{1/p}$$

Part of this work was done while the author visited MSRI. The author thanks the Institute for its hospitality.

where $s_i(T)$, $i = 1, \ldots$ denote the singular values of T. In particular, σ_1 is the nuclear norm, σ_2 is the Hilbert-Schmidt norm and σ_∞ the operator norm. The Schatten trace class of linear mapping on the n-dimensional Euclidean space equipped with the norm σ_p is denoted by S_p^n. More generally, we consider a unitarily invariant norm τ on $L(\ell_2^n)$; it satisfies $\tau(USV) = \tau(S)$ for any $S \in L(\ell_2^n)$ and any isometries U, V on ℓ_2^n. It is associated to a 1-symmetric norm on $\mathbb{R}^n$ and τ is the norm of the n-tuple of singular values. For any Euclidean subspaces E, F of ℓ_2^n, τ induces a unitarily invariant norm on $L(E, F)$ still denoted by τ.

Let $G_{n,k}$ be the Grassmann manifold of the k-dimensional subspaces of $\mathbb{R}^n$. For any subspace $E \in G_{n,k}$ we denote by P_E the orthogonal projection onto E. We will denote by σ_p the metric induced on $G_{n,k}$ by the norm σ_p, when $G_{n,k}$ is considered as a subset of S_p^n. Similarly, we denote by τ the metric induced onto $G_{n,k}$ by the norm τ.

We are mainly interested in estimating $N(B(S_1^n), \sigma_p, \varepsilon)$. The computation of $N(G_{n,k}, \sigma_p, \varepsilon)$ was done in [S1] and is the basic tool in [S2] for solving the finite dimensional basis problem (see also [G1], [G2] and [S3]). For computing the metric entropy of a d-dimensional manifold, we first look for an *atlas*, a collection of charts $(U_i, \varphi_i)_{1 \le i \le N}$ and estimate N. The situation is particularly simple if for each of the charts, φ_i is a bi-Lipschitz correspondence with a d-dimensional ball. Locally, the entropy is computed by volumic method. To estimate the number N of charts in the atlas, we look at the Grassmann manifold in $L(\mathbb{R}^n)$ with the right metric. Such an embedding that does not reflect the dimension of the manifold cannot give directly the right order of magnitude of the entropy, but as we will see, it gives an estimate of the cardinality N of a "good" atlas. The two arguments are combined to give the right order of magnitude for the entropy of $G_{n,k}$. We did not try to give explicit numerical constant and the same letter may be used to denote different constants.

2. Basic Inequalities

Let G be a Gaussian random d-dimensional vector, with mean 0 and the identity as covariance matrix. The following result is the geometric formulation of Sudakov's minoration (See [P]).

LEMMA 1. *There exists a positive constant c such that for any integer $d \ge 1$, any subset A of $\mathbb{R}^d$ and for every $\varepsilon > 0$, we have*

$$\varepsilon \sqrt{\log N(A, |\cdot|, \varepsilon)} \le c \mathbb{E} \sup_{t \in A}(G, t). \tag{2}$$

LEMMA 2. *There exists a positive constant c such that for any integer $n \ge 1$, for any p such that $1 \le p \le \infty$ and for every $\varepsilon > 0$, we have*

$$\log N(B(S_p^n), \sigma_2, \varepsilon) \le c \frac{n^{3-2/p}}{\varepsilon^2}. \tag{3}$$

PROOF. Let A be a subset of the Euclidean space S_2^n equipped with its scalar product given by the trace. Applying the inequality (2) where G is a standard Gaussian matrix whose entries are independent $\mathcal{N}(0,1)$ Gaussian random variables, we get that

$$\varepsilon \sqrt{\log N(A, \sigma_2, \varepsilon)} \leq c \mathbb{E} \sup_{T \in A} \text{trace}(GT).$$

Let q such that $1 \leq q \leq \infty$ and $\frac{1}{p} + \frac{1}{q} = 1$. By the trace duality,

$$|\text{trace}(GT)| \leq \sigma_q(G)\sigma_p(T) \leq \sigma_q(G),$$

for any $T \in B(S_p^n)$. Now it is well known that

$$\mathbb{E}\sigma_q(G) \leq n^{1/q} \mathbb{E}\sigma_\infty(G) \leq \alpha n^{1/q}\sqrt{n}, \tag{4}$$

for some universal constant α (see [MP], Proposition 1.5.). Therefore

$$\sqrt{\log N(B(S_p^n), \sigma_2, \varepsilon)} \leq c\alpha n^{1/q}\sqrt{n}\frac{1}{\varepsilon},$$

which gives the estimate (3). □

The next inequality is in some sense dual to (2). Again G is a Gaussian d-dimensional random vector.

LEMMA 3 (see [PT]). *There exists a positive constant c such that for any integer $d \geq 1$, any norm $\|.\|$ on $\mathbb{R}^d$ and for every $\varepsilon > 0$, we have*

$$\varepsilon\sqrt{\log N(B_2^d, \|.\|, \varepsilon)} \leq c\mathbb{E}\|G\|. \tag{5}$$

LEMMA 4. *There exists a positive constant c such that for any integer $n \geq 1$ and for any q such that $1 \leq q \leq \infty$, we have*

$$\log N(B(S_2^n), \sigma_q, \varepsilon) \leq c \frac{n^{1+2/q}}{\varepsilon^2} \tag{6}$$

for every $\varepsilon > 0$.

PROOF. The proof follows from the formulae (4) and (5) applied with the norm σ_q. □

PROPOSITION 5. *There exists a positive constant c such that for any integer $n \geq 1$, for any p, q such that $1 \leq p \leq \infty$, $2 \leq q \leq \infty$ and for every $\varepsilon > 0$, we have*

$$\log N(B(S_p^n), \sigma_q, \varepsilon) \leq c \frac{n^{(2-1/p-1/q)q'}}{\varepsilon^{q'}}, \tag{7}$$

where $1/q + 1/q' = 1$.

PROOF. We observe that

$$N(B(S_p^n), \sigma_q, \varepsilon) \leq N\left(B(S_p^n), \sigma_2, \frac{\varepsilon}{\theta}\right) N(B(S_2^n), \sigma_q, \theta).$$

Therefore inequalities (3) and (6) give us

$$\log N(B(S_p^n), \sigma_q, \varepsilon) \leq c \left(\frac{n^{3-2/p}\theta^2}{\varepsilon^2} + \frac{n^{1+2/q}}{\theta^2}\right).$$

Optimizing with $\theta^2 = \varepsilon n^{-1+1/p+1/q}$ we arrive at

$$\log N(B(S_p^n), \sigma_q, \varepsilon) \leq 2c \frac{n^{2-1/p+1/q}}{\varepsilon}. \tag{8}$$

In particular for $q = \infty$, we get

$$\log N(B(S_p^n), \sigma_\infty, \varepsilon) \leq 2c \frac{n^{2-1/p}}{\varepsilon}. \tag{9}$$

Let now $2 \leq q \leq \infty$, $\lambda = \frac{2}{q}$, so that $0 \leq \lambda \leq 1$ and $\frac{1}{q} = \lambda \frac{1}{2} + (1-\lambda)\frac{1}{\infty}$. By Hölder's inequality, for every $x, y > 0$, we have

$$N(B(S_p^n), \sigma_q, 2x^\lambda y^{1-\lambda}) \leq N(B(S_p^n), \sigma_2, x) N(B(S_p^n), \sigma_\infty, y).$$

This relation and inequalities (3) and (9) yield

$$\log N(B(S_p^n), \sigma_q, \varepsilon) \leq 8c \left(\frac{n^{3-2/p}}{z^{2/\lambda}} + \frac{n^{2-1/p}}{(\varepsilon/z)^{1/1-\lambda}}\right), \quad z > 0$$

and the optimal choice $z = (q-2)^{(q-2)/q(q-1)} \varepsilon^{1/(q-1)} n^{(1-1/p)(q-2)/q(q-1)}$ gives the estimate (7). $\square$

REMARKS. 1) Estimates (7) and (8) are relevant when $p \leq 2 \leq q$, their accuracy depends on the range of ε, particularly with respect to $1/n^{(1/p-1/q)}$.
2) Note that when $p = 1$, inequality (7) becomes

$$\log N(B(S_1^n), \sigma_q, \varepsilon) \leq c \frac{n}{\varepsilon^{q'}}.$$

3) All the computation of entropy above, could have be done by the same method for the trace classes of operators between two different finite dimensional Euclidean spaces. One can use Chevet inequality [C] to get the relation corresponding to (4).

3. Metric Entropy of the Grassmann Manifold

We consider now the Grassmann manifold $G_{n,k}$ as a subset of S_q^n, which means that $G_{n,k}$ is equipped with the metric $\sigma_q(E, F) = \sigma_q(P_E - P_F)$. In view of evaluating the cardinality of a "good" atlas of the Grassmann manifold, we begin with a first estimate for its entropy.

PROPOSITION 6. *There exists a positive constant c such that for any integers $1 \leq k \leq n$ and for every $\varepsilon > 0$, we have*

$$\log N(G_{n,k}, \sigma_\infty, \varepsilon) \leq c \frac{d}{\varepsilon}, \tag{10}$$

where $d = k(n-k)$.

PROOF. We may suppose that $k \leq n - k$. Since for any rank k orthogonal projection $\sigma_1(P) = k$, inequality (7) with $p = 1$ gives by homogeneity,

$$\log N(G_{n,k}, \sigma_\infty, \varepsilon) \leq \log N(B(S_1^n), \sigma_\infty, \varepsilon k^{-1})$$
$$\leq c \frac{nk}{\varepsilon} \leq 2c \frac{d}{\varepsilon}. \qquad \square$$

Let $E, F \in G_{n,k}$ and $U \in \mathcal{O}_n$ such that $U(F) = E$, then clearly

$$(P_{F^\perp} P_E)^* = (P_{F^\perp} U P_F U^*)^* = U P_F U^* P_{F^\perp} = U P_F P_{E^\perp} U^*.$$

Therefore $P_{F^\perp} P_E$ and $P_F P_{E^\perp}$ have the same singular values. Moreover, since $P_E - P_F = P_{F^\perp} P_E - P_F P_{E^\perp}$ is an orthogonal decomposition, for any unitarily invariant norm τ we have $\tau(P_{F^\perp} P_E) = \tau(P_F P_{E^\perp})$ and

$$\tau(P_{F^\perp} P_E) \leq \tau(E, F) \leq 2\tau(P_{F^\perp} P_E).$$

Denote by $P_{F|E}$ the restriction over E of the orthogonal projection onto F and consider its polar decomposition

$$P_{F|E} = \sum_{i=1}^{k} s_i e_i \otimes f_i, \ s_i \geq 0, \ (e_i, f_j) = \delta_{ij} s_i, 1 \leq i, j \leq k,$$

with $(e_i)_{1 \leq i \leq k}$ an orthonormal basis of E and $(f_i)_{1 \leq i \leq k}$ an orthonormal basis of F. Let $e_i' = (-s_i e_i + f_i)/\sqrt{1 - s_i^2}$ and $f_i' = (-e_i + s_i f_i)/\sqrt{1 - s_i^2}$ when i is such that $s_i \neq 1$. The families (e_i') and (f_i') are respectively orthonormal systems in $E^\perp$ and $F^\perp$ and we have the following polar decomposition:

$$P_E - P_F = P_F P_{E^\perp} - P_E P_{F^\perp} = \sum_{s_i \neq 1} \sqrt{1 - s_i^2} \, e_i' \otimes f_i + \sum_{s_i \neq 1} \sqrt{1 - s_i^2} \, e_i \otimes f_i'.$$

Consequently, for $1 \leq q \leq \infty$

$$\sigma_q(E, F) = (\sigma_q(P_{F^\perp|E}))^q + \sigma_q(P_{F|E^\perp})^q)^{1/q} = \left(2 \sum_{i=1}^{k} (1 - s_i^2)^{q/2} \right)^{1/q}.$$

Note that the Riemannian metric is given by $\sigma_g(E, F) = \left(\sum_{i=1}^{k} \arccos^2 s_i \right)^{1/2}$ and therefore

$$\sigma_2(E, F) \leq \sqrt{2} \sigma_g(E, F) \leq \frac{\pi}{2} \sigma_2(E, F).$$

Let E be a k-dimensional Euclidean subspace of $\mathbb{R}^n$. For any $0 < \rho < 1$, let

$$V_\rho(E) = \{F \in G_{n,k} : \sigma_\infty(E, F) \leq \rho\}.$$

Let us recall a standard chart: $(V_\rho(E), \varphi_E)$ where $\varphi_E : V_\rho(E) \longrightarrow L(E, E^\perp)$ is defined by
$$\varphi_E(F) = P_{E^\perp|F} \circ (P_{E|F})^{-1}.$$
In other words, F is the graph of $u = \varphi_E(F)$ and $F = \{x + u(x) : x \in E\}$. With this notation, we have:

LEMMA 7. *Let $0 < \rho < 1$, $E, F, G \in G_{n,k}$ and $F, G \in V_\rho(E)$, let $u = \varphi_E(F)$ and $v = \varphi_E(G)$, let τ be a unitarily invariant norm on $L(\ell_2^n)$, then we have*

$$\varphi_E(V_\rho(E)) = \left\{ w \in L(E, E^\perp) : \sigma_\infty(w) \leq \frac{\rho}{\sqrt{1-\rho^2}} \right\}, \qquad (11)$$

$$2^{-1}\tau(F, G) \leq \tau(u - v) \leq \frac{2^{1/2}}{1-\rho^2} \tau(F, G). \qquad (12)$$

PROOF. Let $H \in V_\rho(E)$ and $w = \varphi_E(H)$. The relation (11) follows immediately from
$$\sigma_\infty(P_{E^\perp} P_H) = \sigma_\infty(E, H) = \sup_{|x|=1} |w(x)|/\sqrt{1 + |w(x)|^2} = \sigma_\infty(w)/\sqrt{1 + \sigma_\infty(w)^2}.$$

Recall that $F = \{x + u(x) : x \in E\}$ and $G = \{x + v(x) : x \in E\}$. To prove the second relation, let $x, y \in E$ such that $P_G(x + u(x)) = y + v(y)$, then
$$|P_{G^\perp}(x + u(x))| = |x - y + u(x) - v(y)| \geq |x - y|.$$

Therefore
$$\begin{aligned} |u(x) - v(x)|^2 &= |(x + u(x)) - (x + v(x))|^2 \\ &= |P_{G^\perp}(x + u(x))|^2 + |(y - x) + v(y - x)|^2 \\ &\leq (2 + \sigma_\infty(v)^2)|P_{G^\perp}(x + u(x))|^2, \end{aligned}$$

and for every x in E we have
$$|P_{G^\perp}(x + u(x))| \leq |u(x) - v(x)| \leq (2 + \sigma_\infty(v)^2)^{1/2}|P_{G^\perp}(x + u(x))|.$$

The left-hand side inequality means that
$$|P_{G^\perp} P_F(z)| \leq |(u - v)(P_E z)| \quad \text{for any } z \in F.$$

It is well known that if $S, T \in L(\ell_2^n)$ satisfy $|Sx| \leq |Tx|$ for every x then $\tau(S) \leq \tau(T)$. Hence
$$\tau(P_{G^\perp} P_F) \leq \tau(u - v).$$

Applying the same observation to the operators $S = (u - v)P_E$ and
$$T = (2 + \sigma_\infty(v)^2)^{1/2} P_{G^\perp} P_F(P_E + uP_E)$$
and using the right-hand side of the same inequality above, one gets
$$\tau(u - v) \leq (2 + \sigma_\infty(v)^2)^{1/2}(1 + \sigma_\infty(u)^2)^{1/2} \tau(P_{G^\perp} P_F).$$

We conclude using (11) and the relation between $\tau(P_{G^\perp} P_F)$ and $\tau(E, F)$. $\square$

We now give a new proof of a result of Szarek.

PROPOSITION 8 (see [S2]). *For any integers $1 \leq k \leq n$ such that $k \leq n - k$, for any q such that $1 \leq q \leq \infty$ and for every $\varepsilon > 0$, we have*

$$\left(\frac{c}{\varepsilon}\right)^d \leq N(G_{n,k}, \sigma_q, \varepsilon k^{1/q}) \leq \left(\frac{C}{\varepsilon}\right)^d, \tag{13}$$

where we set $d = k(n-k)$ and $c, C > 0$ are universal constants.

PROOF. The relation (11) of lemma 7 shows that if we fix ρ, say $\rho = 1/2$, then φ_E is a bi-Lipschitz correspondence from the "ball" $V_{1/2}(E)$ onto the ball of $L(E, E^\perp)$ of radius $1/\sqrt{3}$ (in the operator norm) and from (12), the Lipschitz constants are universal. Therefore the metric entropy of $V_{1/2}(E)$ for the metric σ_∞ is equivalent to the entropy of a d-dimensional ball of radius $1/\sqrt{3}$ for its own metric. From (1) we get

$$\left(\frac{c_1}{\varepsilon}\right)^d \leq N(V_{1/2}(E), \sigma_\infty, \varepsilon) \leq \left(\frac{c_2}{\varepsilon}\right)^d,$$

for some positive universal constants c_1 and c_2. From inequality (10) of Proposition 6, there is an atlas $(V_{1/2}(E_i), \varphi_{E_i})_{1 \leq i \leq N}$ with

$$\log N \leq \log N(G_{n,k}, \sigma_\infty, 1/2) \leq 2cd.$$

Since for a fixed k-dimensional subspace E,

$$N(G_{n,k}, \sigma_\infty, \varepsilon) \leq N(G_{n,k}, \sigma_\infty, 1/2) N(V_{1/2}(E), \sigma_\infty, \varepsilon),$$

we get

$$\left(\frac{c_1}{\varepsilon}\right)^d \leq N(G_{n,k}, \sigma_\infty, \varepsilon) \leq \left(\frac{c_2 e^{2c}}{\varepsilon}\right)^d. \tag{14}$$

The computation of $\sigma_q(.)$ on $G_{n,k}$ shows that $(2k)^{-1/q} \sigma_q(E, F)$ is an increasing function of $q \in [1, \infty)$ if $E, F \in G_{n,k}$ and so the same is true about $\log N(G_{n,k}, \sigma_q, \varepsilon(2k)^{1/q})$. Therefore the upper bound in (13) is a consequence of (14).

To get a lower bound of the entropy, it is sufficient to look at only one chart, say $(V_{1/2}(E), \varphi_E)$ and for the nuclear norm. Using lemma 7 with $q = 1$, we reduce the problem to a minoration, for the nuclear metric, of the entropy of the unit ball of $L(E, E^\perp)$ with the operator norm. Now we join the method of [S2]; a lower bound for the covering number is obtained by evaluating the ratio of the volume of the operator norm unit ball of $L(E, E^\perp)$ and the volume of the nuclear norm unit ball. This concludes the proof. □

References

[C] S. Chevet, *Séries de variables aléatoires Gaussiennes à valeurs dans $E \otimes_\varepsilon F$. Application aux produits d'espaces de Wiener,* Séminaire Maurey-Schwartz, exposé XIX (1977–1978).

[G1] E. D. Gluskin, *The diameter of the Minkowski compactum is roughly equal to n,* Functional Anal. Appl. **15** (1981), 72–73.

[G2] E. D. Gluskin, *Finite-dimensional analogues of spaces without basis,* Dokl. Akad. Nauk. SSSR **261** (1981), 1046–1050.

[MP] M. Marcus and G. Pisier, *Random Fourier series with applications to harmonic analysis,* Annals of Math. Studies **101** (1981), Princeton Univ. Press.

[MS] V. Milman and G. Schechtmann, *Asymptotic theory of finite dimensional normed spaces,* Springer Lectures Notes **1200** (1986).

[P] G. Pisier, *The volume of convex bodies and Banach space geometry,* Cambridge University Press, 1989.

[PT] A. Pajor and N. Tomczak-Jaegermann, *Subspaces of small codimension of finite-dimensional Banach spaces,* Proc. Amer. Math. Soc. **97** (1986), 637–642.

[S1] S. Szarek, *Metric entropy of homogeneous spaces,* Preprint.

[S2] S. Szarek, *Nets of Grassmann manifold and orthogonal groups,* Proceedings of Banach Spaces Workshop, University of Iowa Press 1982, 169–185.

[S3] S. Szarek, *The finite dimensional basis problem with an appendix on nets of Grassmann manifold,* Acta Math. **151** (1983), 153–179.

ALAIN PAJOR
EQUIPE D'ANALYSE ET MATHÉMATIQUES APPLIQUÉES
UNIVERSITÉ DE MARNE LA VALLÉE
5 BOULEVARD DESCARTES, CHAMPS SUR MARNE
77454 MARNE LA VALLÉE CEDEX 2
FRANCE
   pajor@math.univ-mlv.fr

Curvature of Nonlocal Markov Generators

MICHAEL SCHMUCKENSCHLÄGER

ABSTRACT. Bakry's curvature-dimension condition will be extended to certain nonlocal Markov generators. In particular this gives rise to a possible notion of curvature for graphs.

1. Definition of Curvature

Let (Ω, μ) be a probability space and L a self-adjoint negative but not necessarily bounded operator on $L_2(\mu)$ given by

$$Lf(x) := \int (f(y) - f(x)) K(x,y)\, \mu(dy) \qquad (1)$$

where K is a non negative symmetric kernel. Obviously L remains unchanged if we change K on the diagonal. By $P_t = e^{tL}$ we denote the continuous contraction semigroup on $L_2(\mu)$ with generator L. We will assume that P_t is ergodic and that there exists an algebra $\mathcal{A} \subseteq \bigcap_n \operatorname{dom} L^n$ of bounded functions which is a form core of L. Then the Beurling–Deny condition implies that P_t is a symmetric Markov semigroup, i.e., P_t preserves positivity and extends to a continuous contraction semigroup on $L_p(\mu)$ for all $1 \leq p < \infty$. We will also assume that $\mathcal{A}$ is stable under P_t. On $\mathcal{A} \times \mathcal{A}$ define

$$\Gamma(f,g) := \tfrac{1}{2}(L(fg) - fLg - gLf),$$
$$\Gamma_2(f,g) := \tfrac{1}{2}(L\Gamma(f,g) - \Gamma(f, Lg) - \Gamma(g, Lf)).$$

Following D. Bakry and M. Emery [BE, B] we define the curvature of L at the point $x \in \Omega$ by

$$R(L)(x) := \sup \{r \in \mathbb{R} : \Gamma_2(f,f)(x) \geq r\Gamma(f,f)(x) \text{ for all } f \in \mathcal{A}\},$$

and say that the curvature of L is bounded from below by R if $R(L)(x) \geq R$ for all $x \in \Omega$, i.e., $\Gamma_2(f,f) \geq R\Gamma(f,f)$ for all $f \in \mathcal{A}$. By the definition of R it is clear that $R(\lambda L) = \lambda R(L)$ for any $\lambda > 0$. Let us say a a word about the motivation for this definition. Assume L is the Laplacian on a Riemannian manifold, then

$\Gamma(f,f) = \|\operatorname{grad} f\|^2$ and $\Gamma_2(f,f) = \operatorname{Ric}(\operatorname{grad} f, \operatorname{grad} f) + \|Hf\|^2$, where Ric denotes the Ricci curvature and Hf the Hessian of f. Thus $R(L)$ coincides with the biggest lower bound for the Ricci curvature.

Given Γ we can define a metric d_Γ on Ω by

$$d_\Gamma(x,y) := \sup\{|f(x) - f(y)| : \Gamma(f,f) \leq 1\}.$$

If L is the Laplacian on a Riemannian manifold this is just the metric induced by the Riemannian metric.

From now on we will assume that for all $y \in \Omega$ the function $x \mapsto \sqrt{K(x,y)}$ belongs to the algebra $\mathcal{A}$. In case L is given by (1) we obtain, by putting $\nabla_y f(x) := f(y) - f(x)$ and $d(x) := \int K(x,y)\,\mu(dy)$,

$$\Gamma(f,g)(x) = \tfrac{1}{2} \int \nabla_y f(x) \nabla_y g(x) K(x,y)\,\mu(dy),$$

$$\Gamma_2(f,f)(x) = \tfrac{1}{4} \int \nabla_y f(x)^2 \left(\int K(x,z) K(y,z)\,\mu(dz) + K(x,y)(3d(y) - d(x)) \right) \mu(dy)$$

$$- \tfrac{1}{2} \iint \nabla_y f(x) \nabla_z f(x) K(x,y)(2K(y,z) - K(x,z))\,\mu(dy)\,\mu(dz)$$

For simplicity of notation let us write $\langle f \rangle$ for the mean $\int f\,d\mu$ and $\langle f, g \rangle := \langle fg \rangle$. Suppose that the curvature of L is bounded from below by $R > 0$, then

$$\langle (Lf)^2 \rangle = \langle \Gamma_2(f,f) \rangle \geq R\langle \Gamma(f,f) \rangle,$$

and by Proposition 6.3 in [B] this is equivalent to the spectral gap inequality $\langle f^2 \rangle - \langle f \rangle^2 \leq R^{-1}\langle \Gamma(f,f) \rangle$. Thus $R \leq \lambda_1$, where λ_1 is the spectral gap of $-L$.

Now we are going to check that Ledoux's proof [L1] of the concentration of measure phenomenon on compact Riemannian manifolds still works in the above setting.

1. (Bakry) If $f \in \mathcal{A}$ and if the curvature of L is bounded from below by $R > 0$, then by differentiation of the function $F(s) := P_s \Gamma(P_{t-s}f, P_{t-s}f)$ it is easy to see that $F' \geq 2RF$ and hence, for all f satisfying $\Gamma(f,f) \leq 1$,

$$\Gamma(P_t f, P_t f) \leq e^{-2Rt} P_t \Gamma(f,f) \leq e^{-2Rt}. \tag{2}$$

2. For $f \in \mathcal{A}$ and $\lambda \geq 0$ we have

$$\langle \Gamma(f, e^{\lambda f}) \rangle \leq \lambda \langle e^{\lambda f}, \Gamma(f,f) \rangle. \tag{3}$$

This follows from the elementary inequality $(e^y - e^x)/(y-x) \leq \tfrac{1}{2}(e^y + e^x)$.

3. (Ledoux) For $\lambda > 0$, $f \in \mathcal{A}$ such that $\Gamma(f,f) \leq 1$ and $\langle f \rangle = 0$ define $F(t) := \langle e^{\lambda P_t f} \rangle$, then

$$-F'(t) = -\lambda \langle LP_t f, e^{\lambda P_t f} \rangle = \lambda \langle \Gamma(P_t f, e^{\lambda P_t f}) \rangle$$

$$\leq \lambda \langle \lambda e^{P_t f}, \Gamma(P_t f, P_t f) \rangle \leq \lambda^2 e^{-2Rt} \langle e^{\lambda P_t f} \rangle = \lambda^2 e^{-2Rt} F(t),$$

where we first used (3) and then (2). Thus $(\log F)'(t) \leq -\lambda^2 e^{-2Rt}$ and since $F(\infty) = 1$ we conclude that $F(0) \leq e^{\lambda^2/2R}$, which implies the deviation inequality

$$\mu(f > \varepsilon) \leq e^{-\frac{1}{2}R\varepsilon^2}. \tag{4}$$

If Ω is a finite graph with counting measure μ we define

$$Lf(x) := \sum_{y \sim x}(f(y) - f(x)).$$

$y \sim x$ meaning that y and x are connected by an edge. Suppose that f is 1-Lipschitz with respect to the graph distance, then $\Gamma(f,f)(x) \leq d(x)/2$, where $d(x)$ is the degree of the vertex x. Also, in this case $\Gamma_2(f,f)(x)$ only depends on points whose graph distance to x is at most 2. In this respect the curvature is a local quantity.

Suppose L_1 and L_2 are generators of type (1) on $L_2(\Omega_1, \mu_1)$ and $L_2(\Omega_1, \mu_1)$ respectively. Let P_t^1 and P_t^2 be the corresponding contraction semigroups on $L_2(\Omega_1, \mu_1)$ and $L_2(\Omega_1, \mu_1)$. Then $L := L_1 \otimes 1 + 1 \otimes L_2$ is the generator of $P_t(f \otimes g) := P_t^1 f \otimes P_t^2 g$ and

$$\Gamma^L(f \otimes g, f \otimes g) = f^2 \otimes \Gamma^{L_2}(g,g) + \Gamma^{L_1}(f,f) \otimes g^2,$$
$$\Gamma_2^L(f \otimes g, f \otimes g) = f^2 \otimes \Gamma_2^{L_2}(g,g) + \Gamma_2^{L_1}(f,f) \otimes g^2 + 2\Gamma^{L_1}(f,f) \otimes \Gamma^{L_2}(g,g).$$

For simplicity we assume $L_1 = \cdots = L_n$; we will also drop the superscripts. By induction we obtain, for $F = \bigotimes_{j=1}^n f_j$,

$$\Gamma(F,F) = \sum_j f_1^2 \otimes \cdots \otimes f_{j-1}^2 \otimes \Gamma(f_j, f_j) \otimes f_{j+1}^2 \otimes \cdots \otimes f_n^2$$

$$\Gamma_2(F,F) = \sum_j f_1^2 \otimes \cdots \otimes f_{j-1}^2 \otimes \Gamma_2(f_j, f_j) \otimes f_{j+1}^2 \otimes \cdots \otimes f_n^2$$

$$+ 2\sum_{i<j} f_1^2 \otimes \cdots \otimes f_{i-1}^2 \otimes \Gamma(f_i, f_i) \otimes f_{i+1}^2 \otimes \cdots \otimes f_{j-1}^2$$

$$\otimes \Gamma(f_j, f_j) \otimes f_{j+1}^2 \otimes \cdots \otimes f_n^2.$$

Let $x = (x_1, \ldots, x_n) \in \Omega^n$; put $\widehat{x}_j = (x_1, \ldots, x_{j-1}, x_{j+1}, \ldots, x_n)$ and define $F_{\widehat{x}_j}: \Omega \to \mathbb{R}$ by $F_{\widehat{x}_j}(x_j) = F(x)$, then the terms involving Γ in the second sum can be written as

$$\iint (F_{\widehat{x}_i}(y_i) - F_{\widehat{x}_i}(x_i))^2 (F_{\widehat{x}_j}(y_j) - F_{\widehat{x}_j}(x_j))^2 K(x_i, y_i) K(x_j, y_j) \mu(dy_i) \mu(dy_j).$$

For $x \in \Omega^n$ and $y \in \Omega$, we define $\nabla_y^j F(x) := F_{\widehat{x}_j}(y) - F_{\widehat{x}_j}(x_j)$; then the preceding expression equals

$$\iint \left(\nabla_{y_i}^i \nabla_{y_j}^j F(x)\right)^2 K(x_i, y_i) K(x_j, y_j) \mu(dy_i) \mu(dy_j).$$

Hence, for all $F \in \bigotimes_{j=1}^{n} \mathcal{A}$,

$$\Gamma(F, F)(x) = \sum_j \Gamma(F_{\hat{x}_j}, F_{\hat{x}_j})(x_j),$$

$$\Gamma_2(F, F)(x) = \sum_j \Gamma_2(F_{\hat{x}_j}, F_{\hat{x}_j})(x_j)$$

$$+ 2 \sum_{i<j} \iint \left(\nabla^i_{y_i} \nabla^j_{y_j} F(x)\right)^2 K(x_i, y_i) K(x_j, y_j) \mu(dy_i) \mu(dy_j).$$

We thus have the following analogue to manifolds:

THEOREM 1.1. *Suppose the curvatures of $L_1, \ldots, L_n$ are bounded from below by $R_1, \ldots, R_n$. Then the curvature of*

$$L_1 \otimes 1 \otimes \cdots \otimes 1 + \cdots + 1 \otimes \cdots \otimes 1 \otimes L_n.$$

is bounded from below by $\min_j R_j$.

Finally let us note a somewhat more convenient formula for Γ_2: For each $y \in \Omega$ define $X_y : \mathcal{A} \to \mathcal{A}$ by $X_y f(x) = \sqrt{K(x,y)}(f(y) - f(x))$.

PROPOSITION 1.2. *For all $f, g \in \mathcal{A}$ we have*

$$\Gamma(f, g) = \tfrac{1}{2} \int X_y f X_y g \, \mu(dy),$$

$$\Gamma_2(f, f) = \tfrac{1}{2} \int \Gamma(X_y f, X_y f) + X_y f [L, X_y] f \, \mu(dy),$$

where $[L, X]$ denotes the commutator $LX - XL$.

PROOF. The first formula is just the definition of X_y. As for the second we note that

$$\tfrac{1}{2} L\Gamma(f, f) = \tfrac{1}{4} \int L(X_y f)^2 \, \mu(dy) = \tfrac{1}{2} \int \Gamma(X_y f, X_y f) + X_y f L X_y f \, \mu(dy)$$

and thus the formula follows by the definition of Γ_2. $\square$

2. Curvature of Graphs

Let us consider the trivial example $K(x, y) = 1$. In this case

$$\Gamma(f, f)(x) = \tfrac{1}{2} \int \nabla_y f(x)^2 \, \mu(dy),$$

$$\Gamma_2(f, f)(x) = \tfrac{1}{4}\left(3 \int \nabla_y f(x)^2 \, \mu(dy) - 2\left(\int \nabla_y f(x) \, \mu(dy)\right)^2\right).$$

Choosing $R = \tfrac{1}{2}$ the inequality $\Gamma_2(f, f) \geq R\Gamma(f, f)$ is thus equivalent to

$$\int \nabla_y f(x)^2 \, \mu(dy) \geq \left(\int \nabla_y f(x) \, \mu(dy)\right)^2,$$

i.e., the curvature of L is bounded from below by $\frac{1}{2}$. If Ω is a complete graph of order n then we obtain a slightly larger lower bound for the curvature: $R = 1/2 + 1/n$. In this case the deviation inequality can be obtained much more easily: following M. Ledoux [L2] and using the elementary inequality $(e^x - e^y)/(x-y) \leq \frac{1}{2}(e^x + e^y)$, we get, for all $f \in \mathcal{A}$,

$$\langle fe^f \rangle - \langle e^f \rangle \log \langle e^f \rangle \leq \frac{1}{2} \iint (f(x) - f(y))(e^{f(x)} - e^{f(y)}) \mu(dx)\mu(dy)$$

$$\leq \frac{1}{4} \iint (f(x) - f(y))^2 (e^{f(x)} + e^{f(y)}) \mu(dx)\mu(dy)$$

$$= \langle e^f, \Gamma(f,f) \rangle, \tag{5}$$

which implies (see [L2]) that $\mu(f - \langle f \rangle > \varepsilon) \leq e^{-\varepsilon^2/4}$ provided $\Gamma(f,f) \leq 1$. The latter condition implies that f is bounded: if $\langle f \rangle = 0$, then $|f| \leq \sqrt{2 - \langle f^2 \rangle}$. Ledoux's point is not this deviation inequality but rather the fact that (5) tensorizes easily. In our context this is reflected by the fact that if the curvature of L is bounded from below by R, then so is the curvature of $L \otimes 1 + 1 \otimes L$. In the particular case of the cube $\Omega = \{-1, +1\}$ and the normalized Haar measure μ_1 we get by 4 and Theorem 1.1:

COROLLARY 2.1. *Let $f : \Omega^N \to \mathbb{R}$ be a 1-Lipschitz function with respect to the graph distance. If $\langle f \rangle = 0$, then $\mu_N(f > \varepsilon) \leq e^{-2\varepsilon^2/N}$, where μ_N is the product probability.*

PROOF. Since $\int (\nabla_y f(x))^2 \mu_1(dy) \leq 1/2$ we get $\Gamma(f,f) \leq N/4$. □

More generally:

COROLLARY 2.2. *Let Ω be a complete graph of order n with normalized counting measure μ_1 and Ω^N the product graph with the product measure μ_N. Suppose $f : \Omega^N \to \mathbb{R}$ is a 1-Lipschitz function with respect to the graph distance such that $\langle f \rangle = 0$, then*

$$\mu_N(f > \varepsilon) \leq \exp\left(-\frac{n+2}{2N(n-1)}\varepsilon^2\right).$$

Now suppose $\Omega = \{0, 1, \ldots, n-1\}$ is a finite graph of order n. Any function $f : \Omega \to \mathbb{R}$ can be thought of as a vector $f = (f_0, \ldots, f_{n-1}) \in \mathbb{R}^n$. By μ we denote the counting measure on Ω and by μ_0 the normalized counting measure. For any function $f : \Omega \to \mathbb{R}$ we will also write $\langle f \rangle$ for the mean of f with respect to μ_0. Define $(Lf)_j = \sum_i (f_i - f_j) K_{i,j}$, where $K_{i,j}$ is 1 if and only if $i \sim j$ and put $x_i := f_i - f_0$ and $d_i := \sum_l K_{i,l}$, the degree of the vertex i. Then we obtain

$$\Gamma(f,f)_0 = \frac{1}{2} \sum_{i=1}^{n-1} x_i^2 K_{i,0}$$

and

$$\Gamma_2(f,f)_0 = \tfrac{1}{4}\sum_{i=1}^{n-1} x_i^2\left(\sum_l K_{l,0}K_{i,l} + K_{i,0}(3d_i - d_0 + 2)\right)$$
$$- \sum_{1\leq i<j\leq n-1} x_ix_j(K_{i,j}(K_{i,0} + K_{j,0}) - K_{i,0}K_{j,0}).$$

Define symmetric matrices $A = (a_{i,j})$ and $G = (g_{i,j})$, $1 \leq i,j \leq n-1$ by

$$a_{i,j} = \begin{cases} \tfrac{1}{2}(\sum_l K_{l,0}K_{l,i}) + \tfrac{1}{2}K_{i,0}(3d_i - d_0 + 2) & \text{if } i = j, \\ -K_{i,j}(K_{i,0} + K_{j,0}) + K_{i,0}K_{0,j} & \text{if } i \neq j, \end{cases}$$

and $g_{i,i} = K_{i,0}$ and 0 off the diagonal. Then the curvature R_0 at 0 is bounded from below by

$$\sup\{r \in \mathbb{R} : A - rG \geq 0\}.$$

In this case we conclude by (4) that for all 1-Lipschitz functions $f : \Omega \to \mathbb{R}$

$$\mu_0(f - \langle f \rangle > \varepsilon) \leq e^{-(R/d)\varepsilon^2} \tag{6}$$

where $R = \inf R_i$ and $d = \max d_i$.

The off-diagonal entries of A can take on the values 0, 1 or -1 only:

$$a_{i,j} = \begin{cases} -1 & \text{if } i \sim j \text{ and } (i \sim 0 \text{ or } j \sim 0), \\ 1 & \text{if } i \not\sim j \text{ and } j \sim 0 \text{ and } i \sim 0, \\ 0 & \text{otherwise.} \end{cases}$$

For $\varepsilon > 0$ let $B_\varepsilon(i)$ be the ball $\{j \in \Omega : d(i,j) \leq \varepsilon\}$. Let $I = B_1(0) \setminus \{0\}$ be the set of vertices, which are connected with 0 and put $J = B_2(0) \setminus B_1(0)$. Then, for $i, j \in I$ with $i \neq j$, we get

$$a_{i,i} = \tfrac{1}{2}(c_{i,0}^{(3)} + 3d_i - d_0 + 2) \quad \text{and} \quad a_{i,j} = \begin{cases} -1 & \text{if } i \sim j, \\ +1 & \text{if } i \not\sim j, \end{cases}$$

where $c_{i,0}^{(3)}$ is the number of 3-cycles containing both 0 and i. If $i \neq j$, $i,j \in J$, then

$$a_{i,i} = \tfrac{1}{2}p_{i,0}^{(2)} \quad \text{and} \quad a_{i,j} = 0$$

where $p_{i,0}^{(2)}$ is the number of paths of length 2 joining 0 and i. Finally, if $i \in I$ and $j \in J$, then

$$a_{i,j} = \begin{cases} -1 & \text{if } i \sim j, \\ 0 & \text{if } i \not\sim j. \end{cases}$$

By restricting the vertices to $I \times I$, $J \times J$, $I \times J$ and $J \times I$, we get four submatrices A_{II}, A_{JJ}, A_{IJ} and $A_{JI} = A_{IJ}^t$:

$$A = \begin{pmatrix} A_{II} & A_{IJ} \\ A_{JI} & A_{JJ} \end{pmatrix}.$$

Thus the smallest eigenvalue of A_{II} is a lower bound for the curvature at 0. Since $c_{i,0}^{(3)} \le d_i$, we conclude that

$$R_0 \le \inf\{2d_j - d_0/2 + 1 : j \sim 0\}.$$

Thus the lower bound of the curvature cannot be positive if there exist two connected points i and j such that $d_i \ge 4d_j + 2$. Another more or less obvious upper bound for R involves the diameter D of Ω:

$$D := \sup\{|f_i - f_j| : i, j \in \Omega, f \in \operatorname{Lip}_1(\Omega)\}.$$

Since for all $\varepsilon > 0$ there exists $f \in \operatorname{Lip}_1(\Omega)$ such that $\mu_0(|f - c| > (D/2 - \varepsilon)) \ge 1/n$ for all $c \in [0, D]$, we obtain, by (6), $R \le 4d\log(2n)/D^2$.

Now we turn to a more homogeneous situation: we will assume that for all $i, j \in \Omega$ there exists an isomorphism $h_{i,j}$ from $B_2(i)$ onto $B_2(j)$ such that $h_{i,j}(i) = j$. If follows that each vertex has the same degree d and a lower bound R for the curvature at any point is also a lower bound for the curvature of L. Therefore we will call these graphs, graphs of constant curvature. This situation in particular occurs if there is an underlying group structure that determines the graph: Let $I = \{g_1, \ldots, g_d\} \subseteq G \setminus \{e\}$ be a symmetric subset of a finite group G with neutral element e. Suppose further that $B_1(e) = I \cup \{e\}$ generates G, i.e., $\bigcup_n B_1(e)^n = G$. Two points $x, y \in G$ are connected if there exists a $g_j \in I$ such that $y = g_j x$. Obviously the map $h_{x,y} : B_2(x) \to B_2(y)$ defined by $h_{x,y}(z) := zx^{-1}y$ is an isomorphism.

PROPOSITION 2.3. *Let ∇_j be the operator $\nabla_j f(x) := f(g_j x) - f(x)$. Then the following statements are equivalent.*
1. *The graph distance is a bi-invariant metric.*
2. *For all $g, h \in I$, $ghg^{-1} \in I$.*
3. *For all j the operator ∇_j commutes with L.*

The Ricci curvature of a Lie group with bi-invariant Riemannian metric is always non negative. The following proposition is the analogue of this fact for finite discrete groups.

PROPOSITION 2.4. *Let G be a finite group, $I = \{g_1, \ldots, g_d\}$ a symmetric subset of $G \setminus \{e\}$ such that condition 2 of Proposition 2.3 holds. Then R is a lower bound for the curvature of G if and only if, for all $f : G \to \mathbb{R}$,*

$$\sum_{j,k} (\nabla_j \nabla_k f)^2 \ge 2R \sum_j (\nabla_j f)^2.$$

In particular the curvature of such groups is always non negative.

PROOF. By Proposition 2.3 the commutators vanish and thus the assertion follows from Proposition 1.2. □

A bi-invariant metric on a finite group need not necessarily evolve from a graph structure: The Hilbert–Schmidt metric

$$d_{HS}(\pi_1, \pi_2) := \frac{1}{2}\sqrt{(\pi_1 - \pi_2)(\pi_1 - \pi_2)^*}$$

is a bi-invariant metric on the symmetric group Π_n and $B_1(e) \setminus \{e\}$ is the set of all transpositions, but the metric d determined by $B_1(e) \setminus \{e\}$ is different from d_{HS}.

It's very likely that the curvature of (Π_n, d) is bounded from below by 2. However, this is too small to recover Maurey's deviation inequality for Π_n; see [M] or [MS].

For $n \geq 2$ let $\Omega = \{e_1, e_2, \ldots, e_n, -e_n, \ldots, -e_1\}$ be the set of extreme points of the unit ball of ℓ_1^n. Two points are connected if they are connected by a 1 dimensional face of the unit ball. In this case A is a $(2n-1) \times (2n-1)$ matrix whose diagonal is given by $\{3(n-1), \ldots, 3(n-1), (n-1)\}$. The off diagonal entries $a_{i,j}$ are 1 if $i+j = 2n-1$ and -1 otherwise. The curvature of this graph is bounded from below by n.

The curvature of the icosahedron is bounded from below by $(11 - 3\sqrt{5})/2$.

The curvature of the dodecahedron is bounded from below by 0.

For $n \geq 5$ the curvature of the additive group $\mathbb{Z}_n$ with $I = \{1, n-1\}$ is bounded from below by 0.

Let $(\{1, \ldots, n\}, d)$ be a finite metric space. Putting $K(i,j) := 1/d(i,j)^2$, then $d_\Gamma = d$. Thus the curvature can be defined for any finite metric space.

Acknowledgement

I would like to thank K. Ball and V. Milman for organizing an enjoyable semester in Berkeley. The hospitality of MSRI as well as its support are gratefully acknowledged. Many thanks also to H. Ecker for checking some examples numerically and to V. Milman for drawing my attention to the problem of finding a notion of curvature for graphs.

References

[B] D. Bakry, *L'hypercontractivité et son utilisation en théorie des semigroups*, Lectures in Probability, Ecole d'Eté de Probabilités de Saint-Flour XXII, Lecture Notes in Mathematics **1581**, Springer, 1994, 1–114.

[BE] D. Bakry and M. Emery, *Diffusions hypercontractives*, Séminaire de Probabilités XIX, Lecture Notes in Mathematics **1123**, Springer, 1985, 177–206.

[L1] M. Ledoux, *A heat semigroup approach to concentration on the sphere and on a compact Riemannian manifold*, GAFA **2**:2 (1992), 221-224.

[L2] M. Ledoux, *On Talagrand's deviation inequalities for product measures*, ESAIM Prob. Stat. **1** (1996), 63–87.

[M] B. Maurey. *Construction de suites symétriques*, C. R. Acad. Sci. Paris **288** (1979), 679-681.

[MS] V. Milman and G. Schechtman. *Asymptotic theory of finite dimensional spaces*, Lecture Notes in Mathematics **1200**, Springer, 1986.

MICHAEL SCHMUCKENSCHLÄGER
INSTITUT FÜR MATHEMATIK
JOHANNES KEPLER UNIVERSITÄT LINZ
ALTENBERGER STRASSE 69
A-4040 LINZ
AUSTRIA
schmucki@caddo.bayou.uni-linz.ac.at

An Extremal Property of the Regular Simplex

MICHAEL SCHMUCKENSCHLÄGER

ABSTRACT. If C is a convex body in $\mathbb{R}^n$ such that the ellipsoid of minimal volume containing C—the Löwner ellipsoid—is the euclidean ball B_2^n, then the mean width of C is no smaller than the mean width of a regular simplex inscribed in B_2^n.

1. Introduction and Notation

Suppose that C is a convex body in $\mathbb{R}^n$ such that 0 is an interior point of C, then the mean width $w(C)$ is defined by

$$w(C) := \int_{S^{n-1}} \left(\sup_{y \in C} \langle x, y \rangle - \inf_{y \in C} \langle x, y \rangle \right) \sigma(dx)$$
$$= 2 \int_{S^{n-1}} \sup_{y \in C} |\langle x, y \rangle| \, \sigma(dx) = 2c_n \int_{\mathbb{R}^n} \sup_{y \in C} |\langle x, y \rangle| \, \gamma_n(dx)$$

where c_n is a constant depending only on the dimension, σ the normalized Haar measure on the sphere S^{n-1} and γ_n the n-dimensional standard gaussian measure. Denoting by C^* the polar of C with respect to 0 and by $\|.\|_C$ the gauge of C, we obtain the well known formula

$$w(C) = 2c_n \int_{\mathbb{R}^n} \|x\|_{C^*} \, \gamma_n(dx) =: 2c_n \ell(C^*).$$

The euclidean ball B_2^n is the Löwner ellipsoid of C if and only if B_2^n is the John ellipsoid of C^* i.e., the ellipsoid of maximal volume contained in C^*. Hence, in order to prove that the regular simplex has minimal mean width, it is enough to prove that for all convex bodies K whose John ellipsoid is the euclidean ball, we necessarily have $\ell(K) \geq \ell(T)$, i.e., the ℓ-norm of K is bounded from below by the ℓ-norm of the regular simplex T.

The proof of this inequality will follows closely Keith Ball's proof in [B1], where it is shown that for any convex body K there exists an affine image $\widetilde{K}$ of K for which the isoperimetric quotient $\mathrm{Vol}_{n-1}(\partial \widetilde{K})/\mathrm{Vol}_n(\widetilde{K})^{\frac{n-1}{n}}$ is no larger than the isoperimetric quotient of a regular simplex. Franck Barthe [B] proved a reversed inequality: among convex bodies whose Löwner ellipsoid is the euclidean ball the regular simplex has maximal ℓ-norm.

2. The Proof

The first ingredient of the proof is a well-known theorem of F. John [J]:

THEOREM 2.1. *Let K be a convex body in $\mathbb{R}^n$. Then the euclidean ball B_n^2 is the John ellipsoid of K if and only if there exist unit vectors $u_j \in \partial K$, $1 \leq j \leq m$ and positive numbers c_j such that*

(i) $\sum_{j=1}^m c_j u_j \otimes u_j = id_{\mathbb{R}^n}$ *and*
(ii) $\sum_{j=1}^m c_j u_j = 0$.

The second is an inequality due to Brascamp and Lieb [BL]. We state this inequality in its normalized form, as it was introduced by Ball in [B2].

THEOREM 2.2. *Let u_j, $1 \leq j \leq m$, be a sequence of unit vectors in $\mathbb{R}^n$ and c_j positive numbers such that $\sum_{j=1}^m c_j u_j \otimes u_j = id_{\mathbb{R}^n}$. Then, for all nonnegative integrable functions $f_j : \mathbb{R} \to \mathbb{R}$,*

$$\int_{\mathbb{R}^n} \prod_{j=1}^m f_j(\langle x, u_j \rangle)^{c_j} \, dx \leq \prod_{j=1}^m \left(\int f_j \right)^{c_j}.$$

Equality holds if, for example, the f_j's are identical gaussians or the u_j's form an orthonormal basis.

By John's theorem there exist unit vectors $u_j \in \partial K$ and positive numbers c_j such that

$$\sum_{j=1}^m c_j u_j \otimes u_j = id_{\mathbb{R}^n} \quad \text{and} \quad \sum_{j=1}^m c_j u_j = 0.$$

Putting $v_j := \left(\sqrt{\frac{n}{n+1}} u_j, -\frac{1}{\sqrt{n+1}} \right) \in \mathbb{R}^{n+1}$ and $d_j = \frac{n+1}{n} c_j$ it is easily checked that

$$\sum_{j=1}^m d_j v_j \otimes v_j = id_{\mathbb{R}^{n+1}} \quad \text{and} \quad \sum_{j=1}^m d_j v_j = -\sqrt{n+1}\, \Pr_{n+1} \qquad (1)$$

The first identity implies $\sum d_j \langle z, v_j \rangle^2 = \|z\|_2^2$ and $\sum d_j = n+1$.

For $\alpha \in \mathbb{R}$ let μ be the measure on $\mathbb{R}$ with density

$$\tfrac{1}{\sqrt{2\pi}} \exp(\alpha t \sqrt{n+1} - t^2/2).$$

Then by (1) we obtain

$$\gamma_n \otimes \mu\Big(\bigcap [v_j \leq 0]\Big) = \iint \prod_{j=1}^m I_{(-\infty,0]}(\langle z, v_j \rangle)\, \gamma_n(dx)\, e^{\alpha t \sqrt{n+1}}\, \gamma_1(dt)$$

$$= \int_{\mathbb{R}^{n+1}} I_{(-\infty,0]}(\langle z, v_j \rangle)^{d_j} \exp\Big(-\tfrac{1}{2} \sum_{j=1}^m d_j \langle z, v_j \rangle^2 \Big)$$

$$\times \Big(\tfrac{1}{\sqrt{2\pi}}\Big)^{n+1} \exp\Big(-\alpha \sum_{j=1}^m d_j \langle z, v_j \rangle\Big)\, dz$$

Putting $f(s) = \tfrac{1}{\sqrt{2\pi}} e^{-s^2/2 - \alpha s} I_{(-\infty,0]}(s)$ we conclude by the Brascamp–Lieb inequality that

$$\gamma_n \otimes \mu\Big(\bigcap [v_j \leq 0]\Big) = \int \prod_{j=1}^m f(\langle z, v_j \rangle)^{d_j}\, dz$$

$$\leq \Big(\int f(s)\, ds\Big)^{\sum d_j} = \Big(\int f(s)\, ds\Big)^{n+1},$$

and equality holds if the vectors v_j form an orthonormal basis in $\mathbb{R}^{n+1}$ i.e., if the vectors u_j span a regular simplex. Thus, denoting by u_j^0, $1 \leq j \leq n+1$, the contact points of a regular simplex T and the euclidean ball and by v_j^0 the corresponding unit vectors in $\mathbb{R}^{n+1}$, the above inequality states that

$$\gamma_n \otimes \mu\Big(\bigcap [v_j \leq 0]\Big) \leq \gamma_n \otimes \mu\Big(\bigcap [v_j^0 \leq 0]\Big). \tag{2}$$

On the other hand

$$\bigcap [v_j \leq 0] = \Big\{z = (x, t) \in \mathbb{R}^n \times \mathbb{R} : t \geq 0,\ x \in \tfrac{t}{\sqrt{n}} \widetilde{K}\Big\},$$

where $\widetilde{K} := \bigcap [u_j \leq 1] \supseteq K$. Hence we get, by Fubini's theorem,

$$\gamma_n \otimes \mu\Big(\bigcap [v_j \leq 0]\Big) = \tfrac{1}{\sqrt{2\pi}} \int_0^\infty \gamma_n\Big(\tfrac{t}{\sqrt{n}} \widetilde{K}\Big)\, e^{\alpha t \sqrt{n+1} - t^2/2}\, dt.$$

Now, since $K \subseteq \widetilde{K}$, this implies by (2),

$$\tfrac{1}{\sqrt{2\pi}} \int_0^\infty \gamma_n\Big(\tfrac{t}{\sqrt{n}} K\Big)\, e^{\lambda t - t^2/2}\, dt \leq \tfrac{1}{\sqrt{2\pi}} \int_0^\infty \gamma_n\Big(\tfrac{t}{\sqrt{n}} T\Big)\, e^{\lambda t - t^2/2}\, dt,$$

and therefore

$$\tfrac{1}{\sqrt{2\pi}} \int_0^\infty \gamma_n\Big(\|\cdot\|_K > \tfrac{t}{\sqrt{n}}\Big)\, e^{\lambda t - t^2/2}\, dt \geq \tfrac{1}{\sqrt{2\pi}} \int_0^\infty \gamma_n\Big(\|\cdot\|_T > \tfrac{t}{\sqrt{n}}\Big)\, e^{\lambda t - t^2/2}\, dt.$$

Multiplying both sides by $e^{-\lambda^2/2}$ and integrating over $\lambda \in \mathbb{R}$ we obtain, by Fubini's theorem,

$$\int_0^\infty \gamma_n\Big(\|\cdot\|_K > \tfrac{t}{\sqrt{n}}\Big)\, dt \geq \int_0^\infty \gamma_n\Big(\|\cdot\|_T > \tfrac{t}{\sqrt{n}}\Big)\, dt$$

from which we readily deduce that $\ell(K) \geq \ell(T)$. More generally we get, for each non negative function φ,

$$\int_0^\infty \gamma_n\left(\|\cdot\|_K > \frac{t}{\sqrt{n}}\right) \int_\mathbb{R} \varphi(t-x) e^{-x^2/2}\,dx\,dt$$
$$\geq \int_0^\infty \gamma_n\left(\|\cdot\|_T > \frac{t}{\sqrt{n}}\right) \int_\mathbb{R} \varphi(t-x) e^{-x^2/2}\,dx\,dt.$$

REMARK. If we restrict the problem to convex and symmetric bodies, then we get an inequality for the distribution function (see [SS]): For all convex symmetric bodies B in $\mathbb{R}^n$ whose John ellipsoid is the euclidean ball we have, for all $t > 0$,

$$\gamma_n(\|\cdot\|_B > t) \geq \gamma_n(\|\cdot\|_\infty > t).$$

References

[B] F. Barthe, *Inégalités de Brascamp–Lieb et convexité*, to appear in Comptes Rendus Acad. Sci. Paris.

[BL] H. J. Brascamp and E. H. Lieb, *Best constants in Young's inequality, its converse, and its generalization to more than tree functions*, Advances in Math. **20** (1976), 151–173.

[B1] K. M. Ball, *Volume ratios and a reverse isoperimetric inequality*.

[B2] K. M. Ball, *Volumes of sections of cubes and related problems*, GAFA Seminar, Lecture Notes in Mathematics **1376**, Springer, 1989, 251–260.

[J] F. John, *Extremum problems with inequalities as subsidary conditions*, Courant Aniversary Volume, Interscience, New York, 1948, 187–204.

[SS] G. Schechtman and M. Schmuckenschläger, *A concentration inequality for harmonic measures on the sphere*, GAFA Seminar ed. by J. Lindenstrauss and V. Milman, Operator Theory Advances and Applications **77** (1995), 255–274.

MICHAEL SCHMUCKENSCHLÄGER
INSTITUT FÜR MATHEMATIK
JOHANNES KEPLER UNIVERSITÄT LINZ
ALTENBERGER STRASSE 69
A-4040 LINZ
AUSTRIA
schmucki@caddo.bayou.uni-linz.ac.at

Floating Body, Illumination Body, and Polytopal Approximation

CARSTEN SCHÜTT

ABSTRACT. Let K be a convex body in $\mathbb{R}^d$ and K_t its floating bodies. There is a polytope that satisfies $K_t \subset P_n \subset K$ and has at most n vertices, where

$$n \leq e^{16d} \frac{\operatorname{vol}_d(K \setminus K_t)}{t \operatorname{vol}_d(B_2^d)}.$$

Let K^t be the illumination bodies of K and Q_n a polytope that contains K and has at most n $(d-1)$-dimensional faces. Then

$$\operatorname{vol}_d(K^t \setminus K) \leq cd^4 \operatorname{vol}_d(Q_n \setminus K),$$

where

$$n \leq \frac{c}{dt} \operatorname{vol}_d(K^t \setminus K).$$

1. Introduction

We investigate the approximation of a convex body K in $\mathbb{R}^d$ by a polytope. We measure the approximation by the symmetric difference metric. The symmetric difference metric between two convex bodies K and C is

$$d_S(C, K) = \operatorname{vol}_d((C \setminus K) \cup (K \setminus C)).$$

We study in particular two questions: How well can a convex body K be approximated by a polytope P_n that is contained in K and has at most n vertices and how well can K be approximated by a polytope Q_n that contains K and has at most n $(d-1)$-dimensional faces. Macbeath [Mac] showed that the Euclidean Ball B_2^d is an extremal case: The approximation for any other convex body is better. We have for the Euclidean ball

$$c_1 \, d \operatorname{vol}_d(B_2^d) n^{-\frac{2}{d-1}} \leq d_S(P_n, B_2^d) \leq c_2 \, d \operatorname{vol}_d(B_2^d) n^{-\frac{2}{d-1}}, \tag{1.1}$$

1991 *Mathematics Subject Classification.* 52A22.

This paper was written while the author was visiting MSRI at Berkeley in the spring of 1996.

provided that $n \geq (c_3 \, d)^{(d-1)/2}$. The right hand inequality was first established by Bronshtein and Ivanov [BI] and Dudley [D$_1$,D$_2$]. Gordon, Meyer, and Reisner [GMR$_1$,GMR$_2$] gave a constructive proof for the same inequality. Müller [Mü] showed that random approximation gives the same estimate. Gordon, Reisner, and Schütt [GRS] established the left hand inequality. Gruber [Gr$_2$] obtained an asymptotic formula. If a convex body K in $\mathbb{R}^d$ has a C^2-boundary with everywhere positive curvature, then

$$\inf \{d_S(K, P_n) \mid P_n \subset K \text{ and } P_n \text{ has at most } n \text{ vertices}\}$$

is asymptotically the same as

$$\tfrac{1}{2}\mathrm{del}_{d-1} \left(\int_{\partial K} \kappa(x)^{\frac{1}{d+1}} \, d\mu(x) \right)^{\frac{d+1}{d-1}} \left(\frac{1}{n} \right)^{\frac{2}{d-1}},$$

where del_{d-1} is a constant that is connected with Delone triangulations. In this paper we are not concerned with asymptotic estimates, but with uniform.

$\mathrm{Int}(M)$ denotes the interior of a set M. $H(x, \xi)$ denotes the hyperplane that contains x and is orthogonal to ξ. $H^+(x,\xi)$ denotes the halfspace that contains the vector $x - \xi$, and $H^-(x, \xi)$ the halfspace containing $x + \xi$. $e_i, i = 1, \ldots, d$ denotes the unit vector basis in $\mathbb{R}^d$. $[A, B]$ is the convex hull of the sets A and B. The convex floating body K_t of a convex body K is the intersection of all halfspaces whose defining hyperplanes cut off a set of volume t from K.

The illumination body K^t of a convex body K is [W]

$$\{x \in \mathbb{R}^d \mid \mathrm{vol}_d([x, K] \setminus K) \leq t\}.$$

K^t is a convex body. It is enough to show this for polytopes. Let F_i denote the faces of a polytope P, ξ_i the outer normal and x_i an element of F_i. Then

$$\mathrm{vol}_d([x, P] \setminus P) = \frac{1}{d} \sum_{i=1}^n \max\{0, \langle \xi_i, x - x_i \rangle\} \mathrm{vol}_{d-1}(F_i).$$

The right-hand side is a convex function.

2. The Floating Body

THEOREM 2.1. *Let K be a convex body in $\mathbb{R}^d$. Then, for every t satisfying $0 \leq t \leq \tfrac{1}{4} e^{-5} \mathrm{vol}_d(K)$, there exist $n \in \mathbb{N}$ with*

$$n \leq e^{16d} \frac{\mathrm{vol}_d(K \setminus K_t)}{t \, \mathrm{vol}_d(B_2^d)}$$

and a polytope P_n that has n vertices and such that

$$K_t \subset P_n \subset K.$$

We want to see what kind of asymptotic estimate we get for bodies with smooth boundary from Theorem 2.1. We have [SW]

$$\operatorname{vol}_d(K \setminus K_t) \sim t^{\frac{2}{d+1}} \frac{1}{2} \left(\frac{d+1}{\operatorname{vol}_{d-1}(B_2^{d-1})} \right)^{\frac{2}{d+1}} \int_{\partial K} \kappa(x)^{\frac{1}{d+1}} d\mu(x)$$

$$\sim t^{\frac{2}{d+1}} d \int_{\partial K} \kappa(x)^{\frac{1}{d+1}} d\mu(x).$$

Since
$$n \sim d^{\frac{d}{2}} \frac{1}{t} \operatorname{vol}_d(K \setminus K_t),$$

we get

$$\operatorname{vol}_d(K \setminus K_t) \sim d \left(d^{\frac{d}{2}} \frac{1}{n} \operatorname{vol}_d(K \setminus K_t) \right)^{\frac{2}{d+1}} \int_{\partial K} \kappa(x)^{\frac{1}{d+1}} d\mu(x),$$

$$\operatorname{vol}_d(K \setminus K_t)^{\frac{d-1}{d+1}} \sim d^2 n^{-\frac{2}{d+1}} \int_{\partial K} \kappa(x)^{\frac{1}{d+1}} d\mu(x).$$

Thus we get

$$\operatorname{vol}_d(K \setminus P_n) \leq \operatorname{vol}_d(K \setminus K_t) \sim d^2 n^{-\frac{2}{d-1}} \left(\int_{\partial K} \kappa(x)^{\frac{1}{d+1}} d\mu(x) \right)^{\frac{d+1}{d-1}}.$$

When K is the Euclidean ball we get

$$\operatorname{vol}_d(B_2^d \setminus P_n) \leq c d^2 n^{-\frac{2}{d-1}} \operatorname{vol}_d(B_2^d),$$

where c is an absolute constant. If one compares this to the optimal result (1.1) one sees that there is an additional factor d.

The volume difference $\operatorname{vol}_d(P) - \operatorname{vol}_d(P_t)$ for a polytope P is of a much smaller order than for a convex body with smooth boundary. In fact, we have [S] that it is of the order $t |\ln t|^{d-1}$. In [S] this has been used to get estimates for approximation of convex bodies by polytopes.

The same result as in Theorem 2.1 holds if we fix the number of (d-1)-dimensional faces instead of the number of vertices. This follows from the same proof as for Theorem 2.1 and also from the economic cap covering for floating bodies [BL, Theorem 6]. I. Bárány showed us a proof for Theorem 2.1 using the economic cap covering. The constants are not as good as in Theorem 2.1.

The following lemmata are not new. They have usually been formulated for symmetric, convex bodies [B,H,MP]. Lemma 2.2 is due to Grünbaum [Grü].

LEMMA 2.2. *Let K be a convex body in $\mathbb{R}^d$ and let $H(cg(K), \xi)$ be the hyperplane passing through the center of gravity $cg(K)$ of K and being orthogonal to ξ. Then we have, for all $\xi \in \partial B_2^d$:*

(i) $(1 - \frac{1}{d+1})^d \operatorname{vol}_d(K) \leq \operatorname{vol}_d(K \cap H^+(cg(K), \xi)) \leq (1 - (1 - \frac{1}{d+1})^d) \operatorname{vol}_d(K)$.

(ii) *For all hyperplanes H in $\mathbb{R}^d$ that are parallel to $H(cg(K), \xi)$,*

$$\left(1 - \frac{1}{d+1}\right)^{d-1} \text{vol}_{d-1}(K \cap H) \leq \text{vol}_{d-1}(K \cap H(cg(K), \xi)).$$

The sequence $(1 - \frac{1}{d+1})^d$, $d = 2, 3, \ldots$ is monotonely decreasing. Indeed, by Bernoulli's inequality we have $1 - \frac{1}{d} \leq (1 - \frac{1}{d^2})^d$, or $\frac{d-1}{d} \leq (\frac{d^2-1}{d^2})^d$. Therefore we get $(\frac{d}{d+1})^d \leq (\frac{d-1}{d})^{d-1}$, which implies $(1 - \frac{1}{d+1})^d \leq (1 - \frac{1}{d})^{d-1}$.

Therefore we get for the inequalities (i)

$$\frac{1}{e} \text{vol}_d(K) \leq \text{vol}_d(K \cap H^+(cg(K), \xi)) \leq \left(1 - \frac{1}{e}\right) \text{vol}_d(K). \tag{2.1}$$

By the preceding calculations, $(1 + \frac{1}{d})^d$ is a monotonely increasing sequence. Thus $(1 + \frac{1}{d})^{d-1} < e$. For (ii) we get

$$\text{vol}_{d-1}(K \cap H) \leq e \, \text{vol}_{d-1}(K \cap H(cg(K), \xi)). \tag{2.2}$$

PROOF. (i) We can reduce the inequality to the case that K is a cone with a Euclidean ball of dimension $d - 1$ as base. To see this we perform a Schwarz symmetrization parallel to $H(cg(K), \xi)$ and denote the symmetrized body by $S(K)$. The Schwarz symmetrization replaces a section parallel to $H(cg(K), \xi)$ by a $(d-1)$-dimensional Euclidean sphere of the same $(d-1)$-dimensional volume. This does not change the volume of K and $K \cap H^+(cg(K), \xi)$ and the center of gravity $cg(K)$ is still an element of $H(cg(K), \xi)$. Now we consider the cone

$$[z, S(K) \cap H(cg(K), \xi)]$$

such that

$$\text{vol}_d([z, S(K) \cap H(cg(K), \xi)]) = \text{vol}_d(K \cap H^-(cg(K), \xi))$$

and such that z lies on the axis of symmetry of $S(K)$ and in $H^-(cg(K), \xi)$. See Figure 1.

The set

$$\tilde{K} = (K \cap H^+(cg(K), \xi)) \cup [z, S(K) \cap H(cg(K), \xi)]$$

is a convex set such that $\text{vol}_d(K) = \text{vol}_d(\tilde{K})$ and such that the center of gravity $cg(\tilde{K})$ of $\tilde{K}$ is contained in $[z, S(K) \cap H(cg(K), \xi)]$. Thus

$$\text{vol}_d(\tilde{K} \cap H^+(cg(\tilde{K}), \xi)) \geq \text{vol}_d(\tilde{K} \cap H^+(cg(K), \xi)) = \text{vol}_d(K \cap H^+(cg(K), \xi)).$$

We apply a similar argument to the set $S(K) \cap H^+(cg(K), \xi)$ and show that we may assume that $S(K)$ is a cone with z as its vertex. Thus we may assume that

$$K = [(0, \ldots, 0, 1), \{(x_1, \ldots, x_{d-1}, 0) \mid \sum_{i=1}^{d-1} |x_i|^2 \leq 1\}] \quad \text{and} \quad \xi = (0, \ldots, 0, 1).$$

Then

$$\text{vol}_d(K) = \frac{1}{d} \text{vol}_{d-1}(B_2^{d-1})$$

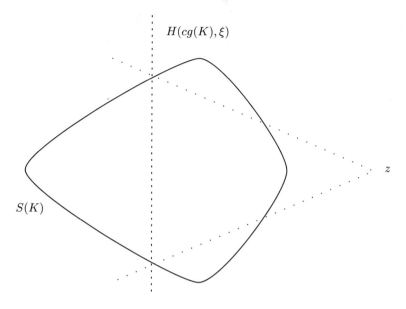

Figure 1.

and

$$\frac{1}{\operatorname{vol}_d(K)} \int_K x_d\, dx_d = d\int_0^1 t(1-t)^{d-1}\, dt = d\int_0^1 (1-s)s^{d-1}\, ds = \frac{1}{d+1}.$$

We obtain

$$\operatorname{vol}_d(K \cap H^-(cg(K),(0,\ldots,0,1))) = \left(1 - \frac{1}{d+1}\right)^d \operatorname{vol}_d(K).$$

(ii) Let H be a hyperplane parallel to $H(cg(K),\xi)$ and such that $\operatorname{vol}_{d-1}(K\cap H) > \operatorname{vol}_{d-1}(K \cap H(cg(K),\xi))$. Otherwise there is nothing to prove. We apply a Schwarz symmetrization parallel to $H(cg(K),\xi)$ to K. The symmetrized body is denoted by $S(K)$. Let z be the element of the axis of symmetry of $S(K)$ such that

$$[z, S(K) \cap H] \cap H(cg(K),\xi) = S(K) \cap H(cg(K),\xi).$$

Since $\operatorname{vol}_{d-1}(K \cap H) > \operatorname{vol}_{d-1}(K \cap H(cg(K),\xi))$ there is such a z. We may assume that $H^+(cg(K),\xi)$ is the half-space containing z. Then

$$[z, S(K) \cap H] \cap H^-(cg(K),\xi) \subset S(K) \cap H^-(cg(K),\xi),$$
$$[z, S(K) \cap H] \cap H^+(cg(K),\xi) \supset S(K) \cap H^+(cg(K),\xi).$$

Therefore

$$cg([z, S(K) \cap H]) \in H^+(cg(K),\xi).$$

Therefore, if h_{cg} denotes the distance of z to $H(cg(K), \xi)$ and h the distance of z to H, we get as in the proof of (i) that

$$h_{cg} \geq h\left(1 - \frac{1}{d+1}\right).$$

Thus we get

$$\begin{aligned}
\mathrm{vol}_{d-1}(K \cap H(cg(K), \xi)) &= \mathrm{vol}_{d-1}(S(K) \cap H(cg(K), \xi)) \\
&\geq (1 - \frac{1}{d+1})^{d-1} \mathrm{vol}_{d-1}(S(K) \cap H) \\
&= (1 - \frac{1}{d+1})^{d-1} \mathrm{vol}_{d-1}(K \cap H).
\end{aligned}$$ □

LEMMA 2.3. *Let K be a convex body in $\mathbb{R}^d$ and let $\Theta(\xi)$ be the infimum of all positive numbers t such that*

$$\mathrm{vol}_{d-1}(K \cap H(cg(K), \xi)) \geq e\, \mathrm{vol}_{d-1}(K \cap H(cg(K) + t\xi, \xi)).$$

Then

$$\frac{1}{2e^3} \mathrm{vol}_d(K) \leq \Theta(\xi) \mathrm{vol}_{d-1}(K \cap H(cg(K), \xi)) \leq e\, \mathrm{vol}_d(K).$$

PROOF. The right hand inequality follows from Fubini's theorem and Brunn–Minkowski's theorem. Now we verify the left hand inequality. We consider first the case in which, for all t such that $t > \Theta(\xi)$,

$$K \cap H(cg(K) + t\xi, \xi) = \emptyset.$$

Then, by (2.1) and (2.2),

$$\begin{aligned}
\frac{1}{e} \mathrm{vol}_d(K) &\leq \mathrm{vol}_d(K \cap H^+(cg(K), \xi)) \\
&= \int_0^{\Theta(\xi)} \mathrm{vol}_{d-1}(K \cap H(cg(K) + t\xi, \xi))\, dt \\
&\leq e\, \Theta(\xi) \mathrm{vol}_{d-1}(K \cap H(cg(K), \xi)).
\end{aligned}$$

If, for some t such that $t > \Theta(\xi)$, we have $K \cap H(cg(K) + t\xi, \xi) \neq \emptyset$, then, by continuity,

$$\mathrm{vol}_{d-1}(K \cap H(cg(K), \xi)) = e\, \mathrm{vol}_{d-1}(K \cap H(cg(K) + \Theta(\xi)\xi, \xi)).$$

We perform a Schwarz symmetrization parallel to $H(cg(K), \xi)$. We consider the cone

$$[z, S(K) \cap H(cg(K), \xi)]$$

such that z is an element of the axis of symmetry of $S(K)$ and such that

$$[z, S(K) \cap H(cg(K), \xi)] \cap H(cg(K) + \Theta(\xi)\xi, \xi) = S(K) \cap H(cg(K) + \Theta(\xi)\xi, \xi).$$

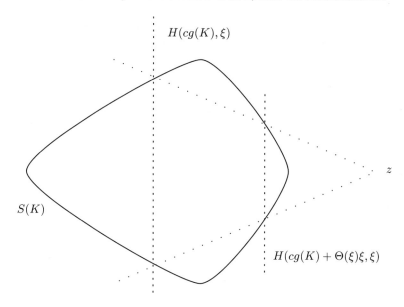

Figure 2.

Let $H^+(cg(K),\xi)$ and $H^+(cg(K)+\Theta(\xi)\xi,\xi)$ be the half-spaces that contain z. Then, by convexity,

$$[z, S(K) \cap H(cg(K),\xi)] \cap H^+(cg(K)+\Theta(\xi)\xi,\xi)$$
$$\supset S(K) \cap H^+(cg(K)+\Theta(\xi)\xi,\xi). \quad (2.3)$$

We get by (2.1)

$$\frac{1}{e}\operatorname{vol}_d(K) \leq \operatorname{vol}_d(K \cap H^+(cg(K),\xi))$$
$$= \operatorname{vol}_d(K \cap H^+(cg(K),\xi) \cap H^-(cg(K)+\Theta(\xi)\xi,\xi))$$
$$+ \operatorname{vol}_d(K \cap H^+(cg(K)+\Theta(\xi)\xi,\xi))$$
$$= \operatorname{vol}_d(S(K) \cap H^+(cg(K),\xi) \cap H^-(cg(K)+\Theta(\xi)\xi,\xi))$$
$$+ \operatorname{vol}_d(S(K) \cap H^+(cg(K)+\Theta(\xi)\xi,\xi)).$$

By the hypothesis of the lemma we have, for all s with $0 \leq s \leq \Theta(\xi)$,

$$\operatorname{vol}_{d-1}(K \cap H(cg(K),\xi)) \leq e\operatorname{vol}_{d-1}(K \cap H(cg(K)+s\xi,\xi)).$$

Using this and (2.2) we estimate the first summand. The second summand is estimated by using (2.3). Thus the above expression is not greater than

$$e^2 \operatorname{vol}_d([z, S(K) \cap H(cg(K),\xi)] \cap H^-(cg(K)+\Theta(\xi)\xi,\xi))$$
$$+ \operatorname{vol}_d([z, S(K) \cap H(cg(K),\xi)] \cap H^+(cg(K)+\Theta(\xi)\xi,\xi)).$$

This is the volume of a cone with the base $S(K) \cap H(cg(K), \xi)$. By an elementary computation for the volume of a cone we get that the latter expression is smaller than
$$2e^2 \operatorname{vol}_d([z, S(K) \cap H(cg(K), \xi)] \cap H^-(cg(K) + \Theta(\xi)\xi, \xi)).$$
Since in a cone the base has the greatest surface area, the above expression is smaller than
$$2e^2 \Theta(\xi) \operatorname{vol}_{d-1}(K \cap H(cg(K), \xi)). \qquad \square$$

LEMMA 2.4. *Let K be a convex body in $\mathbb{R}^d$. Then there is a linear transform T with $\det(T) = 1$ so that, for all $\xi \in \partial B_2^d$,*
$$\int_{T(K)} |\langle x, \xi \rangle|^2 \, dx = \frac{1}{d} \int_{T(K)} \sum_{i=1}^d |\langle x, e_i \rangle|^2 \, dx.$$

We say that a convex body is in an isotropic position if the linear transform T in Lemma 2.4 can be chosen to be the identity. See [B,H].

PROOF. We claim that there is a orthogonal transform U such that, for all $i, j = 1, \ldots, d$ with $i \neq j$,
$$\int_{U(K)} \langle x, e_i \rangle \langle x, e_j \rangle \, dx = 0.$$
Clearly, the matrix
$$\left(\int_K \langle x, e_i \rangle \langle x, e_j \rangle \, dx \right)_{i,j=1}^d$$
is symmetric. Therefore there is an orthogonal $d \times d$-matrix U so that
$$U \left(\int_K \langle x, e_i \rangle \langle x, e_j \rangle \, dx \right)_{i,j=1}^d U^t$$
is a diagonal matrix. We have
$$U \left(\int_K \langle x, e_i \rangle \langle x, e_j \rangle \, dx \right)_{i,j=1}^d U^t = \left(\int_K \sum_{i,j=1}^d u_{l,i} \langle x, e_i \rangle \langle x, e_j \rangle u_{k,j} \, dx \right)_{l,k=1}^d$$
$$= \left(\int_K \langle x, U^t(e_l) \rangle \langle x, U^t(e_k) \rangle \, dx \right)_{l,k=1}^d$$
$$= \left(\int_{U(K)} \langle y, e_l \rangle \langle y, e_k \rangle \, dy \right)_{l,k=1}^d.$$
So the latter matrix is a diagonal matrix. All the diagonal elements are strictly positive. This argument is repeated with a diagonal matrix so that the diagonal

elements turn out to be equal. Therefore there is a matrix T with $\det T = 1$ such that

$$\int_{T(K)} \langle x, e_i \rangle \langle x, e_j \rangle \, dx = \begin{cases} 0 & \text{if } i \neq j, \\ \dfrac{1}{d} \displaystyle\int_{T(K)} \sum_{j=1}^{d} |\langle x, e_j \rangle|^2 \, dx & \text{if } i = j. \end{cases}$$

From this the lemma follows. $\square$

LEMMA 2.5. *Let K be a convex body in $\mathbb{R}^d$ that is in an isotropic position and whose center of gravity is at the origin. Then, for all $\xi \in \partial B_2^d$,*

$$\frac{1}{24e^{10}} \operatorname{vol}_d(K)^3 \leq \operatorname{vol}_{d-1}(K \cap H(cg(K), \xi))^2 \frac{1}{d} \int_K \sum_{i=1}^{d} |\langle x, e_i \rangle|^2 \, dx$$

$$\leq 6 \, e^3 \operatorname{vol}_d(K)^3.$$

PROOF. By Lemma 2.4 we have, for all $\xi \in \partial B_2^d$,

$$\frac{1}{d} \int_K \sum_{i=1}^{d} |\langle x, e_i \rangle|^2 \, dx = \int_K |\langle x, \xi \rangle|^2 \, dx.$$

By Fubini's theorem, this equals

$$\int_{-\infty}^{\infty} t^2 \operatorname{vol}_{d-1}(K \cap H(t\xi, \xi)) \, dt \geq \int_0^{\Theta(\xi)} t^2 \operatorname{vol}_{d-1}(K \cap H(t\xi, \xi)) \, dt,$$

where $\Theta(\xi)$ is as defined in Lemma 2.3. By the definition of $\Theta(\xi)$ the above expression is greater than

$$\frac{1}{e} \operatorname{vol}_{d-1}(K \cap H(cg(K), \xi)) \int_0^{\Theta(\xi)} t^2 \, dt \geq \frac{1}{3e} \Theta(\xi)^3 \operatorname{vol}_{d-1}(K \cap H(cg(K), \xi)).$$

By Lemma 2.3 this is greater than

$$\frac{1}{24e^{10}} \frac{\operatorname{vol}_d(K)^3}{\operatorname{vol}_{d-1}(K \cap H(cg(K), \xi))^2}.$$

Now we show the right hand inequality. By Lemma 2.4 we have

$$\frac{1}{d} \int_K \sum_{i=1}^{d} |\langle x, e_i \rangle|^2 \, dx = \int_K |\langle x, \xi \rangle|^2 \, dx = \int_{-\infty}^{\infty} t^2 \operatorname{vol}_{d-1}(K \cap H(t\xi, \xi)) \, dt$$

$$= \int_0^{\Theta(\xi)} t^2 \operatorname{vol}_{d-1}(K \cap H(t\xi, \xi)) \, dt + \int_{\Theta(\xi)}^{\infty} t^2 \operatorname{vol}_{d-1}(K \cap H(t\xi, \xi)) \, dt$$

$$+ \int_{\Theta(-\xi)}^{0} t^2 \operatorname{vol}_{d-1}(K \cap H(t\xi, \xi)) \, dt + \int_{-\infty}^{\Theta(-\xi)} t^2 \operatorname{vol}_{d-1}(K \cap H(t\xi, \xi)) \, dt.$$

By (2.2) this is not greater than

$$\frac{e}{3}\Theta(\xi)^3 \operatorname{vol}_{d-1}(K \cap H(cg(K),\xi)) + \int_{\Theta(\xi)}^{\infty} t^2 \operatorname{vol}_{d-1}(K \cap H(t\xi,\xi))\, dt$$

$$+ \frac{e}{3}\Theta(-\xi)^3 \operatorname{vol}_{d-1}(K \cap H(cg(K),\xi)) + \int_{-\infty}^{\Theta(-\xi)} t^2 \operatorname{vol}_{d-1}(K \cap H(t\xi,\xi))\, dt.$$

The integrals can be estimated by

$$2\,\Theta(\xi)^3 \operatorname{vol}_{d-1}(K \cap H(cg(K),\xi)) \quad \text{and} \quad 2\,\Theta(-\xi)^3 \operatorname{vol}_{d-1}(K \cap H(cg(K),\xi)),$$

respectively. We treat here only the case ξ; the case $-\xi$ is treated in the same way. If the integral equals 0, there is nothing to show. If the integral does not equal 0, we have

$$\operatorname{vol}_{d-1}(K \cap H(cg(K),\xi)) = e \operatorname{vol}_{d-1}(K \cap H(cg(K) + \Theta(\xi)\xi,\xi)).$$

We consider the Schwarz symmetrization $S(K)$ of K with respect to the plane $H(cg(K),\xi)$. We consider the cone C that is generated by the Euclidean spheres $S(K) \cap H(cg(K),\xi)$ and $S(K) \cap H(cg(K) + \Theta(\xi)\xi,\xi)$. We

$$S(K) \cap H^+(cg(K) + \Theta(\xi)\xi,\xi) \subset C$$

and the height of C equals

$$\frac{\Theta(\xi)}{1 - e^{-\frac{1}{d-1}}}.$$

Since $(1 + \frac{1}{d-1})^{d-1} < e$, we have $1 - e^{-\frac{1}{d-1}} > \frac{1}{d}$. Thus the height of the cone C is less than $d\,\Theta(\xi)$. Thus, for all t with $\Theta(\xi) \le t \le d\,\Theta(\xi)$,

$$\operatorname{vol}_{d-1}(K \cap H(cg(K) + t\xi,\xi)) \le \left(1 - \frac{t}{d\Theta(\xi)}\right)^{d-1} \operatorname{vol}_{d-1}(K \cap H(cg(K),\xi)).$$

Now we get

$$\int_{\Theta(\xi)}^{\infty} t^2 \operatorname{vol}_{d-1}(K \cap H(t\xi,\xi))\, dt$$

$$\le \int_{\Theta(\xi)}^{d\,\Theta(\xi)} t^2 \left(1 - \frac{t}{d\Theta(\xi)}\right)^{d-1} \operatorname{vol}_{d-1}(K \cap H(cg(K),\xi))\, dt$$

$$\le \operatorname{vol}_{d-1}(K \cap H(cg(K),\xi))(d\,\Theta(\xi))^3 \int_0^1 s^2(1-s)^{d-1}\, ds$$

$$= \operatorname{vol}_{d-1}(K \cap H(cg(K),\xi))(d\,\Theta(\xi))^3 \frac{2}{d(d+1)(d+2)}$$

$$\le 2 \operatorname{vol}_{d-1}(K \cap H(cg(K),\xi))\Theta(\xi)^3.$$

Therefore

$$\frac{1}{d}\int_K \sum_{i=1}^d |\langle x,e_i\rangle|^2\, dx \le \left(\frac{e}{3}+2\right)(\Theta(\xi)^3 + \Theta(-\xi)^3) \operatorname{vol}_{d-1}(K \cap H(cg(K),\xi)).$$

Now we apply Lemma 2.3 and get

$$2(\frac{e}{3}+2)e^3 \frac{\operatorname{vol}_d(K)^3}{\operatorname{vol}_{d-1}(K \cap H(cg(K),\xi))^2}. \qquad \square$$

LEMMA 2.6. *Let K be a convex body in $\mathbb{R}^d$ such that the origin is an element of K. Then*

$$\frac{1}{d}\int_K \sum_{i=1}^d |\langle x, e_i\rangle|^2\, dx \geq \frac{d^{\frac{2}{d}}}{d+2}\operatorname{vol}_{d-1}(\partial B_2^d)^{-\frac{2}{d}}\operatorname{vol}_d(K)^{\frac{d+2}{d}}.$$

PROOF. Let $r(\xi)$ be the distance of the origin to the boundary of K in direction ξ. By passing to spherical coordinates we get

$$\frac{1}{d}\int_K \sum_{i=1}^d |\langle x, e_i\rangle|^2\, dx = \frac{1}{d}\int_{\partial B_2^d}\int_0^{r(\xi)} \rho^{d+1}\, d\rho\, d\xi = \frac{1}{d(d+2)}\int_{\partial B_2^d} r(\xi)^{d+2}\, d\xi$$

By Hölder's inequality, this expression is greater than

$$\frac{\operatorname{vol}_{d-1}(\partial B_2^d)}{d(d+2)}\left(\frac{1}{\operatorname{vol}_{d-1}(\partial B_2^d)}\int_{\partial B_2^d} r(\xi)^d\, d\xi\right)^{\frac{d+2}{d}}$$

$$= \frac{d^{\frac{2}{d}}}{d+2}\operatorname{vol}_{d-1}(\partial B_2^d)^{-\frac{2}{d}}\operatorname{vol}_d(K)^{\frac{d+2}{d}}. \qquad \square$$

The following lemma can be found in [MP]. It is formulated there for the case of symmetric convex bodies.

LEMMA 2.7. *Let K be a convex body in $\mathbb{R}^d$ such that the origin coincides with the center of gravity of K and such that K is in an isotropic position. Then*

$$B_2^d(cg(K), \frac{1}{24e^5\sqrt{\pi}}\operatorname{vol}_d(K)^{\frac{1}{d}}) \subset K_{\frac{1}{4e^4}\operatorname{vol}_d(K)}.$$

An affine transform can put a convex body into this position.

PROOF. As in Lemma 2.3, let $\Theta(\xi)$ be the infimum of all numbers t such that

$$\operatorname{vol}_{d-1}(K \cap H(cg(K),\xi)) \geq e\operatorname{vol}_{d-1}(K \cap H(cg(K)+t\xi,\xi)).$$

By Lemma 2.3,

$$\Theta(\xi) \geq \frac{1}{2e^3}\frac{\operatorname{vol}_d(K)}{\operatorname{vol}_{d-1}(K \cap H(cg(K),\xi))}.$$

By Lemma 2.5 we get

$$\Theta(\xi) \geq \frac{1}{2e^3\sqrt{6}e^{\frac{3}{2}}}\left(\frac{1}{\operatorname{vol}_d(K)}\frac{1}{d}\int_K \sum_{i=1}^d |\langle x, e_i\rangle|^2\, dx\right)^{\frac{1}{2}}.$$

We have

$$\operatorname{vol}_d(B_2^d) = \frac{\pi^{\frac{d}{2}}}{\Gamma(\frac{d}{2}+1)} \leq \frac{\pi^{(d-1)/2}(2e)^{\frac{d}{2}}}{d^{\frac{d+1}{2}}},$$

and thus
$$\mathrm{vol}_d(B_2^d)^{\frac{1}{d}} \leq \sqrt{\frac{2\pi e}{d}}.$$

Therefore, by Lemma 2.6,
$$\Theta(\xi) \geq \frac{1}{2e^3\sqrt{6}e^{\frac{3}{2}}} \frac{d^{\frac{1}{d}}}{\sqrt{d+2}} \left(\frac{\mathrm{vol}_d(K)}{\mathrm{vol}_{d-1}(\partial B_2^d)}\right)^{\frac{1}{d}} \geq \frac{1}{12e^5\sqrt{\pi}} \mathrm{vol}_d(K)^{\frac{1}{d}}.$$

On the other hand,
$$\mathrm{vol}_d(K \cap H^-(cg(K) + \tfrac{1}{2}\Theta(\xi)\xi, \xi)) \geq \int_{\frac{1}{2}\Theta(\xi)}^{\Theta(\xi)} \mathrm{vol}_{d-1}(K \cap H(cg(K) + t\xi, \xi))\, dt,$$

where $H^-(cg(K) + \tfrac{1}{2}\Theta(\xi)\xi, \xi)$ is the half-space not containing the origin. By the definition of $\Theta(\xi)$ this expression is greater than
$$\frac{\Theta(\xi)}{2e} \mathrm{vol}_{d-1}(K \cap H(cg(K), \xi)).$$

By Lemma 2.3 we get that this is greater than
$$\frac{1}{4e^4} \mathrm{vol}_d(K).$$

Therefore, every hyperplane that has distance
$$\frac{1}{24e^5\sqrt{\pi}} \mathrm{vol}_d(K)^{\frac{1}{d}}$$

from the center of gravity cuts off a set of volume greater than $\frac{1}{4e^4}\mathrm{vol}_d(K)$. $\square$

PROOF OF THEOREM 2.1. We are choosing the vertices $x_1, \ldots, x_n \in \partial K$ of the polytope P_n. $N(x_k)$ denotes the normal to ∂K at x_k. x_1 is chosen arbitrarily. Having chosen $x_1, \ldots, x_{k-1}$ we choose x_k such that
$$\{x_1, \ldots, x_{k-1}\} \cap \mathrm{Int}(K \cap H^-(x_k - \Delta_k N(x_k), N(x_k)) = \varnothing,$$

where Δ_k is determined by
$$\mathrm{vol}_d(K \cap H^-(x_k - \Delta_k N(x_k), N(x_k))) = t.$$

If the normal at x_k is not unique it suffices that just one of the normals satisfies the condition. It could be that the hyperplane $H(x_k - \Delta_k N(x_k), N(x_k))$ is not tangential to the floating body K_t, but this does not affect the computation. We claim that this process terminates for some n with
$$n \leq e^{16d} \frac{\mathrm{vol}_d(K \setminus K_t)}{t\, \mathrm{vol}_d(B_2^d)}. \tag{2.4}$$

This claim proves the theorem: If we cannot choose another x_{n+1}, then there is no cap of volume t that does not contain an element of the polytope $P_n = [x_1, \ldots, x_n]$. By the theorem of Hahn–Banach we get $K_t \subset P_n$. We show now

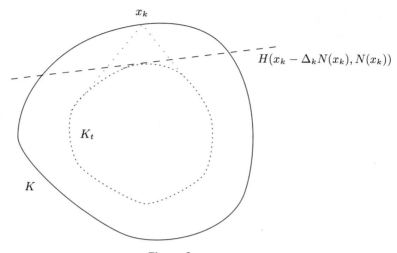

Figure 3.

the claim. We assume that we manage to choose points $x_1, \ldots, x_n$ where n is to big that (2.4) does not hold. We put

$$S_n = K \cap H^-(x_n - \Delta_n N(x_n), N(x_n)) \tag{2.5}$$

and

$$S_k = K \cap \left(\bigcap_{i=k+1}^{n} H^+(x_i - \Delta_i N(x_i), N(x_i)) \right) \cap H^-(x_k - \Delta_k N(x_k), N(x_k))$$

for $k = 1, \ldots, n-1$. For $k \neq l$, we have

$$\mathrm{vol}_d(S_k \cap S_l) = 0.$$

Let $k < l < n$. Then

$$S_k \cap S_l = K \cap \left(\bigcap_{i=k+1}^{n} H^+(x_i - \Delta_i N(x_i), N(x_i)) \right) \cap H^-(x_k - \Delta_k N(x_k), N(x_k))$$

$$\cap K \cap \left(\bigcap_{i=l+1}^{n} H^+(x_i - \Delta_i N(x_i), N(x_i)) \right) \cap H^-(x_l - \Delta_l N(x_l), N(x_l))$$

$$\subset H^+(x_l - \Delta_l N(x_l), N(x_l)) \cap H^-(x_l - \Delta_l N(x_l), N(x_l))$$

$$= H(x_l - \Delta_l N(x_l), N(x_l)).$$

Thus we have

$$\mathrm{vol}_d(S_k \cap S_l) \leq \mathrm{vol}_d(H(x_l - \Delta_l N(x_l), N(x_l))) = 0. \tag{2.6}$$

The case $k < l = n$ is shown in the same way. We have, for $k = 1, \ldots, n-1$,

$$S_k = K \cap \left(\bigcap_{i=k+1}^{n} H^+(x_i - \Delta_i N(x_i), N(x_i)) \right) \cap H^-(x_k - \Delta_k N(x_k), N(x_k))$$
$$\supset [x_k, K_t] \cap H^-(x_k - \Delta_k N(x_k), N(x_k))$$
$$\supset [x_k, (K \cap H^-(x_k - \tilde{\Delta}_k N(x_k), N(x_k))_t] \cap H^-(x_k - \Delta_k N(x_k), N(x_k)),$$

where $\tilde{\Delta}_k$ is determined by

$$\text{vol}_d(K \cap H^-(x_k - \tilde{\Delta}_k N(x_k), N(x_k))) = 4e^4 t.$$

By Lemma 2.7 there is an ellipsoid $\mathcal{E}$ contained in

$$(K \cap H^-(x_k - \tilde{\Delta}_k N(x_k), N(x_k)))_t$$

whose center is $cg(K \cap H^-(x_k - \tilde{\Delta}_k N(x_k), N(x_k)))$ and that has volume

$$\text{vol}_d(\mathcal{E}) = \frac{4e^4}{(24e^5\sqrt{\pi})^d} \, t \, \text{vol}_d(B_2^d)$$

Since $(K \cap H^-(x_k - \tilde{\Delta}_k N(x_k), N(x_k)))_t$ is contained in K_t, $\mathcal{E}$ is contained in K_t. Thus

$$S_k \supset [x_k, \mathcal{E}] \cap H^-(x_k - \Delta_k N(x_k), N(x_k)).$$

We claim now that $[x_k, \mathcal{E}] \cap H^-(x_k - \Delta_k N(x_k), N(x_k))$ contains an ellipsoid $\tilde{\mathcal{E}}$ such that

$$\text{vol}_d(\tilde{\mathcal{E}}) = \frac{4e^4}{(24e^5\sqrt{\pi})^d} \frac{1}{(4e^5)^d} \, t \, \text{vol}_d(B_2^d),$$

and consequently

$$\text{vol}_d(S_k) \geq \frac{4e^4}{(24e^5\sqrt{\pi})^d} \frac{1}{(4e^5)^d} \, t \, \text{vol}_d(B_2^d) = \frac{4e^4}{(96e^{10}\sqrt{\pi})^d} \, t \, \text{vol}_d(B_2^d). \quad (2.7)$$

For this we have to see that $\tilde{\Delta}_k \leq 4e^5 \Delta_k$. By the assumption $t \leq \frac{1}{4} e^{-5} \text{vol}_d(K)$ we get

$$\text{vol}_d(K \cap H^-(x_k - \tilde{\Delta}_k N(x_k), N(x_k))) \leq \frac{1}{e} \, \text{vol}_d(K).$$

Therefore, by (2.1), $cg(K) \in H^+(x_k - \tilde{\Delta}_k N(x_k), N(x_k))$. We consider two cases. If

$$\text{vol}_{d-1}(K \cap H(x_k - \tilde{\Delta}_k N(x_k), N(x_k))) \leq \text{vol}_{d-1}(K \cap H(x_k - \Delta_k N(x_k), N(x_k))),$$

the theorem of Brunn–Minkowski implies that, for all s in the range $\Delta_k \leq s \leq \tilde{\Delta}_k$, we have

$$\text{vol}_{d-1}(K \cap H(cg(K), N(x_k))) \leq \text{vol}_{d-1}(K \cap H(x_k - \tilde{\Delta}_k N(x_k), N(x_k)))$$
$$\leq \text{vol}_{d-1}(K \cap H(x_k - sN(x_k), N(x_k))). \quad (2.8)$$

We get, by (2.2),
$$\Delta_k \geq \frac{t}{e\ \mathrm{vol}_{d-1}(K \cap H(cg(K), N(x_k)))}.$$

By (2.8),

$(\tilde{\Delta}_k - \Delta_k)\,\mathrm{vol}_{d-1}(K \cap H(cg(K), N(x_k)))$
$\leq \mathrm{vol}_d(K \cap H^-(x_k - \tilde{\Delta}_k N(x_k), N(x_k))) - \mathrm{vol}_d(K \cap H^-(x_k - \Delta_k N(x_k), N(x_k))).$

This implies
$$\tilde{\Delta}_k - \Delta_k \leq \frac{(4e^4 - 1)t}{\mathrm{vol}_{d-1}(K \cap H(cg(K), N(x_k)))}.$$

Therefore
$$\tilde{\Delta}_k \leq \frac{(4e^4 - 1)t}{\mathrm{vol}_{d-1}(K \cap H(cg(K), N(x_k)))} + \Delta_k \leq 4e^5\,\Delta_k.$$

If
$$\mathrm{vol}_{d-1}(K \cap H(x_k - \Delta_k N(x_k), N(x_k))) \leq \mathrm{vol}_{d-1}(K \cap H(x_k - \tilde{\Delta}_k N(x_k), N(x_k))),$$

the theorem of Brunn–Minkowski implies that, for all u in the range $0 \leq u \leq \Delta_k$, and all s in the range $\Delta_k \leq s \leq \tilde{\Delta}_k$, we have

$\mathrm{vol}_{d-1}(K \cap H(x_k - uN(x_k), N(x_k))) \leq \mathrm{vol}_{d-1}(K \cap H(x_k - \Delta_k N(x_k), N(x_k)))$
$\leq \mathrm{vol}_{d-1}(K \cap H(x_k - sN(x_k), N(x_k))).$

We get
$$\Delta_k \geq \frac{t}{\mathrm{vol}_{d-1}(K \cap H(x_k - \Delta_k N(x_k), N(x_k)))}$$

and
$$\tilde{\Delta}_k - \Delta_k \leq \frac{(4e^4 - 1)t}{\mathrm{vol}_{d-1}(K \cap H(x_k - \Delta_k N(x_k), N(x_k)))}.$$

Therefore
$$\tilde{\Delta}_k \leq \frac{(4e^4 - 1)t}{\mathrm{vol}_{d-1}(K \cap H(x_k - \Delta_k N(x_k), N(x_k)))} + \Delta_k \leq 4e^4 \Delta_k.$$

We have verified (2.7). From (2.6) and (2.7) we get

$$\mathrm{vol}_d(K \setminus K_t) \geq \mathrm{vol}_d\left(\bigcup_{k=1}^n S_k\right) = \sum_{k=1}^n \mathrm{vol}_d(S_k) \geq n\frac{4e^4}{(96e^{10}\sqrt{\pi})^d}\,t\,\mathrm{vol}_d(B_2^d).$$

Thus we get the desired equation (2.4):
$$\mathrm{vol}_d(K \setminus K_t) \geq e^{-16d} n\,t\,\mathrm{vol}_d(B_2^d). \qquad \square$$

3. The Illumination Body

THEOREM 3.1. *Let K be a convex body in $\mathbb{R}^d$ such that*

$$\frac{1}{c_1} B_2^d \subset K \subset c_2 B_2^d.$$

Let $0 \leq t \leq (5c_1c_2)^{-d-1} \operatorname{vol}_d(K)$ and let $n \in \mathbb{N}$ be such that

$$(\frac{128}{7}\pi)^{(d-1)/2} \leq n \leq \frac{1}{32\, edt} \operatorname{vol}_d(K^t \setminus K).$$

Then we have, for every polytope P_n that contains K and has at most n $(d-1)$-dimensional faces,

$$\operatorname{vol}_d(K^t \setminus K) \leq 10^7 \, d^2 (c_1 c_2)^{2+\frac{1}{d-1}} \operatorname{vol}_d(P_n \setminus K).$$

We want to see what this result means for bodies with a smooth boundary. We have the asymptotic formula [W]

$$\lim_{t \to 0} \frac{\operatorname{vol}_d(K^t) - \operatorname{vol}_d(K)}{t^{\frac{2}{d+1}}} = \frac{1}{2} \left(\frac{d(d+1)}{\operatorname{vol}_{d-1}(B_2^{d-1})} \right)^{\frac{2}{d+1}} \int_{\partial K} \kappa(x)^{\frac{1}{d+1}} d\mu(x).$$

Thus

$$\operatorname{vol}_d(K^t) - \operatorname{vol}_d(K) \sim t^{\frac{2}{d+1}} d \int_{\partial K} \kappa(x)^{\frac{1}{d+1}} d\mu(x).$$

And by the theorem we have

$$n \sim \frac{1}{dt} \operatorname{vol}_d(K^t \setminus K).$$

Thus

$$\operatorname{vol}_d(K^t) - \operatorname{vol}_d(K) \sim d \left(\frac{1}{dn} \operatorname{vol}_d(K^t \setminus K) \right)^{\frac{2}{d+1}} \int_{\partial K} \kappa(x)^{\frac{1}{d+1}} d\mu(x),$$

or

$$\operatorname{vol}_d(K^t \setminus K)^{\frac{d-1}{d+1}} \sim d \left(\frac{1}{dn} \right)^{\frac{2}{d+1}} \int_{\partial K} \kappa(x)^{\frac{1}{d+1}} d\mu(x),$$

$$\operatorname{vol}_d(K^t \setminus K) \sim d \left(\frac{1}{n} \right)^{\frac{2}{d-1}} \left(\int_{\partial K} \kappa(x)^{\frac{1}{d+1}} d\mu(x) \right)^{\frac{d+1}{d-1}}.$$

By Theorem 3.1 we now get

$$\operatorname{vol}_d(P_n \setminus K) \gtrsim \frac{1}{d} \left(\frac{1}{c_1 c_2} \right)^{1+\frac{d}{d+1}} \left(\frac{1}{n} \right)^{\frac{2}{d-1}} \left(\int_{\partial K} \kappa(x)^{\frac{1}{d+1}} d\mu(x) \right)^{\frac{d+1}{d-1}}.$$

By a theorem of F. John [J] we have $c_1 c_2 \leq d$.

The following lemma is due to Bronshtein and Ivanov [BI] and Dudley [D_1, D_2]. It can also be found in [GRS].

LEMMA 3.2. *For all dimensions d, $d \geq 2$, and all natural numbers n, $n \geq 2d$, there is a polytope Q_n that has n vertices and is contained in the Euclidean ball B_2^d such that*

$$d_H(Q_n, B_2^d) \leq \frac{16}{7} \left(\frac{\mathrm{vol}_{d-1}(\partial B_2^d)}{\mathrm{vol}_{d-1}(B_2^{d-1})} \right)^{\frac{2}{d-1}} n^{-\frac{2}{d-1}}.$$

We have

$$\mathrm{vol}_{d-1}(\partial B_2^d) = d \, \mathrm{vol}_d(B_2^d) = d \frac{\pi^{\frac{d}{2}}}{\Gamma(\frac{d}{2}+1)}$$

$$= d\sqrt{\pi} \frac{\Gamma(\frac{d-1}{2}+1)}{\Gamma(\frac{d}{2}+1)} \mathrm{vol}_{d-1}(B_2^{d-1}) \leq d\sqrt{\pi} \, \mathrm{vol}_{d-1}(B_2^{d-1}). \quad (3.1)$$

Since $d^{\frac{2}{d-1}} \leq 4$ and $(1-t)^d \geq 1 - dt$, (3.1) yields

$$d_H(B_2^d, Q_n) \leq \frac{16}{7} \left(\frac{d\sqrt{\pi}}{n} \right)^{\frac{2}{d-1}} \leq \frac{64}{7} \pi n^{-\frac{2}{d-1}}. \quad (3.2)$$

PROOF OF THEOREM 3.1. We denote the $(d-1)$-dimensional faces of P_n by F_i, for $i = 1, \ldots, n$, and the cones generated by the origin and a face F_i by C_i, for $i = 1, \ldots, n$. Take $x_i \in F_i$ and let ξ_i, with $\|\xi_i\|_2 = 1$, be orthogonal to F_i and pointing to the outside of P_n. Then $H(x_i, \xi_i)$ is the hyperplane containing F_i and $H^+(x_i, \xi_i)$ the halfspace containing P_n. See Figure 4.

We may assume that the hyperplanes $H(x_i, \xi_i)$, $i = 1, \ldots, n$, are supporting hyperplanes of K. Otherwise we can choose a polytope of lesser volume. Let Δ_i be the height of the set

$$K^t \cap H^-(x_i, \xi_i) \cap C_i,$$

that is, the smallest number s such that

$$K^t \cap H^-(x_i, \xi_i) \cap C_i \subset H^+(x_i + s\xi_i, \xi_i).$$

Let z_i be a point in $\partial K^t \cap C_i$ where the height Δ_i is attained. We may assume that $B_2^d \subset K \subset cB_2^d$ where $c = c_1 c_2$. Also we may assume that

$$P_n \subset 2cB_2^d \quad (3.3)$$

if we allow twice as many faces. This follows from (3.2): There is a polytope Q_k such that $\frac{1}{2}B_2^d \subset Q_k \subset B_2^d$ and the number of vertices k is smaller than $(\frac{128}{7}\pi)^{(d-1)/2}$. Thus Q_k^* satisfies $B_2^d \subset Q_k^* \subset 2B_2^d$ and has at most $(\frac{128}{7}\pi)^{(d-1)/2}$ $(d-1)$-dimensional faces. As the new polytope P_n we choose the intersection of cQ_k^* with the original polytope P_n. Since we have by assumption that n is greater than $(\frac{128}{7}\pi)^{(d-1)/2}$ the new polytope has at most

$$\frac{1}{16 \, edt} \mathrm{vol}_d(K^t \setminus K). \quad (3.4)$$

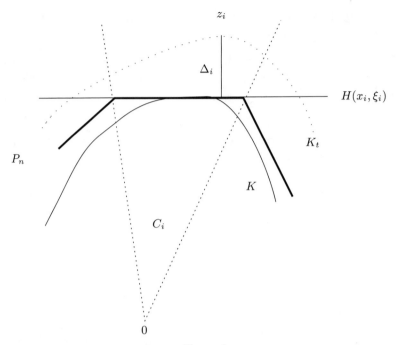

Figure 4.

$(d-1)$-dimensional faces.

We show first that for t with $0 \leq t \leq (5cd)^{-d-1} \operatorname{vol}_d(K)$ and all i, $i = 1, \ldots, n$ we have

$$\Delta_i \leq \frac{1}{d} \tag{3.5}$$

Assume that there is a face F_i with $\Delta_i > \frac{1}{d}$. Consider the smallest infinite cone D_i having z_i as vertex and containing K. Since $H(x_i, \xi_i)$ is a supporting hyperplane to K and $K \subset c\, B_2^d$ we have

$$K \subset D_i \cap H^+(x_i, \xi_i) \cap H^-(x_i - 2c\xi_i, \xi_i)$$

and

$$D_i \cap H^-(x_i, \xi) = [z_i, K] \cap H^-(x_i, \xi)$$

We have

$$t = \operatorname{vol}_d([z_i, K] \setminus K) \geq \operatorname{vol}_d([z_i, K] \cap H^-(x_i, \xi_i)) = \operatorname{vol}_d(D_i \cap H^-(x_i, \xi_i)) =$$

$$\frac{1}{d} \Delta_i \operatorname{vol}_{d-1}(D_i \cap H(x_i, \xi_i)) \geq \frac{1}{d^2} \operatorname{vol}_{d-1}(D_i \cap H(x_i, \xi_i))$$

Thus

$$\operatorname{vol}_{d-1}(D_i \cap H(x_i, \xi_i)) \leq d^2 t \tag{3.6}$$

Since (3.5) does not hold we have

$$\text{vol}_{d-1}(D_i \cap H(x_i - 2c\xi_i, \xi_i)) = \left(\frac{2c + \Delta_i}{\Delta_i}\right)^{d-1} \text{vol}_{d-1}(D_i \cap H(x_i, \xi_i))$$
$$\leq (2cd + 1)^{d-1} \text{vol}_{d-1}(D_i \cap H(x_i, \xi_i)).$$

By (3.6) we get

$$\text{vol}_{d-1}(D_i \cap H(x_i - 2c\xi_i, \xi_i)) \leq (2cd+1)^{d-1} d^2 t \leq (3cd)^{d-1} d^2 t.$$

Thus

$$\text{vol}_d(K) \leq \text{vol}_d(D_i \cap H^+(x_i, \xi_i) \cap H^-(x_i - 2c\xi_i, \xi_i))$$
$$\leq 2c(3cd)^{d-1} d^2 t \leq (3cd)^{d+1} t,$$

and we conclude that

$$t \geq (3cd)^{-d-1} \text{vol}_d(K).$$

This is a contradiction to the assumption on t in the hypothesis of the theorem. Thus we have shown (3.5). We consider now two cases: All those heights Δ_i that are smaller than $2dt/\text{vol}_{d-1}(F_i)$ and those that are greater. We may assume that Δ_i, $i = 1, \ldots, k$ are smaller than $2dt/\text{vol}_{d-1}(F_i)$ and Δ_i, $i = k+1, \ldots, n$ are strictly greater. We have

$$\text{vol}_d((K^t \setminus P_n) \cap C_i) = \int_0^{\Delta_i} \text{vol}_{d-1}((K^t \setminus P_n) \cap C_i \cap H(x_i + s\xi_i, \xi_i)) \, ds.$$

Since $B_2^d \subset K \subset P_n$ we get

$$\text{vol}_d((K^t \setminus P_n) \cap C_i) \leq \int_0^{\Delta_i} \text{vol}_{d-1}(F_i)(1+s)^{d-1} \, ds \leq \Delta_i (1+\Delta_i)^{d-1} \text{vol}_{d-1}(F_i).$$

By (3.5) we get

$$\text{vol}_d((K^t \setminus P_n) \cap C_i) \leq \Delta_i \left(1 + \frac{1}{d}\right)^{d-1} \text{vol}_{d-1}(F_i).$$

For $i = 1, \ldots, k$ we get

$$\text{vol}_d((K^t \setminus P_n) \cap C_i) \leq \frac{2dt}{\text{vol}_{d-1}(F_i)} \left(1 + \frac{1}{d}\right)^{d-1} \text{vol}_{d-1}(F_i) \leq 2edt.$$

Thus

$$\text{vol}_d\left((K^t \setminus P_n) \cap \bigcup_{i=1}^k C_i\right) \leq 2kedt \leq 2nedt.$$

By (3.4) we get

$$\text{vol}_d\left((K^t \setminus P_n) \cap \bigcup_{i=1}^k C_i\right) \leq \tfrac{1}{8} \text{vol}_d(K^t \setminus K). \tag{3.7}$$

Now we consider the other faces. For $i = k+1, \ldots, n$, we have

$$\Delta_i \geq \frac{2dt}{\text{vol}_{d-1}(F_i)}. \tag{3.8}$$

We show that, for $i = k+1, \ldots, n$, we have

$$\Delta_i \leq 5c \left(\frac{5c \, \text{vol}_{d-1}(F_i)}{2d \, \text{vol}_d(K)} \right)^{\frac{1}{d-1}}. \tag{3.9}$$

Suppose that there is a face F_i so that (3.9) does not hold. Then

$$t = \text{vol}_d([z_i, K] \setminus K) \geq \text{vol}_d([z_i, K] \cap H^-(x_i, \xi_i)) = \frac{\Delta_i}{d} \text{vol}_{d-1}([z_i, K] \cap H(x_i, \xi_i)).$$

Therefore we get, by (3.8),

$$\text{vol}_{d-1}([z_i, K] \cap H(x_i, \xi_i)) \leq \frac{dt}{\Delta_i} \leq \frac{1}{2} \text{vol}_{d-1}(F_i). \tag{3.10}$$

Since $K \subseteq B_2^d$ we have

$$K \subset D_i \cap H^+(x_i, \xi_i) \cap H^-(x_i - 2c\xi_i, \xi_i).$$

Thus

$$\text{vol}_d(K) \leq \text{vol}_d(D_i \cap H^-(x_i - 2c\xi_i, \xi_i)).$$

The cone $D_i \cap H^-(x_i - 2c\xi_i, \xi_i)$ has a height equal to $2c + \Delta_i$. Therefore

$$\text{vol}_d(K) \leq \frac{1}{d}(2c + \Delta_i) \left(\frac{2c + \Delta_i}{\Delta_i} \right)^{d-1} \text{vol}_{d-1}(D_i \cap H(x_i, \xi_i)).$$

By (3.5) we have $\Delta_i \leq 1$. Therefore we get

$$\text{vol}_d(K) \leq \frac{3c}{d} \left(\frac{3c}{\Delta_i} \right)^{d-1} \text{vol}_{d-1}(D_i \cap H(x_i, \xi_i))$$

$$= \frac{3c}{d} \left(\frac{3c}{\Delta_i} \right)^{d-1} \text{vol}_{d-1}([z_i, K] \cap H(x_i, \xi_i)).$$

By (3.10) we get

$$\text{vol}_d(K) \leq \frac{3c}{2d} \left(\frac{3c}{\Delta_i} \right)^{d-1} \text{vol}_{d-1}(F_i),$$

which implies (3.9).

Let y_i be the unique point

$$y_i = [0, z_i] \cap H(x_i, \xi_i).$$

We want to make sure that $y_i \in F_i \cap [z_i, K]$. This holds since $z_i \in C_i \cap H^-(x_i, \xi_i)$ and $\Delta_i > 0$. Since $y_i \in F_i$ we have

$$\text{vol}_{d-1}(F_i) = \frac{\text{vol}_{d-1}(B_2^{d-1})}{\text{vol}_{d-2}(\partial B_2^{d-1})} \int_{\partial B_2^{d-1}} r_i(\eta)^{d-1} \, d\mu(\eta),$$

where $r_i(\eta)$ is the distance of y_i to the boundary ∂F_i in direction η, $\eta \in \partial B_2^{d-1}$, and, since $y_i \in F_i \cap [z_i, K]$, we have

$$\text{vol}_{d-1}(F_i \cap [z_i, K]) = \frac{\text{vol}_{d-1}(B_2^{d-1})}{\text{vol}_{d-2}(\partial B_2^{d-1})} \int_{\partial B_2^{d-1}} \rho_i(\eta)^{d-1} \, d\mu(\eta),$$

where $\rho_i(\eta)$ is the distance of y_i to the boundary $\partial(F_i \cap [z_i, K])$. Consider the set
$$A_i = \{\eta \mid (1 - \tfrac{1}{4d})r_i(\eta) \le \rho_i(\eta) \}.$$
We show that
$$\tfrac{1}{4} \operatorname{vol}_{d-1}(F_i) \le \frac{\operatorname{vol}_{d-1}(B_2^{d-1})}{\operatorname{vol}_{d-2}(\partial B_2^{d-1})} \int_{A_i^c} r_i(\eta)^{d-1} - \rho_i(\eta)^{d-1}\, d\mu(\eta) \qquad (3.11)$$
We have
$$\frac{\operatorname{vol}_{d-1}(B_2^{d-1})}{\operatorname{vol}_{d-2}(\partial B_2^{d-1})} \int_{A_i} r_i(\eta)^{d-1} - \rho_i(\eta)^{d-1}\, d\mu(\eta)$$
$$\le \frac{\operatorname{vol}_{d-1}(B_2^{d-1})}{\operatorname{vol}_{d-2}(\partial B_2^{d-1})} \int_{A_i} r_i(\eta)^{d-1}(1 - (1 - \tfrac{1}{4d})^{d-1})\, d\mu(\eta)$$
$$\le \frac{1}{4} \frac{\operatorname{vol}_{d-1}(B_2^{d-1})}{\operatorname{vol}_{d-2}(\partial B_2^{d-1})} \int_{A_i} r_i(\eta)^{d-1}\, d\mu(\eta) \le \tfrac{1}{4} \operatorname{vol}_{d-1}(F_i).$$

Therefore
$$\frac{\operatorname{vol}_{d-1}(B_2^{d-1})}{\operatorname{vol}_{d-2}(\partial B_2^{d-1})} \int_{A_i^c} r_i(\eta)^{d-1} - \rho_i(\eta)^{d-1}\, d\mu(\eta)$$
$$\ge \frac{\operatorname{vol}_{d-1}(B_2^{d-1})}{\operatorname{vol}_{d-2}(\partial B_2^{d-1})} \int_{\partial B_2^{d-1}} r_i(\eta)^{d-1} - \rho_i(\eta)^{d-1}\, d\mu(\eta)$$
$$- \frac{\operatorname{vol}_{d-1}(B_2^{d-1})}{\operatorname{vol}_{d-2}(\partial B_2^{d-1})} \int_{A_i} r_i(\eta)^{d-1} - \rho_i(\eta)^{d-1}\, d\mu(\eta)$$
$$\ge \operatorname{vol}_{d-1}(F_i) - \operatorname{vol}_{d-1}(F_i \cap [z_i, K]) - \tfrac{1}{4}\operatorname{vol}_{d-1}(F_i).$$

By (3.10) this is greater than $\tfrac{1}{4}\operatorname{vol}_{d-1}(F_i)$. This implies
$$\tfrac{1}{4}\operatorname{vol}_{d-1}(F_i) \le \frac{\operatorname{vol}_{d-1}(B_2^{d-1})}{\operatorname{vol}_{d-2}(\partial B_2^{d-1})} \int_{A_i^c} r_i(\eta)^{d-1} - \rho_i(\eta)^{d-1}\, d\mu(\eta).$$
Thus we have established (3.11).

We shall show that
$$\operatorname{vol}_d((K^t \setminus P_n) \cap C_i) \le 10^6 \, ed^2 c^{2 + \frac{1}{d-1}} \operatorname{vol}_d((P_n \setminus K) \cap C_i). \qquad (3.12)$$
We have
$$\operatorname{vol}_d(D_i^c \cap H^+(x_i, \xi_i) \cap C_i) \le \operatorname{vol}_d((P_n \setminus K) \cap C_i).$$
Compare Figure 5. Therefore, if we want to verify (3.12) it is enough to show that
$$\operatorname{vol}_d((K^t \setminus P_n) \cap C_i) \le 10^6 \, ed^2 c^{2 + \frac{1}{d-1}} \operatorname{vol}_d(D_i^c \cap H^+(x_i, \xi_i) \cap C_i).$$
We may assume that y_i and z_i are orthogonal to $H(x_i, \xi_i)$. This is accomplished by a linear, volume preserving map: Any vector orthogonal to ξ_i is mapped onto itself and y_i is mapped to $\langle \xi_i, y_i \rangle \xi_i$. See Figure 6.

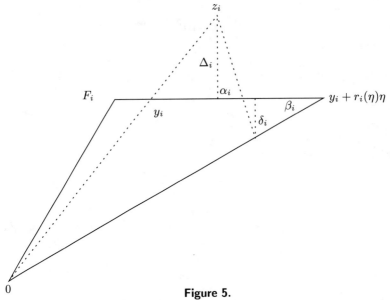

Figure 5.

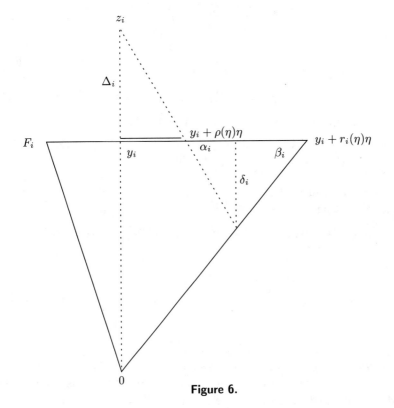

Figure 6.

Let $w_i(\eta) \in D_i^c \cap H^+(x_i, \xi_i) \cap C_i$ such that $w_i(\eta)$ is an element of the 2-dimensional subspace containing 0, y_i, and $y_i + \eta$. Let $\delta_i(\eta)$ be the distance of $w_i(\eta)$ to the plane $H(x_i, \xi_i)$. Then

$$\frac{1}{d} \frac{\mathrm{vol}_{d-1}(B_2^{d-1})}{\mathrm{vol}_{d-2}(\partial B_2^{d-1})} \int_{A_i^c} (r_i(\eta)^{d-1} - \rho_i(\eta)^{d-1}) \delta_i(\eta) \, d\mu(\eta)$$

$$\leq \mathrm{vol}_d(D_i^c \cap H^+(x_i, \xi_i) \cap C_i).$$

Thus, in order to verify (3.12), it suffices to show that

$$\mathrm{vol}_d((K^t \setminus P_n) \cap C_i)$$
$$\leq 10^6 \, ed^2 c^{2 + \frac{1}{d-1}} \frac{1}{d} \frac{\mathrm{vol}_{d-1}(B_2^{d-1})}{\mathrm{vol}_{d-2}(\partial B_2^{d-1})} \int_{A_i^c} (r_i(\eta)^{d-1} - \rho_i(\eta)^{d-1}) \delta_i(\eta) \, d\mu(\eta). \quad (3.13)$$

In order to do this we shall show that for all $i = k+1, \ldots, n$ and all $\eta \in A_i^c$ there is $w_i(\eta)$ such that the distance $\delta_i(\eta)$ of w_i from $H(x_i, \xi_i)$ satisfies

$$\frac{\Delta_i}{\delta_i} \leq \begin{cases} 32dc & \text{if } 0 \leq \alpha_i \leq \frac{\pi}{4}, \\ \frac{160 \, dc^2}{r_i} \left(\frac{5c \, \mathrm{vol}_{d-1}(F_i)}{2d \, \mathrm{vol}_d(K)} \right)^{\frac{1}{d-1}} & \text{if } \frac{\pi}{4} \leq \alpha_i \leq \frac{\pi}{2}. \end{cases} \quad (3.14)$$

The angles $\alpha_i(\eta)$ and $\beta_i(\eta)$ are given in Figure 6. We have for all $\eta \in A_i^c$

$$\delta_i = (r_i - \rho_i) \frac{\sin(\alpha_i) \sin(\beta_i)}{\sin(\pi - \alpha_i - \beta_i)}, \quad (3.15)$$

$$\Delta_i = \rho_i \tan \alpha_i,$$

with $0 \leq \alpha_i, \beta_i \leq \frac{\pi}{2}$. Thus we get

$$\frac{\Delta_i}{\delta_i} \leq \frac{\rho_i}{r_i - \rho_i} \frac{\sin(\pi - \alpha_i - \beta_i)}{\cos(\alpha_i) \sin(\beta_i)} \leq \frac{\rho_i}{(r_i - \rho_i) \cos(\alpha_i) \sin(\beta_i)}.$$

By (3.11) we have $\rho_i \leq (1 - \frac{1}{4d}) r_i$. Therefore

$$\frac{\Delta_i}{\delta_i} \leq \frac{4d}{\cos(\alpha_i) \sin(\beta_i)}.$$

Since $B_2^d \subset K \subset P_n \subset 2c \, B_2^d$ we have $\tan \beta_i \geq \frac{1}{4c}$: Here we have to take into account that we applied a transform to K mapping y_i to $\langle \xi_i, y_i \rangle \xi_i$. That leaves the distance of F_i to the origin unchanged and $r_i(\eta)$ is less than $4c$. If $\beta_i \geq \frac{\pi}{4}$ we have $\sin \beta_i \geq \frac{1}{\sqrt{2}}$. If $\beta_i \leq \frac{\pi}{4}$ then $\frac{1}{4c} \leq \tan \beta_i = \frac{\sin \beta_i}{\cos \beta_i} \leq \sqrt{2} \sin \beta_i$. Therefore we get

$$\frac{\Delta_i}{\delta_i} \leq \frac{16\sqrt{2} \, dc}{\cos \alpha_i}.$$

Therefore we get, for all $0 \leq \alpha_i \leq \frac{\pi}{4}$,

$$\frac{\Delta_i}{\delta_i} \leq 32dc.$$

By (3.9) and (3.15) we get

$$\frac{\Delta_i}{\delta_i} \leq \frac{1}{r_i - \rho_i} \frac{\sin(\pi - \alpha_i - \beta_i)}{\sin(\alpha_i)\sin(\beta_i)} 5c \left(\frac{5c \, \text{vol}_{d-1}(F_i)}{2d \, \text{vol}_d(K)} \right)^{\frac{1}{d-1}}.$$

We proceed as in the estimate above and obtain

$$\frac{\Delta_i}{\delta_i} \leq \frac{16\sqrt{2} \, dc}{r_i} \frac{5c}{\sin(\alpha_i)} \left(\frac{5c \, \text{vol}_{d-1}(F_i)}{2d \, \text{vol}_d(K)} \right)^{\frac{1}{d-1}}.$$

Thus we get for $\frac{\pi}{4} \leq \alpha_i \leq \frac{\pi}{2}$

$$\frac{\Delta_i}{\delta_i} \leq \frac{32 \, dc}{r_i} 5c \left(\frac{5c \, \text{vol}_{d-1}(F_i)}{2d \, \text{vol}_d(K)} \right)^{\frac{1}{d-1}}.$$

We verify now (3.13). By the definition of A_i we get

$$\frac{\text{vol}_{d-1}(B_2^{d-1})}{\text{vol}_{d-2}(\partial B_2^{d-1})} \int_{A_i^c} (r_i(\eta)^{d-1} - \rho_i(\eta)^{d-1})\delta_i(\eta) \, d\mu(\eta)$$

$$\geq (1 - e^{-\frac{1}{8}}) \frac{\text{vol}_{d-1}(B_2^{d-1})}{\text{vol}_{d-2}(\partial B_2^{d-1})} \int_{A_i^c} r_i(\eta)^{d-1} \delta_i(\eta) \, d\mu(\eta).$$

We get by (3.14) that the last expression is greater than

$$\frac{1}{320dc} \Delta_i \frac{\text{vol}_{d-1}(B_2^{d-1})}{\text{vol}_{d-2}(\partial B_2^{d-1})}$$

$$\times \left(\int_{\substack{A_i^c \\ \alpha_i \leq \frac{\pi}{4}}} r_i^{d-1} \, d\mu + \frac{1}{5c} \left(\frac{2d \, \text{vol}_d(K)}{5c \, \text{vol}_{d-1}(F_i)} \right)^{\frac{1}{d-1}} \int_{\substack{A_i^c \\ \alpha_i > \frac{\pi}{4}}} r_i^d \, d\mu \right).$$

By (3.11) we get that either

$$\frac{\text{vol}_{d-1}(B_2^{d-1})}{\text{vol}_{d-2}(\partial B_2^{d-1})} \int_{\substack{A_i^c \\ \alpha_i \leq \frac{\pi}{4}}} r_i^{d-1} \, d\mu \geq \tfrac{1}{8} \, \text{vol}_{d-1}(F_i)$$

or

$$\frac{\text{vol}_{d-1}(B_2^{d-1})}{\text{vol}_{d-2}(\partial B_2^{d-1})} \int_{\substack{A_i^c \\ \alpha_i > \frac{\pi}{4}}} r_i^{d-1} \, d\mu \geq \tfrac{1}{8} \, \text{vol}_{d-1}(F_i).$$

In the first case we get for the above estimate

$$\frac{\text{vol}_{d-1}(B_2^{d-1})}{\text{vol}_{d-2}(\partial B_2^{d-1})} \int_{A_i^c} (r_i(\eta)^{d-1} - \rho_i(\eta)^{d-1})\delta_i(\eta) \, d\mu(\eta)$$

$$\geq \frac{\Delta_i}{2560dc} \, \text{vol}_{d-1}(F_i) \geq \frac{1}{2560edc} \, \text{vol}_d((K^t \setminus P_n) \cap C_i).$$

The last inequality is obtained by using (3.5): Since $B_2^d \subset K$ we have, for all hyperplanes H that are parallel to F_i,

$$\text{vol}_{d-1}(K^t \cap H \cap C_i) \leq (1 + \Delta_i)^{d-1} \, \text{vol}_{d-1}(F_i).$$

By (3.5) we get $\text{vol}_{d-1}(K^t \cap H \cap C_i) \leq e\, \text{vol}_{d-1}(F_i)$. In the second case we have

$$\frac{\text{vol}_{d-1}(B_2^{d-1})}{\text{vol}_{d-2}(\partial B_2^{d-1})} \int_{A_i^c} (r_i(\eta)^{d-1} - \rho_i(\eta)^{d-1}) \delta_i(\eta)\, d\mu(\eta)$$

$$\geq \frac{1}{5c} \left(\frac{2d\, \text{vol}_d(K)}{5c\, \text{vol}_{d-1}(F_i)} \right)^{\frac{1}{d-1}} \frac{1}{320dc} \Delta_i \frac{\text{vol}_{d-1}(B_2^{d-1})}{\text{vol}_{d-2}(\partial B_2^{d-1})} \int_{\substack{A_i^c \\ \alpha_i > \frac{\pi}{4}}} r_i^d\, d\mu$$

$$\geq \frac{1}{5c} \left(\frac{2d\, \text{vol}_d(K)}{5c\, \text{vol}_{d-1}(F_i)} \right)^{\frac{1}{d-1}} \frac{1}{320dc} \Delta_i \frac{\text{vol}_{d-1}(B_2^{d-1})}{(\text{vol}_{d-2}(\partial B_2^{d-1}))^{\frac{d}{d-1}}} \left(\int_{\substack{A_i^c \\ \alpha_i > \frac{\pi}{4}}} r_i^{d-1}\, d\mu \right)^{\frac{d}{d-1}}$$

$$\geq \frac{1}{5c} \left(\frac{2d\, \text{vol}_d(K)}{5c\, \text{vol}_{d-1}(F_i)} \right)^{\frac{1}{d-1}} \frac{\Delta_i}{320dc} \text{vol}_{d-1}(B_2^{d-1})^{-\frac{1}{d-1}} (\tfrac{1}{8} \text{vol}_{d-1}(F_i))^{\frac{d}{d-1}}$$

$$= \frac{1}{5c} \left(\frac{d\, \text{vol}_d(K)}{20c\, \text{vol}_{d-1}(B_2^{d-1})} \right)^{\frac{1}{d-1}} \frac{\Delta_i}{2560dc} \text{vol}_{d-1}(F_i)$$

$$\geq \frac{1}{5c} \left(\frac{d\, \text{vol}_d(K)}{20c\, \text{vol}_{d-1}(B_2^{d-1})} \right)^{\frac{1}{d-1}} \frac{1}{2560edc} \text{vol}_d((K^t \setminus P_n) \cap C_i).$$

Since $B_2^d \subset K$ we get

$$\frac{\text{vol}_{d-1}(B_2^{d-1})}{\text{vol}_{d-2}(\partial B_2^{d-1})} \int_{A_i^c} (r_i(\eta)^{d-1} - \rho_i(\eta)^{d-1}) \delta_i(\eta)\, d\mu(\eta)$$

$$\geq \frac{1}{5c} \left(\frac{d\, \text{vol}_d(B_2^d)}{20c\, \text{vol}_{d-1}(B_2^{d-1})} \right)^{\frac{1}{d-1}} \frac{1}{2560edc} \text{vol}_d((K^t \setminus P_n) \cap C_i)$$

$$\geq \frac{1}{5c} \left(\frac{1}{20c} \right)^{\frac{1}{d-1}} \frac{1}{2560edc} \text{vol}_d((K^t \setminus P_n) \cap C_i)$$

$$\geq (10^6\, edc^{2+\frac{1}{d-1}})^{-1} \text{vol}_d((K^t \setminus P_n) \cap C_i).$$

The second case gives a weaker estimate. Therefore we get for both cases

$$\text{vol}_d((K^t \setminus P_n) \cap C_i)$$
$$\leq 10^6\, edc^{2+\frac{1}{d-1}} \frac{\text{vol}_{d-1}(B_2^{d-1})}{\text{vol}_{d-2}(\partial B_2^{d-1})} \int_{A_i^c} (r_i(\eta)^{d-1} - \rho_i(\eta)^{d-1}) \delta_i\, d\mu(\eta).$$

Thus we have verified (3.13) and thereby also (3.12). By (3.12) we get

$$\text{vol}_d\left((K^t \setminus P_n) \cap \bigcup_{i=k+1}^n C_i \right) \leq 10^6\, ed^2 c^{2+\frac{1}{d-1}} \text{vol}_d\left(\left(\bigcup_{i=k+1}^n C_i \right) \cap (P_n \setminus K) \right)$$

$$\leq 10^6\, ed^2 c^{2+\frac{1}{d-1}} \text{vol}_d((P_n \setminus K)). \qquad (3.16)$$

If the assertion of the theorem does not hold we have

$$\mathrm{vol}_d((P_n \setminus K)) \leq \frac{\mathrm{vol}_d(K^t \setminus K)}{10^7 \ ed^2 c^{2+\frac{1}{d-1}}}. \tag{3.17}$$

Thus we get

$$\mathrm{vol}_d\left((K^t \setminus P_n) \cap \bigcup_{i=k+1}^{n} C_i\right) \leq \tfrac{1}{10} \mathrm{vol}_d(K^t \setminus K).$$

Together with (3.7) we obtain

$$\mathrm{vol}_d(K^t \setminus P_n) \leq \tfrac{1}{4} \mathrm{vol}_d(K^t \setminus K) \leq \tfrac{1}{4}\{\mathrm{vol}_d(K^t \setminus P_n) + \mathrm{vol}_d(P_n \setminus K)\}. \tag{3.18}$$

By (3.17) we have

$$\mathrm{vol}_d(P_n \setminus K) \leq \frac{\mathrm{vol}_d(K^t \setminus K)}{10^7 \ ed^2 c^{2+\frac{1}{d-1}}}$$
$$\leq \tfrac{1}{2} \mathrm{vol}_d(K^t \setminus K) \leq \tfrac{1}{2} \mathrm{vol}_d(K^t \setminus P_n) + \tfrac{1}{2} \mathrm{vol}_d(P_n \setminus K).$$

This implies

$$\mathrm{vol}_d(P_n \setminus K) \leq \mathrm{vol}_d(K^t \setminus P_n).$$

Together with (3.18) we get now the contradiction

$$\mathrm{vol}_d(K^t \setminus P_n) \leq \tfrac{1}{2} \mathrm{vol}_d(K^t \setminus P_n). \qquad \square$$

References

[B] K. Ball, "Logarithmically concave functions and sections of convex sets in $\mathbb{R}^n$", *Studia Math.* **88** (1988), 69–84.

[BL] I. Bárány and D. G. Larman, "Convex bodies, economic cap covering, random polytopes", *Mathematika* **35** (1988), 274–291.

[BI] E. M. Bronshtein and L. D. Ivanov, "The approximation of convex sets by polyhedra", *Siberian Math. J.* **16** (1975), 1110–1112.

[D_1] R. Dudley, "Metric entropy of some classes of sets with differentiable boundaries", *J. of Approximation Theory* **10** (1974), 227–236.

[D_2] R. Dudley, "Correction to 'Metric entropy of some classes of sets with differentiable boundaries'", *J. of Approximation Theory* **26** (1979), 192–193.

[F–T] L. Fejes Toth, "Über zwei Maximumsaufgaben bei Polyedern", *Tohoku Math. J.* **46** (1940), 79–83.

[GMR_1] Y. Gordon, M. Meyer, and S. Reisner, "Volume approximation of convex bodies by polytopes—a constructive method", *Studia Math.* (1994), **111** 81–95.

[GMR_2] Y. Gordon, M. Meyer and S. Reisner, "Constructing a polytope to approximate a convex body", *Geometriae Dedicata* **57** (1995), 217–222.

[GRS] Y. Gordon, S. Reisner, and C. Schütt, "Umbrellas and polytopal approximation of the Euclidean ball", *J. of Approximation Theory* **90** (1997), 9–22.

[Gr_1] P. M. Gruber, "Volume approximation of convex bodies by inscribed polytopes", *Math. Annalen* **281** (1988), 292–245.

[Gr₂] P. M. Gruber, "Asymptotic estimates for best and stepwise approximation of convex bodies II", *Forum Mathematicum* (1993), **5** 521–538.

[GK] P. M. Gruber and P. Kenderov, "Approximation of convex bodies by polytopes", *Rend. Circolo Mat. Palermo* **31** (1982), 195–225.

[Grü] B. Grünbaum, "Partitions of mass-distributions and of convex bodies by hyperplanes, *Pacific J. Math.* **10** (1960), 1257–1261.

[H] D. Hensley, "Slicing convex bodies-bounds for slice area in terms of the body's covariance", *Proc. Amer. Math. Soc.* **79** (1980), 619–625.

[J] F. John, "Extremum problems with inequalities as subsidiary conditions", R. Courant Anniversary Volume, Interscience, New York, 1948, 187–204.

[Mac] A. M. Macbeath, "An extremal property of the hypersphere", *Proc. Cambridge Phil. Soc.* **47** (1951), 245–247.

[MP] V. Milman and A. Pajor, "Isotropic position and inertia ellipsoids and zonoids of the unit ball of a normed n-dimensional space", *Geometric Aspects of Functional Analysis* (GAFA) 1987–88, edited by J. Lindenstrauss and V. D. Milman, Springer, 1989, 64–104.

[Mü] J. S. Müller, "Approximation of the ball by random polytopes", *J. Approximation Theory* **63** (1990), 198–209.

[R] C. A. Rogers, *Packing and covering*, Cambridge University Press, 1964.

[S] C. Schütt, "The convex floating body and polyhedral approximation", *Israel J. Math.* **73** (1991), 65–77.

[SW] C. Schütt and E. Werner, "The convex floating body", *Math. Scandinavica* **66** (1990), 275–290.

[W] E. Werner, "Illumination bodies and the affine surface area", *Studia Math.* **110** (1994), 257–269.

CARSTEN SCHÜTT
MATHEMATISCHES SEMINAR
CHRISTIAN ALBRECHTS UNIVERSITÄT
D-24098 KIEL
GERMANY
CarstenSchuett@compuserve.com

A Note on the M^*-Limiting Convolution Body

ANTONIS TSOLOMITIS

ABSTRACT. We introduce the mixed convolution bodies of two convex symmetric bodies. We prove that if the boundary of a body K is smooth enough then as δ tends to 1 the δ-M^*-convolution body of K with itself tends to a multiple of the Euclidean ball after proper normalization. On the other hand we show that the δ-M^*-convolution body of the n-dimensional cube is homothetic to the unit ball of ℓ_1^n.

1. Introduction

Throughout this note K and L denote convex symmetric bodies in $\mathbb{R}^n$. Our notation will be the standard notation that can be found, for example, in [2] and [4]. For $1 \leq m \leq n$, $V_m(K)$ denotes the m-th mixed volume of K (i.e., mixing m copies of K with $n-m$ copies of the Euclidean ball $\mathcal{B}_n$ of radius one in $\mathbb{R}^n$). Thus if $m = n$ then $V_n(K) = \text{vol}_n(K)$ and if $m = 1$ then $V_1(K) = w(K)$ the mean width of K.

For $0 < \delta < 1$ we define the m-th mixed δ-convolution body of the convex symmetric bodies K and L in $\mathbb{R}^n$:

DEFINITION. The m-th mixed δ-convolution body of K and L is defined to be the set
$$C_m(\delta; K, L) = \{x \in \mathbb{R}^n : V_m(K \cap (x+L)) \geq \delta V_m(K)\}.$$

It is a consequence of the Brunn–Minkowski inequality for mixed volumes that these bodies are convex.

If we write $h(u)$ for the support function of K in the direction $u \in \mathbb{S}^{n-1}$, we have
$$w(K) = 2M_K^* = 2\int_{\mathbb{S}^{n-1}} h(u)\, d\nu(u), \tag{1.1}$$

where ν is the Lebesgue measure of $\mathbb{R}^n$ restricted on $\mathbb{S}^{n-1}$ and normalized so that $\nu(\mathbb{S}^{n-1}) = 1$. In this note we study the limiting behavior of $C_1(\delta; K, K)$

Work partially supported by an NSF grant.

(which we will abbreviate with $C_1(\delta)$) as δ tends to 1 and K has a C_+^2 boundary. For simplicity we will call $C_1(\delta)$ the δ-M^*-convolution body of K.

We are looking for suitable $\alpha \in \mathbb{R}$ so that the limit

$$\lim_{\delta \to 1^-} \frac{C_1(\delta)}{(1-\delta)^\alpha}$$

exists (convergence in the Hausdorff distance). In this case we call the limiting body "the limiting M^*-convolution body of K".

We prove that for a convex symmetric body K in $\mathbb{R}^n$ with C_+^2 boundary the limiting M^*-convolution body of K is homothetic to the Euclidean ball. We also get a sharp estimate (sharp with respect to the dimension n) of the rate of the convergence of the δ-M^*-convolution body of K to its limit. By C_+^2 we mean that the boundary of K is C^2 and that the principal curvatures of $\mathrm{bd}(K)$ at every point are all positive.

We also show that some smoothness condition on the boundary of K is necessary for this result to be true, by proving that the limiting M^*-convolution body of the n-dimensional cube is homothetic to the unit ball of ℓ_1^n.

2. The Case Where the Boundary of K Is a C_+^2 Manifold

THEOREM 2.1. *Let K be a convex symmetric body in $\mathbb{R}^n$ so that $\mathrm{bd}(K)$ is a C_+^2 manifold. Then for all $x \in \mathbb{S}^{n-1}$ we have*

$$\left| \|x\|_{\frac{C_1(\delta)}{1-\delta}} - \frac{c_n}{M_K^*} \right| \leq C \frac{c_n}{M_K^*} \left(M_K^* n(1-\delta) \right)^2, \tag{2.1}$$

where $c_n = \int_{\mathbb{S}^{n-1}} |\langle x, u \rangle| \, d\nu(u) \sim 1/\sqrt{n}$ and C is a constant independent of the dimension n. In particular,

$$\lim_{\delta \to 1^-} \frac{C_1(\delta)}{1-\delta} = \frac{M_K^*}{c_n} \mathcal{B}_n.$$

Moreover the estimate (2.1) is sharp with respect to the dimension n.

By "sharp" with respect to the dimension n we mean that there are examples (for instance the n-dimensional Euclidean ball) for which the inequality (2.1) holds true if "$\leq$" is replaced with "$\geq$" and the constant C changes by a (universal) constant factor.

Before we proceed with the proof we will need to collect some standard notation which can be found in [4]. We write $p : \mathrm{bd}(K) \to \mathbb{S}^{n-1}$ for the Gauss map $p(x) = N(x)$ where $N(x)$ denotes the unit normal vector of $\mathrm{bd}(K)$ at x. W_x denotes the Weingarten map, that is, the differential of p at the point $x \in \mathrm{bd}(K)$. W^{-1} is the reverse Weingarten map and the eigenvalues of W_x and W_u^{-1} are respectively the principal curvatures and principal radii of curvatures of the manifold $\mathrm{bd}(K)$ at $x \in \mathrm{bd}(K)$ and $u \in \mathbb{S}^{n-1}$. We write $\|W\|$ and $\|W^{-1}\|$

for the quantities $\sup_{x\in \mathrm{bd}(K)} \|W_x\|$ and $\sup_{u\in \mathbb{S}^{n-1}} \|W_u^{-1}\|$, respectively. These quantities are finite since the manifold $\mathrm{bd}(K)$ is assumed to be C_+^2.

For $\lambda \in \mathbb{R}$ and $x \in \mathbb{S}^{n-1}$ we write K_λ for the set $K \cap (\lambda x + K)$. $p_\lambda^{-1} : \mathbb{S}^{n-1} \to \mathrm{bd}(K_\lambda)$ is the reverse Gauss map, that is, the affine hyperplane $p_\lambda^{-1}(u) + [u]^\perp$ is tangent to K_λ at $p_\lambda^{-1}(u)$. The normal cone of K_λ at x is denoted by $N(K_\lambda, x)$ and similarly for K. The normal cone is a convex set (see [4]). Finally h_λ will denote the support function of K_λ.

PROOF. Without loss of generality we may assume that both the $\mathrm{bd}(K)$ and $\mathbb{S}^{n-1}$ are equipped with an atlas whose charts are functions which are Lipschitz, their inverses are Lipschitz and they all have the same Lipschitz constant $c > 0$.

Let $x \in \mathbb{S}^{n-1}$ and $\lambda = 1/\|x\|_{C_1(\delta)}$; hence $\lambda x \in \mathrm{bd}\,(C_1(\delta))$ and

$$M^*_{K_\lambda} = \delta M^*_K.$$

We estimate now $M^*_{K_\lambda}$. Let $u \in \mathbb{S}^{n-1}$. We need to compare $h_\lambda(u)$ and $h(u)$. Set $Y_\lambda = \mathrm{bd}(K) \cap \mathrm{bd}(\lambda x + K)$.

Case 1. $p_\lambda^{-1}(u) \notin Y_\lambda$.

In this case it is easy to see that

$$h_\lambda(u) = h(u) - |\langle \lambda x, u\rangle|.$$

Case 2. $p_\lambda^{-1}(u) \in Y_\lambda$.

Let $y_\lambda = p_\lambda^{-1}(u)$ and $y'_\lambda = y_\lambda - \lambda x \in \mathrm{bd}(K)$. The set $N(K_\lambda, y_\lambda) \cap \mathbb{S}^{n-1}$ defines a curve γ which we assume to be parametrized on $[0,1]$ with $\gamma(0) = N(K, y_\lambda)$ and $\gamma(1) = N(K, y'_\lambda)$. We use the inverse of the Gauss map p to map the curve γ to a curve $\tilde{\gamma}$ on $\mathrm{bd}(K)$ by setting $\tilde{\gamma} = p^{-1}\gamma$. The end points of $\tilde{\gamma}$ are y_λ (label it with A) and y'_λ (label it with B). Since $u \in \gamma$ we conclude that the point $p^{-1}(u)$ belongs to the curve $\tilde{\gamma}$ (label this point by Γ). Thus we get

$$0 \leq h(u) - h_\lambda(u) = |\langle \vec{A\Gamma}, u\rangle|.$$

It is not difficult to see that the cosine of the angle of the vectors $\vec{A\Gamma}$ and u is less than the largest principal curvature of $\mathrm{bd}(K)$ at Γ times $|\vec{A\Gamma}|$, the length of the vector $\vec{A\Gamma}$. Consequently we can write

$$0 \leq h(u) - h_\lambda(u) \leq \|W\|\,|\vec{A\Gamma}|^2.$$

In addition we have

$$|\vec{A\Gamma}| \leq \mathrm{length}\,(\tilde{\gamma}|_A^\Gamma) \leq \mathrm{length}\,(\tilde{\gamma}|_A^B) = \int_0^1 |d_t\tilde{\gamma}|\,dt = \int_0^1 |d_t p^{-1}\gamma|\,dt$$

$$\leq \|W^{-1}\|\mathrm{length}(\gamma) \leq \frac{2}{\pi}\|W^{-1}\|\,|p(y_\lambda) - p(y'_\lambda)|,$$

where $|\cdot|$ is the standard Euclidean norm. Without loss of generality we can assume that the points y_λ and y'_λ belong to the same chart at y_λ. Let φ be the chart mapping $\mathbb{R}^{n-1}$ to a neighborhood of y_λ on $\mathrm{bd}(K)$ and ψ the chart mapping

$\mathbb{R}^{n-1}$ on $\mathbb{S}^{n-1}$. We assume, as we may, that the graph of γ is contained in the range of the chart ψ. It is now clear from the above series of inequalities that

$$|\vec{A\Gamma}| \le c_0 \|W^{-1}\| \, |\psi^{-1}p\varphi(t) - \psi^{-1}p\varphi(s)|,$$

where t and s are points in $\mathbb{R}^{n-1}$ such that $\varphi(t) = y_\lambda$ and $\varphi(s) = y'_\lambda$ and $c_0 > 0$ is a universal constant. Now the mean value theorem for curves gives

$$|\vec{A\Gamma}| \le C\|W^{-1}\|\|W\| \, |t-s| \le C\|W^{-1}\|\|W\| \, |y_\lambda - y'_\lambda| = C\|W^{-1}\|\|W\|\lambda,$$

where C may denote a different constant every time it appears. Thus we have

$$0 \le h(u) - h_\lambda(u) \le C\|W\| \left(\|W^{-1}\|\|W\|\right)^2 \lambda^2.$$

Consequently,

$$\int_{\mathbb{S}^{n-1}\setminus p_\lambda(Y_\lambda)} (h(u) - |\langle \lambda x, u\rangle|) \, d\nu(u) + \int_{p_\lambda(Y_\lambda)} (h(u) - C\lambda^2) \, d\nu(u)$$

$$\le M^*_{K_\lambda} = \delta M^*_K \le$$

$$\int_{\mathbb{S}^{n-1}\setminus p_\lambda(Y_\lambda)} (h(u) - |\langle \lambda x, u\rangle|) \, d\nu(u) + \int_{p_\lambda(Y_\lambda)} h(u) \, d\nu(u),$$

where C now depends on $\|W\|$ and $\|W^{-1}\|$.

Rearranging and using c_n for the quantity $\int_{\mathbb{S}^{n-1}} |\langle x, u\rangle| \, d\nu(u)$ and the fact that $\lambda = 1/\|x\|_{C_1(\delta)}$ we get

$$\left| \|x\|_{\frac{C_1(\delta)}{1-\delta}} - \frac{c_n}{M^*_K} \right| \le \frac{c_n}{M^*_K} \left(\frac{\int_{p_\lambda(Y_\lambda)} |\langle x, u\rangle| \, d\nu(u)}{c_n} + C\lambda \frac{\mu(p_\lambda(Y_\lambda))}{c_n} \right).$$

We observe now that for $u \in p_\lambda(Y_\lambda), |\langle x, u\rangle| \le \text{length}(\gamma)/2 \le \|W\|\lambda$. Using this in the last inequality and the fact that $p_\lambda(Y_\lambda)$ is a band around an equator of $\mathbb{S}^{n-1}$ of width at most $\text{length}(\gamma)/2$ we get

$$\left| \|x\|_{\frac{C_1(\delta)}{1-\delta}} - \frac{c_n}{M^*_K} \right| \le \frac{c_n}{M^*_K} Cn\lambda^2 \le \frac{c_n}{M^*_K} Cn \frac{(1-\delta)^2}{\|x\|^2_{\frac{C_1(\delta)}{1-\delta}}}.$$

Our final task is to get rid of the norm that appears on the right side of the latter inequality. Set

$$T = \frac{\|x\|_{C_1(\delta)/1-\delta}}{c_n/M^*_K}.$$

We have shown that

$$T^2|T-1| \le C \frac{M^*_K}{c_n} n(1-\delta)^2.$$

If $T \ge 1$ then we can just drop the factor T^2 and we are done. If $T < 1$ we write $T^2|T-1|$ as $(1-(1-T))^2(1-T)$ and we consider the function

$$f(x) = (1-x)^2 x : (-\infty, \tfrac{1}{3}) \to \mathbb{R}.$$

This function is strictly increasing thus invertible on its range, that is, f^{-1} is well defined and increasing in $(-\infty, \frac{4}{27})$. Consequently, if

$$C\frac{M_K^*}{c_n}n(1-\delta)^2 \leq \frac{4}{27}, \tag{2.2}$$

we conclude that

$$0 \leq 1 - T \leq f^{-1}\left(C\frac{M_K^*}{c_n}n(1-\delta)^2\right) \leq C\frac{M_K^*}{c_n}n(1-\delta)^2.$$

The last inequality is true since the derivative of f^{-1} at zero is 1. Observe also that the convergence is "essentially realized" after (2.2) is satisfied. □

We now proceed to show that some smoothness conditions on the boundary of K are necessary, by proving that the limiting M^*-convolution body of the n-dimensional cube is homothetic to the unit ball of ℓ_1^n. In fact we show that the δ-M^*-convolution body of the cube is already homothetic to the unit ball of ℓ_1^n.

EXAMPLE 2.2. Let $P = [-1,1]^n$. Then for $0 < \delta < 1$ we have

$$C_1(P) = \frac{C_1(\delta; P, P)}{1-\delta} = n^{3/2}\operatorname{vol}_{n-1}(\mathbb{S}^{n-1})B_{\ell_1^n}.$$

PROOF. Let $x = \sum_{j=1}^n x_j e_j$ where $x_j \geq 0$ for all $j = 1, 2, \ldots, n$ and e_j is the standard basis of $\mathbb{R}^n$. Let $\lambda > 0$ be such that $\lambda x \in \operatorname{bd}(C_1(\delta))$. Then

$$P \cap (\lambda x + P) = \left\{y \in \mathbb{R}^n : y = \sum_{j=1}^n y_i e_i, -1 + \lambda x_i \leq y_i \leq 1\right\}.$$

The vertices of $P_\lambda = P \cap (\lambda x + P)$ are the points $\sum_{j=1}^n \alpha_j e_j$ where α_j is either 1 or $-1 + \lambda x$ for all j. Without loss of generality we can assume that $-1 + \lambda x_j < 0$ for all the indices j. Put $\operatorname{sign}\alpha_j = \alpha_j/|\alpha_j|$ when $\alpha_j \neq 0$ and $\operatorname{sign} 0 = 0$. Fix a sequence of α_j's so that the point $v = \sum_{j=1}^n \alpha_j e_j$ is a vertex of P_λ. Clearly,

$$N(P_\lambda, v) = N\left(P, \sum_{j=1}^n (\operatorname{sign}\alpha_j)e_j\right).$$

If $u \in \mathbb{S}^{n-1} \cap N(P_\lambda, v)$ then

$$h_\lambda(u) = h(u) - \left|\left\langle\sum_{j=1}^n (\alpha_j - \operatorname{sign}\alpha_j)e_j, u\right\rangle\right|.$$

If $\operatorname{sign}\alpha_j = 1$ then $\alpha_j - \operatorname{sign}\alpha_j = 0$ otherwise $\alpha_j - \operatorname{sign}\alpha_j = \lambda x$.
Let $\mathcal{A} \subseteq \{1, 2, \ldots, n\}$. Consider the "$\mathcal{A}$-orthant"

$$\mathcal{O}_\mathcal{A} = \{y \in \mathbb{R}^n : \langle y, e_j\rangle < 0, \text{ if } j \in \mathcal{A} \text{ and } \langle y, e_j\rangle \geq 0 \text{ if } j \notin \mathcal{A}\}.$$

Then $\mathcal{O}_\mathcal{A} = N\left(P, \sum_{j=1}^n (\text{sign } \alpha_j) e_j\right)$ if and only if $\text{sign } \alpha_j = 1$ exactly for every $j \notin \mathcal{A}$. Thus we get

$$h_\lambda(u) = h(u) - \left|\left\langle \sum_{j \in \mathcal{A}} \lambda x_j e_j, u \right\rangle\right|,$$

for all $u \in \mathcal{O}_\mathcal{A} \cap \mathbb{S}^{n-1}$. Hence using the facts $M_{P_\lambda}^* = \delta M_P^*$ and $\lambda = 1/\|x\|_{C_1(\delta)}$ we get

$$\|x\|_{\frac{C_1(\delta)}{1-\delta}} = -\frac{1}{M_P^*} \sum_{\mathcal{A} \subseteq \{1,2,\ldots,n\}} \sum_{j \in \mathcal{A}} x_j \int_{\mathcal{O}_\mathcal{A} \cap \mathbb{S}^{n-1}} \langle e_j, u \rangle \, d\nu(u),$$

which gives the result since

$$\int_{\mathcal{O}_\mathcal{A} \cap \mathbb{S}^{n-1}} \langle e_j, u \rangle \, d\nu(u) = \frac{1}{2^{n-1}} \int_{\mathbb{S}^{n-1}} |\langle e_1, u \rangle| \, d\nu(u). \qquad \square$$

Acknowledgement

We want to thank Professor V. D. Milman for his encouragement and his guidance in this research and for suggesting the study of mixed convolution bodies.

References

[1] K. Kiener, "Extremalität von Ellipsoiden und die Faltungsungleichung von Sobolev", *Arch. Math.* **46** (1986), 162–168.

[2] V. Milman and G. Schechtmann, *Asymptotic theory of finite dimensional normed spaces*, Lecture Notes in Math. **1200**, Springer, 1986.

[3] M. Schmuckenschläger, "The distribution function of the convolution square of a convex symmetric body in $\mathbb{R}^n$", *Israel Journal of Mathematics* **78** (1992), 309–334.

[4] R. Schneider, *Convex bodies: The Brunn–Minkowski theory*, Encyclopedia of Mathematics and its Applications **44**, Cambridge University Press, 1993.

[5] A. Tsolomitis, Convolution bodies and their limiting behavior, *Duke Math. J.* **87**:1 (1997), 181–203.

ANTONIS TSOLOMITIS
THE OHIO STATE UNIVERSITY
DEPARTMENT OF MATHEMATICS
231 W.18TH AVENUE
COLUMBUS, OH 43210
UNITED STATES OF AMERICA
    atsol@eexi.gr
    Current address: University of Crete, Department of Mathematics, 71409 Heraklion, Crete, Greece